Everyday Mathematics®

The University of Chicago School Mathematics Project

STUDENT REFERENCE BOOK

McGraw Hill Education

The University of Chicago School Mathematics Project

Max Bell, Director, *Everyday Mathematics* First Edition
James McBride, Director, *Everyday Mathematics* Second Edition
Andy Isaacs, Director, *Everyday Mathematics* Third, CCSS, and Fourth Editions
Amy Dillard, Associate Director, *Everyday Mathematics* Third Edition
Rachel Malpass McCall, Associate Director, *Everyday Mathematics* CCSS and Fourth Editions
Mary Ellen Dairyko, Associate Director, *Everyday Mathematics* Fourth Edition

Authors
Max Bell
Jean Bell
John Bretzlauf
Amy Dillard
James Flanders
Robert Hartfield
Andy Isaacs
Catherine Randall Kelso
James McBride
Kathleen Pitvorec
Peter Saecker

Writers
Lisa J. Bernstein
Andy Carter
Jeanne Di Domenico
Lila K.S. Goldstein
Jesch Reyes
Elizabet Spaepen
Judith S. Zawojewski

Digital Development Team
Carla Agard-Strickland, Leader
John Benson
Gregory Berns-Leone
Scott Steketee

Technical Art
Diana Barrie, Senior Artist
Cherry Inthalangsy

UCSMP Editorial
Elizabeth Olin
Kristen Pasmore
Molly Potnick

Contributors
Rosalie A. DeFino
Lance Campbell
Kathryn Flores

www.everydaymath.com

Copyright © McGraw-Hill Education

All rights reserved. No part of this publication may be reproduced or distributed in any form or by any means, or stored in a database or retrieval system, without the prior written consent of McGraw-Hill Education, including, but not limited to, network storage or transmission, or broadcast for distance learning.

Send all inquiries to:
McGraw-Hill Education
8787 Orion Place
Columbus, OH 43240

ISBN: 978-0-02-138360-3
MHID: 0-02-138360-X

Printed in the United States of America.

1 2 3 4 5 6 7 8 9 QVS 20 19 18 17 16 15

About the *Student Reference Book* xi
How to Use the *Student Reference Book* xii

Standards for Mathematical Practice — 1

Mathematical Practices 2
Problem Solving: Make Sense and Keep Trying. 4
Create and Make Sense of Representations. 9
Make Conjectures and Arguments. 13
Create and Use Mathematical Models 17
Choose Tools to Solve Problems 20
Be Precise and Accurate 23
Use Structures to Solve Problems. 26
Create and Explain Rules and Shortcuts 29
A Problem-Solving Diagram. 32
Problem Solving and the Mathematical Practices 33

Ratios and Proportional Relationships — 37

Ratios 38
Equivalent Ratios 41
Comparing Ratios 42
Using Ratio/Rate Tables to Solve Problems 43
Using Arrays and Tape Diagrams to Solve Ratio Problems . . . 45
Unit Ratios 49
Solve Problems Using Ratio/Rate Graphs 51
Estimating Solutions Using Ratio/Rate Graphs. 53
Percents. 54
Percents, Fractions, and Decimals 56
Using Unit Percents to Solve Problems 59
Using Percents to Find a Part 60
Finding a Percent. 61
Using Percents to Find the Whole 64
Ratios in Systems of Measurement. 65
Unit Conversions. 68
Multi-Step Unit Conversions 70
Using Ratios to Describe Size Changes 72

Scale Models 81

Contents

The Number System — 87

Uses of Numbers 88
Kinds of Numbers 89
The Real Number Line 91
Absolute Value 92
Number Lines 94
Coordinate Grids 95
Place Value for Whole Numbers 97
Exponential Notation 98
Powers of 10 99
Expanded Form 100
Comparing Numbers and Amounts 101
Factors of a Counting Number 102
Divisibility. 103
Prime and Composite Numbers. 104
Greatest Common Factor 105
Least Common Multiples 106
Decimals 107
Extending Place Value to Decimals 109
Comparing Decimals 111
Comparing Decimals on a Number Line 112
Estimation 113
Rounding 115
Addition and Subtraction of Decimals 117
Addition Methods 118
Subtraction Methods 121
Units and Precision in Decimal Addition and Subtraction 128
Multiplying by Powers of 10 .129
Extended Multiplication Facts 132
Multiplication Methods. 133
Extended Division Facts 139
Dividing Decimals by Powers of 10 140
Interpreting Remainders in Division 141
Division Methods. 143

Dividing Decimals by Decimals 145
U.S. Traditional Long Division: Single-Digit Divisors 147
U.S. Traditional Long Division: Multidigit Divisors 149
U.S. Traditional Long Division: Decimal Dividends 151
U.S. Traditional Long Division: Decimal Divisors 153
U.S. Traditional Long Division: Decimal Divisors and Dividends   154
Fractions 155
Reading and Writing Fractions 156
Meanings of Fractions 158
Uses of Fractions 159
Equivalent Fractions 160
Renaming Fractions Greater Than One 161
Renaming Mixed Numbers 162
Renaming Fractions as Decimals 164
Comparing Fractions 168
Finding Fractions Between Fractions 169
Fraction-Stick Chart 171
Common Denominators 173
Using Unit Fractions to Solve Problems 174
Estimating with Fractions 176
Adding and Subtracting Fractions 179
Adding Mixed Numbers 180
Subtracting Mixed Numbers 181
Finding a Fraction of a Number 184
Finding a Fraction of a Fraction 185
Predicting the Size of Products 186
Multiplying Fractions and Whole Numbers 188
Multiplying Fractions 189
Multiplying Mixed Numbers 190
Division of Fractions 193
Division of Fractions and Mixed Numbers 196

Expressions and Equations   197

Algebra 198
Expressions 200
Order of Operations 203

Contents

The Distributive Property	204
Equivalent Expressions	206
Number Sentences	207
Inequalities	210
Formulas	212
Using Trial and Error to Solve Equations	214
Using Bar Models to Solve Equations	216
Pan-Balance Problems and Solving Equations	217
Using Inverse Operations to Solve Equations	219
"What's My Rule?" Problems	221
Rules, Tables, and Graphs	222
Independent and Dependent Variables	223
Representing Patterns with Algebra	225
How to Balance a Mobile	226
Solving Problems with Computer Spreadsheets and Formulas	228
Properties of Numbers and Operations	231

Geometry — 233

What is Geometry?	234
Angles	236
Line Segments, Rays, Lines, and Angles	237
Polygons	238
Triangles	240
Triangle Hierarchy	241
Quadrilaterals	242
Quadrilateral Hierarchy	243
Geometric Solids	244
Polyhedrons	246
Pyramids	247
Prisms	248
Regular Polyhedrons	249
Length	250
Perimeter	251
Area	252

Area of a Rectangle 253

Area of a Parallelogram 254

Area of a Triangle 255

Finding Area by Composing and Decomposing Shapes 257

Volume 258

Volume of Geometric Solids 259

Volume of a Rectangular or Triangular Prism 260

Volume of a Rectangular or Triangular Pyramid 261

Volume of a Rectangular Prism with Fractional Edge Lengths 262

Introduction to Surface Area 263

Calculating Surface Area 264

Coordinate Grid 265

Plotting Polygons on a Coordinate Grid 267

Congruent Figures 268

Similar Figures 269

Line Symmetry 270

Reflections 271

Paper Folding 273

Statistics 279

Statistical Questions 280

Collecting Data 281

Organizing Data 283

Statistical Landmarks: Measures of Center 284

Measures of Center: The Mean (or Average) 285

Choosing a Measure of Center 289

Statistical Landmarks: Measures of Spread and Variability . . 291

Data Representations: Bar Graphs 295

Data Representations: Histograms 296

Data Representations: Box Plots 300

Reading and Analyzing Graphs 303

The Shape of the Data Distribution 305

Persuasive Graphs 308

The U.S. Census 309

Contents

Games 315

Games 316
Absolute Value Sprint 318
Algebra Election 320
Build-It 322
Daring Division 323
Divisibility Dash 324
Doggone Decimal 325
Factor Captor 326
First to 100 327
Fraction Action, Fraction Friction 328
Fraction Capture 329
Fraction Top-It 330
Fraction/Whole Number Top-It 331
Getting to One 332
Hidden Treasure 333
High-Number Toss (Decimal Version) 334
Landmark Shark 335
Mixed-Number Spin 337
Multiplication Bull's Eye 338
Multiplication Wrestling (Mixed-Number Version) 339
Name That Number 340
Percent Spin 341
Polygon Capture 343
Ratio Comparison 344
Ratio Dominoes 345
Ratio Memory Match 346
Solution Search 347
Spoon Scramble 348
Top-It Games 349

Real-World Data — 351

- Introduction 352
- United States: Utah National Parks 353
- United States: State Facts 354
- United States: Native American Population in 2010 356
- United States: Foreign-Born Population over Time 357
- World: Largest Countries by Land Area 358
- World: Tallest and Largest 359
- World: Most-Visited Museums 360
- Sports: Boston Marathon Records 361
- Sports: Tour de France Records .362
- Sports: Olympic Medal Counts 363
- Sports: International Athletes in the NBA and MLB 364
- Science and Nature: Biodiversity 366
- Science and Nature: Tallest and Deepest 368
- Science and Nature: Electricity Consumption 370
- Data Sources 371

Tables and Charts — 373

Appendix — 379

Glossary — 411

Answer Key — 439

Index — 453

About the *Student Reference Book*

The *Student Reference Book* is a helpful guide to review math concepts and skills, a resource to find the meaning of math terms, and interesting reading when you want to learn about a new topic in mathematics.

A reference book is organized to help readers find information quickly and easily. Dictionaries, encyclopedias, atlases, and cookbooks are examples of reference books. Reference books are not like novels and biographies, which you often read in order from beginning to end. When you read reference books, you look for specific information at the time you need it. Then you read just the pages you need at that time.

You can use this *Student Reference Book* to look up and review information on topics in mathematics. It includes the following informaton:

- A **table of contents** that lists the topics and gives an overview of how the book is organized

- Essays describing how to use **mathematical practices** to solve problems and show mathematical thinking

- Essays on **mathematical content,** such as ratios and proportions, the number system, expressions and equations, geometry, and statistics

- A collection of tables, charts, diagrams, and maps that includes **real-world data**

- A collection of **photo essays** that show in words and pictures some of the ways that scientists, artists, engineers, and others have used mathematics throughout history or how they use it today

- Directions on how to play **mathematical games** to practice your math skills

- A set of **tables** and **charts** that summarize information, such as a place-value chart, rules for the order of operations, tables of equivalent values, and a chart of personal measurement references

- An **appendix** that includes directions on how to use a **calculator** to perform various mathematical operations

- A **glossary** of mathematical terms consisting of definitions and some illustrations

- An **answer key** for the **Check Your Understanding** problems

- An **index** to help you locate topics quickly

- **Videos** and **interactive problems** available through the electronic version of this book in the Student Learning Center

How to Use the Student Reference Book

As you work in class or at home, you can use the *Student Reference Book* to help you solve problems. For example, when you don't remember the meaning of a word or aren't sure what method to use, you can use the *Student Reference Book* as a tool.

You can look in the **table of contents** or the **index** to find pages that give a brief explanation of the topic. The explanation will often include definitions of important math words and show examples of problems with step-by-step sample solutions.

While reading the text, you can take notes that include words, pictures, and diagrams to help you understand what you are reading. Work through the examples and try to follow each step.

At the end of many of the essays, you will find problems in **Check Your Understanding** boxes. Solve these problems and then check the **answer key** at the back of the book. These exercises will help you make sure that you understand the information you have been reading. Make sense of the problems by comparing your answers with those in the answer key. If necessary, work backward from the sample answers to revise your work.

The **Standards for Mathematical Practice** section includes interesting problems that you can solve. The discussions illustrate how sixth-grade students use the practices to solve these problems.

The world of mathematics is a very interesting and exciting place. Read the **photo essays**, explore **real-world data**, or review topics learned in class. The *Student Reference Book* is a great place to continue your investigation of math topics and ideas.

Once you are familiar with the overall structure of the *Student Reference Book*, you can use it to read about different mathematical concepts. As you follow your interests, you will find that your skills as an independent reader and problem-solver will improve.

Standards for Mathematical Practice

Standards for Mathematical Practice

Mathematical Practices

Mathematicians, scientists, engineers, and others who use mathematics develop ways of working that help them solve problems. These ways of doing mathematics are called **mathematical practices.** When you solve problems, you are using mathematical practices.

Look at each picture below. In what ways might these students be thinking and working as mathematicians?

The next page lists the Standards for Mathematical Practice you will use in *Everyday Mathematics*. Below each standard is a list of goals that can help you understand what it means to use the practices. On the following pages of the *Student Reference Book*, you will see how some sixth-grade students use mathematical practices as they solve problems and reason about mathematics. As you read, think about the ways you can develop these practices to become a more powerful problem solver.

Standards for Mathematical Practice

Mathematical Practice 1: Make sense of problems and persevere in solving them.
- GMP1.1 Make sense of your problem.
- GMP1.2 Reflect on your thinking as you solve your problem.
- GMP1.3 Keep trying when your problem is hard.
- GMP1.4 Check whether your answer makes sense.
- GMP1.5 Solve problems in more than one way.
- GMP1.6 Compare the strategies you and others use.

Mathematical Practice 2: Reason abstractly and quantitatively.
- GMP2.1 Create mathematical representations using numbers, words, pictures, symbols, gestures, tables, graphs, and concrete objects.
- GMP2.2 Make sense of the representations you and others use.
- GMP2.3 Make connections between representations.

Mathematical Practice 3: Construct viable arguments and critique the reasoning of others.
- GMP3.1 Make mathematical conjectures and arguments.
- GMP3.2 Make sense of others' mathematical thinking.

Mathematical Practice 4: Model with mathematics.
- GMP4.1 Model real-world situations using graphs, drawings, tables, symbols, numbers, diagrams, and other representations.
- GMP4.2 Use mathematical models to solve problems and answer questions.

Mathematical Practice 5: Use appropriate tools strategically.
- GMP5.1 Choose appropriate tools.
- GMP5.2 Use tools effectively and make sense of your results.

Mathematical Practice 6: Attend to precision.
- GMP6.1 Explain your mathematical thinking clearly and precisely.
- GMP6.2 Use an appropriate level of precision for your problem.
- GMP6.3 Use clear labels, units, and mathematical language.
- GMP6.4 Think about accuracy and efficiency when you count, measure, and calculate.

Mathematical Practice 7: Look for and make use of structure.
- GMP7.1 Look for mathematical structures such as categories, patterns, and properties.
- GMP7.2 Use structures to solve problems and answer questions.

Mathematical Practice 8: Look for and express regularity in repeated reasoning.
- GMP8.1 Create and justify rules, shortcuts, and generalizations.

Standards for Mathematical Practice

Problem Solving: Make Sense and Keep Trying

When you solve a problem, you make sense of the problem, keep trying when the problem is hard, and reflect on your thinking as you work. When you compare your strategies to the strategies of others, you learn to solve problems in more than one way. Comparing strategies also helps you check your work.

Think about this problem:

Choose one value for x and one value for y so that both inequalities are true.

$$x + y < 20 \quad \text{and} \quad x + y \geq 10$$

Find all possible combinations of x and y that work from the options below.

Possible Values of x	Possible Values of y
15.5 11.25 5.75	4.5 8.75 4.25

Ben and London start thinking about the problem together.

How would you start to solve this problem?

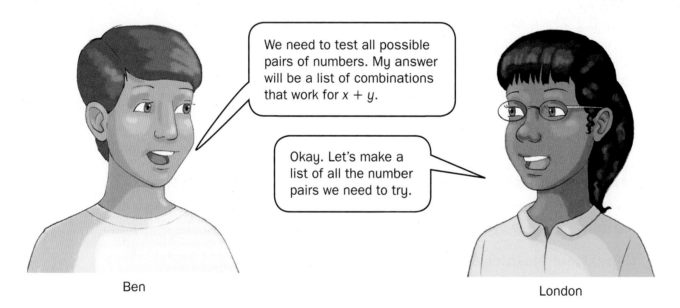

We need to test all possible pairs of numbers. My answer will be a list of combinations that work for $x + y$.

Okay. Let's make a list of all the number pairs we need to try.

Ben

London

Ben and London make a list. They find 9 possible number pairs and write them down in a list of 9 addition problems.

GMP1.1 Make sense of your problem.

Ben and London made sense of the problem by listing all possible combinations of $x + y$.

Standards for Mathematical Practice

Here is Ben and London's list:

	x + y	
15.5 + 4.5	11.25 + 4.5	5.75 + 4.5
15.5 + 8.75	11.25 + 8.75	5.75 + 8.75
15.5 + 4.25	11.25 + 4.25	5.75 + 4.25

London adds the first expression, 15.5 + 4.5, in her head by adding the whole-number parts and then decimal parts: 15 + 4 = 19 and 0.5 + 0.5 = 1. The total is 20.

She tells Ben, "Since the sum is 20, we should circle 15.5 + 4.5 on our list to show this pair works for both inequalities."

Ben agrees that the sum is 20, but argues that 15.5 + 4.5 doesn't work because the sum needs to be *less than* 20, but not *equal to* 20. Ben crosses out 15.5 + 4.5 to show that this combination does not work.

GMP1.2 Reflect on your thinking as you solve your problem.

When Ben reflected on the sum and the meaning of the expression $x + y < 20$, he realized that the sum of 20 should not be included as a solution.

London writes the next pair of numbers on her paper.

$$\begin{array}{r} 15.5 \\ +\ 8.75 \\ \hline \end{array}$$

She puts her pencil down before she adds. She says, "Look at the whole numbers. I can see 15 + 8 is more than 20. Even without adding the decimal parts, I know this pair doesn't work."

London crosses out 15.5 + 8.75. She and Ben know that two pairs of numbers do not work so far.

	x + y	
~~15.5 + 4.5~~	11.25 + 4.5	5.75 + 4.5
~~15.5 + 8.75~~	11.25 + 8.75	5.75 + 8.75
15.5 + 4.25	11.25 + 4.25	5.75 + 4.25

London looks for other pairs of numbers that have a whole-number sum greater than 20. She does not find any.

five SRB 5

Standards for Mathematical Practice

Ben notices a strategy for adding the whole-number parts and estimating the decimal parts to see whether the total sum is between 10 and 20.

Look at 11.25 + 4.5. I can see that 11 + 4 is 15, and each decimal part is less than 1. That means the sum of the decimal parts is less than 2. So, the total sum is less than 17. This pair works. Are there any more like this?

Ben

London thinks about Ben's strategy and uses it to find two other expressions that have a sum between 10 and 20.

11.25 + **4**.25: The sum of the whole numbers is 11 + 4 = 15. The sum of the decimal parts is less than 2. So, this pair works because the total sum is between 15 and 17.

5.75 + **8**.75: The sum of the whole numbers is 5 + 8 = 13. The sum of the decimal parts is less than 2. So, this pair works because the total sum is between 13 and 15.

London circles the three pairs of numbers that they know work.

	$x + y$	
~~15.5 + 4.5~~	(11.25 + 4.5)	5.75 + 4.5
~~15.5 + 8.75~~	11.25 + 8.75	(5.75 + 8.75)
15.5 + 4.25	(11.25 + 4.25)	5.75 + 4.25

GMP1.3 Keep trying when your problem is hard.

London and Ben tried and applied multiple strategies as they worked on the problem. They first found a strategy to identify pairs by adding just the whole-number parts to see whether their sum was greater than 20. When that strategy did not help find any other pairs of numbers, London and Ben revised the strategy to look for other pairs between 10 and 20.

Ben and London use mental arithmetic to test another pair of numbers by first adding the whole-number parts and then adding the decimal parts.

Ben: Look at 15.5 + 4.25. I know that 15 + 4 = 19. To find the sum of the decimal parts, think about the 5 tenths in 15.5 as $\frac{1}{2}$, and the 25 hundredths in 4.25 as $\frac{1}{4}$. The sum of the decimal parts is the fraction $\frac{3}{4}$. This pair works because the total sum is $19\frac{3}{4}$, which is less than 20 but greater than 10.

London: I got the same answer, but I thought about money. 15.5 is like $15.50 and 4.25 is like $4.25. That's $19 whole dollars, and then you have 2 quarters plus 1 quarter. Since 3 quarters is 75¢, the sum is $19.75. The total is less than $20.

Ben: Our strategies for thinking about the decimal parts are similar. The decimal 0.5 is the same as $\frac{1}{2}$, and 2 quarters is $\frac{1}{2}$ of a dollar. The decimal 0.25 is the same as $\frac{1}{4}$, and 1 quarter is $\frac{1}{4}$ of a dollar. The fraction total $\frac{3}{4}$ is like 3 quarters of a dollar, or 75¢. That's why we got the same answer.

GMP1.5 Solve problems in more than one way.
Ben and London solved the problem using mental math in different ways. Ben thought about the decimals as fractions, while London thought about the decimals as quarters. They used two different ways to solve the same problem.

GMP1.6 Compare the strategies you and others use.
Ben compared his strategy with London's to explain why they got the same result.

Ben circles 15.5 + 4.25.

	x + y	
15.5 + 4.5	⟨11.25 + 4.5⟩	5.75 + 4.5
15.5 + 8.75	11.25 + 8.75	⟨5.75 + 8.75⟩
⟨15.5 + 4.25⟩	⟨11.25 + 4.25⟩	5.75 + 4.25

seven

Standards for Mathematical Practice

Ben and London work independently on the remaining three pairs of numbers.

Name: Ben	Name: London
$x + y < 20$ and $x + y \geq 10$	$x + y < 20$ and $x + y \geq 10$
~~11.25~~ 5.75 ~~+ 8.75~~ + 4.5 ~~20.00~~ (10.25)	11.25 + 8.75 = ? 5.75 + 4.5 = ? 11 + 8 = 19 5 + 4 = 9 0.25 + 0.75 = 1 0.75 + 0.5 = 1.25 19 + 1 = 20 9 + 1.25 = 10.25 So, ~~11.25 + 8.75 = 20.~~ (So, 5.75 + 4.5 = 10.25)
~~5.75~~ ~~+ 4.25~~ ~~10.00~~	5.75 + 4.25 = ? 5 + 4 = 9 0.75 + 0.25 = 1 9 + 1 = 10 (So, 5.75 + 4.25 = 10)

London and Ben compare their results and see that they disagree about 5.75 + 4.25. They agree that the sum is 10, but London says to circle it and Ben says to cross it out.

London explains her reasoning, "The sum of 10 works because the original problem says that the sum has to be greater than *or equal* to 10."

Ben looks back at his work after thinking about London's explanation. He changes his mind, "Oh, I was thinking that the sum had to be greater than 10. I see that the inequality says the sum can also be equal to 10. It *does* work. Good thing we checked with each other."

GMP1.4 Check whether your answer makes sense.
Even though London and Ben found the same sum for 5.75 + 4.25, at first they disagreed about whether the combination worked for both inequalities. By checking with London and looking back at the original problem, Ben realized he made a mistake and corrected his error.

London and Ben show their final solution.

	$x + y$	
~~15.5 + 4.5~~	(11.25 + 4.5)	(5.75 + 4.5)
~~15.5 + 8.75~~	~~11.25 + 8.75~~	(5.75 + 8.75)
(15.5 + 4.25)	(11.25 + 4.25)	(5.75 + 4.25)

Mathematical Practice 1: Make sense of problems and persevere in solving them.

Standards for Mathematical Practice

Create and Make Sense of Representations

You can use mathematical **representations** to solve problems. Mathematical representations can be numbers, words, pictures, symbols, gestures, tables, graphs, or real objects.

Think about this problem: $1\frac{1}{2} * 2\frac{2}{3} = ?$

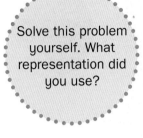

Solve this problem yourself. What representation did you use?

Erik knows that multiplication can be represented by the area of a rectangle. He measures and draws a rectangle that is $1\frac{1}{2}$ inches wide and $2\frac{2}{3}$ inches long. To make the multiplication easier, Erik partitions the width of $1\frac{1}{2}$ inches into two parts: 1 inch and $\frac{1}{2}$ inch. He does the same for the length: 2 inches and $\frac{2}{3}$ inch.

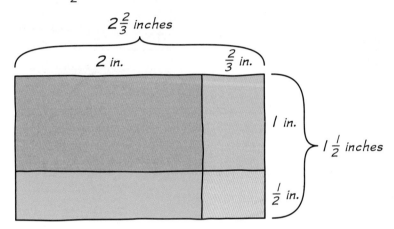

After partitioning the sides into parts, Erik writes an **expression** in each small rectangle to find its area. These expressions represent the partial products for the multiplication problem.

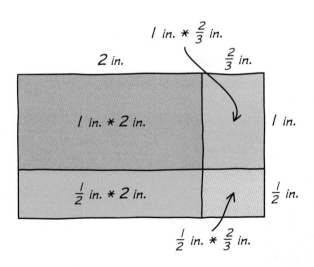

GMP2.1 Create mathematical representations using numbers, words, pictures, symbols, gestures, tables, graphs, and concrete objects.
Erik created an area model using a partial-products diagram to represent the problem and then identified the partial products he would need to calculate.

SRB
nine 9

Standards for Mathematical Practice

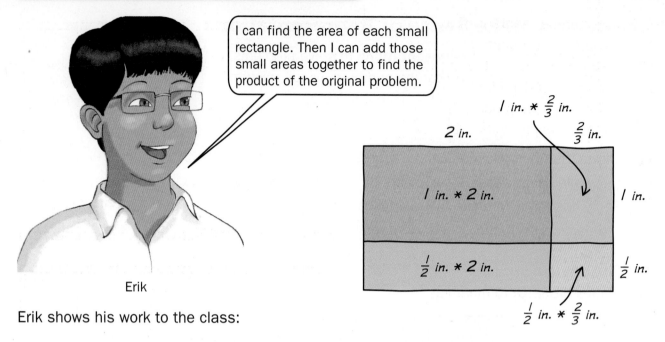

"I can find the area of each small rectangle. Then I can add those small areas together to find the product of the original problem."

Erik

Erik shows his work to the class:

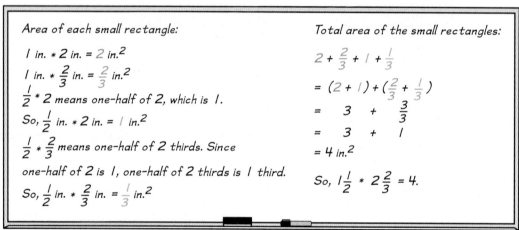

Ms. Kim points out that Erik used the Associative and Commutative Properties to regroup the addends to make finding the total area easier.

GMP2.2 Make sense of the representations you and others use.

Erik made sense of the area representation when he figured out that he needed to add the areas of each small rectangle to find the area of the whole rectangle and when he realized the area of the whole rectangle would be the same as the product of the original problem.

Standards for Mathematical Practice

Ella uses number sentences to represent the problem. She breaks the factors into easier parts. Then she uses the Distributive Property. Ella first shows how she breaks the products apart:

$$1\tfrac{1}{2} * 2\tfrac{2}{3} = ?$$

$1\tfrac{1}{2}$ is the same as $1 + \tfrac{1}{2}$.

$2\tfrac{2}{3}$ is the same as $2 + \tfrac{2}{3}$.

$1\tfrac{1}{2} * 2\tfrac{2}{3}$ is the same as $(1 + \tfrac{1}{2}) * (2 + \tfrac{2}{3})$.

Then Ella shows how she uses the Distributive Property to multiply:

$$(1 + \tfrac{1}{2}) * (2 + \tfrac{2}{3})$$
$$= [1 * (2 + \tfrac{2}{3})] + [\tfrac{1}{2} * (2 + \tfrac{2}{3})]$$
$$= (1 * 2) + (1 * \tfrac{2}{3}) + (\tfrac{1}{2} * 2) + (\tfrac{1}{2} * \tfrac{2}{3})$$

Ella

> I see that all of my partial products are the same as Erik's in the rectangles. I can find the answer to each partial product using mental math.
>
> Just like Erik, I can use the Associative and Commutative Properties to regroup the addends to find the total of the partial products.

Standards for Mathematical Practice

Ella finishes her work and gets the same final product as Erik.

$$= (1 * 2) + (1 * \tfrac{2}{3}) + (\tfrac{1}{2} * 2) + (\tfrac{1}{2} * \tfrac{2}{3})$$
$$= 2 + \tfrac{2}{3} + 1 + \tfrac{2}{6}$$
$$= (2 + 1) + (\tfrac{2}{3} + \tfrac{2}{6})$$
$$= 3 + (\tfrac{4}{6} + \tfrac{2}{6})$$
$$= 3 + \tfrac{6}{6}$$
$$= 3 + 1$$
$$= 4$$

GMP2.3 Make connections between representations.

Ella made a connection between her representation and Erik's representation when she noticed that their different representations resulted in the same partial products and similar calculations.

When you make sense of and solve a problem using representations, you are reasoning abstractly. When you think about numbers and amounts in the representations, you are reasoning quantitatively.

Mathematical Practice 2: Reason abstractly and quantitatively.

Check Your Understanding

Use what you learned about making mathematical representations to solve this problem using two different representations:

$$2\tfrac{1}{2} + 3\tfrac{1}{4} = ?$$

How are your representations similar? How are they different?

Standards for Mathematical Practice

Make Conjectures and Arguments

A **conjecture** is a statement that might be true. In mathematics, conjectures are not simply guesses. They are claims based on information or mathematical thinking. **Arguments** in mathematics use mathematical reasoning to show whether a conjecture is true or false. Mathematical arguments can use words, pictures, symbols, or other representations.

A sixth-grade class is looking for a rule they can use to find the area of any triangle.

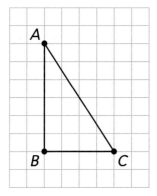 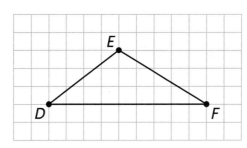 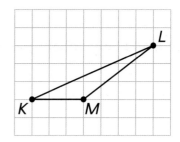

The side length of 1 square represents 1 cm.
One square represents 1 cm².

Ashley looks at triangle *ABC* first. She writes this conjecture:

To find the area of a triangle, multiply the lengths of two of the sides, then divide the product by 2.

She supports her conjecture with an argument.

Ashley

Look at triangle *ABC*. One side is 6 cm long and another side is 4 cm long. I can draw a rectangle around the triangle. The whole rectangle's area is 6 cm ∗ 4 cm = 24 cm².

Triangle *ABC* covers exactly half of the rectangle, so its area is 24 cm² ÷ 2 = 12 cm². That's the product of two of its sides divided by 2.

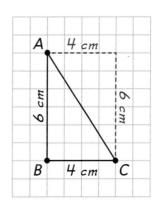

thirteen 13

Standards for Mathematical Practice

Caleb disagrees with Ashley's conjecture. He explains why and shows his work.

Caleb

I tried your rule with triangle DEF, but it doesn't work. I measured two sides, 5 cm and 9 cm, and then multiplied to get 45 cm². When I divided by 2, I got an area of $22\frac{1}{2}$ cm². But when I made a quick estimate of the number of square centimeters inside the triangle, I counted about 14 square centimeters. The answer I got using your conjecture doesn't make sense. For your conjecture to be true, it must be true for all triangles.

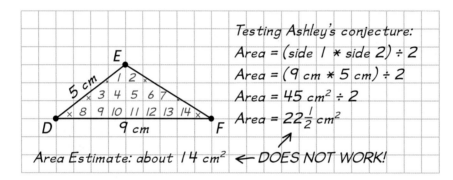

GMP3.1 Make mathematical conjectures and arguments.

Ashley made a conjecture and supported it with one example that worked. Caleb showed that Ashley's conjecture was false by giving an example that didn't work with her conjecture. Caleb reminded Ashley that for a conjecture to be true, it must always be true.

Ashley thinks about Caleb's argument and says, "You're right. Another problem is that if you use different sides, you get different areas. I have to revise my rule. Let me think. For triangle ABC, the 2 sides I multiplied together make a right angle. They were also the length and width of the rectangle I drew."

Ashley writes a new conjecture:

To find the area of a triangle, multiply the lengths of two sides of a triangle that form a right angle. Then divide the product by 2.

Caleb points out that not all triangles have a right angle, so Ashley's conjecture still doesn't work for triangle DEF.

Standards for Mathematical Practice

Ashley draws a line to divide triangle *DEF* into two right triangles. Ms. Kim shows the class the names for the parts of triangle *DEF* so they can use them in their conjecture.

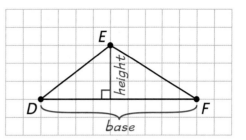

Ms. Kim

Ashley, the line you drew is called the *height* of the triangle. The height is a line segment drawn from one vertex of a triangle to the opposite side, so that the height makes a right angle with the opposite side. The opposite side is called the *base*.

Caleb shows Ashley how he uses the two triangles she made to find the area of triangle *DEF*.

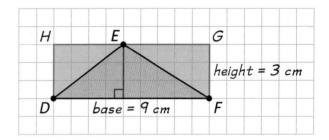

Caleb

Triangle *DEF* is divided into 2 small right triangles that I shaded red. Each small red triangle is half of the rectangle drawn around it, so the area of the whole red triangle *DEF* is half of the whole rectangle *DHGF*.

I can see that in my head. You could cut out the 2 green triangles and flip them so they cover the area of the whole red triangle exactly.

Ashley

Caleb finds the area of the whole rectangle by multiplying the length (9 cm) by the width (3 cm). Then to find the area of triangle DEF, he divides the area of the rectangle by 2.

The area of triangle DEF is:
(9 cm * 3 cm) / 2 = 27 cm² / 2 = $13\frac{1}{2}$ cm²

Ashley and Caleb see that the base of the triangle is the same as the length of the large rectangle and that the height of the triangle is the same as the width of the large rectangle. They realize they can use that information to improve Ashley's conjecture even more, so that it will work for triangles with or without right angles. They write a new conjecture:

To find the area of a triangle, multiply its height by its base, and then divide the product by 2.

GMP3.2 Make sense of others' mathematical thinking.

Caleb made sense of Ashley's drawing when he drew a rectangle around it to help find the area of the triangle. When Ashley made sense of Caleb's explanation of the revised drawing, they figured out a different conjecture that would work for all of the triangles.

Mathematical Practice 3: Construct viable arguments and critique the reasoning of others.

Check Your Understanding

Ms. Kim shows the class how to find the height of triangle KLM. Ashley and Caleb use their new conjecture to find the area of triangle KLM.

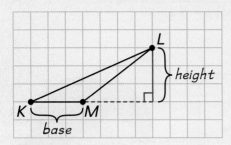

Area = (base * height) / 2

Area = (3 cm * 3 cm) / 2

Area = 9 cm² / 2 = $4\frac{1}{2}$ cm²

1. Ashley and Caleb decide to find the area of △KLM by drawing a rectangle around it. They subtract the areas of the two green triangles from the area of the rectangle to find the area of △KLM. Complete Ashley and Caleb's work to find the area of △KLM.

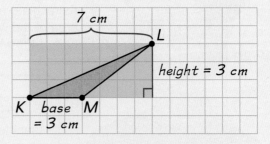

2. Does the area you found in Problem 1 match the answer for the area that Ashley and Caleb found using their conjecture? Does their conjecture work for △KLM?

Standards for Mathematical Practice

Create and Use Mathematical Models

Mathematical **models** represent situations or objects in the real world. Models can include graphs, drawings, tables, symbols, number models, diagrams, or words.

You use mathematical models when you think about a real-world situation, create a model for it, and use the model to answer a question. If the model doesn't help you find a meaningful solution to the problem, you should revise the model or create a new one to better answer the question or represent the real world.

Think about this problem:

Erik's older brother needs to save $1,200 in one year for a trip with his baseball team. He works for their neighbors. Each month, the neighbors pay Erik's brother $20 for babysitting and $8 for every hour that he does yard work.

How many hours of yard work does Erik's brother need to do each month in order to save $1,200 in one year?

How would you solve this problem?

London starts with a table to show how much Erik's brother can earn in one month.

hours of yard work	0	1	2	3	4	5
money earned	$20	$28	$36	$44	$52	$60

↙ $20 from babysitting with no yard work

London then writes ordered pairs using the numbers in the table and makes a graph.

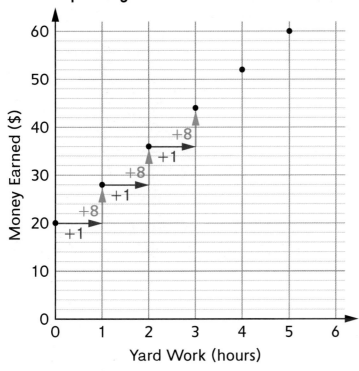

Money Earned in One Month, Depending on Hours of Yard Work Done

(hours, dollars)

(0, 20)
(1, 28)
(2, 36)
(3, 44)
(4, 52)
(5, 60)

seventeen 17

Standards for Mathematical Practice

London explains, "My graph shows that the amount of money Erik's brother earns in a month depends on the number of hours of yard work he does. But it doesn't show how many hours he needs to work to make $1,200. Maybe there is another way to think about this problem."

GMP4.1 Model real-world situations using graphs, drawings, tables, symbols, numbers, diagrams, and other representations.

London used a table and a graph to model the relationship between the number of hours of yard work Erik's brother does each month and the amount of money he earns. After looking at her representations, she realized she needed to try a different approach to solve the problem.

Camila and Amir use London's table and graph to write a number model that shows how much money Erik's brother can earn each month.

h = the number of hours of yard work
money earned in one month = 20 + (8 * h)

Erik's brother gets $20 from babysitting. That's the first column in the table and the first point on the graph. Then he gets $8 multiplied by the number of hours of yard work he does that month. On the graph, the money earned goes up $8 every time he works another hour. In my number model, the variable h represents the number of hours he does yard work.

Camila

Let's see if the number model works for one of the numbers in the table like in a "What's My Rule?" problem. I'll try $h = 5$ hours.

money earned in one month = 20 + (8 * h)
money earned = 20 + (8 * h)
= 20 + (8 * 5)
= 20 + 40 = 60
The money earned when $h = 5$ hours is $60.

Amir

Camila and Amir agree that the number model works. When Amir uses 5 in the number model for h, the money earned is $60. That matches the last column in London's table.

Standards for Mathematical Practice

Amir uses the number model to find the number of hours Erik's brother needs to work to make $1,200. He first figures out that 1,200 is the same as 12 hundreds, and there are 12 months in a year. That means Erik's brother needs to make $100 per month, because 12 ∗ 100 is 1,200.

money earned in one month = $20 + (8 * h)$
$100 = 20 + (8 * h)$
$100 - 20 = 20 - 20 + (8 * h)$
$80 = 8 * h$
Since $8 * 10 = 80$, $h = 10$.

Camila

Amir's calculations show $h = 10$, or 10 hours. In addition to babysitting, Erik's brother needs to do 10 hours of yard work each month to reach his goal of $1,200.

I don't think Erik's brother has to work exactly 10 hours every month. Some months he might have more time to work extra hours, and other months it might be too cold to work in the yard. He just needs to work an average of 10 hours per month in one year.

London

GMP4.2 Use mathematical models to solve problems and answer questions.

Amir used Camila's number model to figure out the number of hours Erik's brother needs to do yard work each month. London thought about the answer in terms of the real-world situation. She helped to clarify that the 10 hours of yard work per month might actually vary across the months.

Mathematical Practice 4: Model with mathematics.

Standards for Mathematical Practice

Choose Tools to Solve Problems

You can use a variety of **tools** to solve problems in mathematics. Some examples include rulers, diagrams, tables, graphs, calculators, spreadsheets, equations, and procedures.

Consider this problem:

Ben's parents drive him and his older sister, Julia, to school every day. They leave at the same time each morning. About half of the time, their mother drives them, and she takes the route that is the shortest distance. The other half of the time, their father drives them. He takes roads where he can drive faster, even though the distance is longer. Ben and Julia want to figure out whose route is better.

Julia: Let's tell Mom to use Dad's route. His way seems faster.

Ben: I like Mom's route better. Even though it seems to take longer, we are never late when she drives.

Ben and Julia decide to figure out whose route is better by keeping track of the time it takes to get to school for a month. They time each trip from the moment the car starts moving to the time they stop in front of the school. They record the time to the nearest minute on a calendar.

March

Sun	Mon	Tue	Wed	Thu	Fri	Sat
			1: 12	2: 10	3: 10	4
5	6: 11	7: 9	8: 12	9: 13	10: 17	11
12	13: 13	14: 12	15: 9	16: 13	17: 12	18
19	20: 11	21: 10	22: 12	23: 11	24: 11	25
26	27: 13	28: 18	29: 12	30: 13	31: 12	

KEY: ■ = Mom's Route (minutes)
■ = Dad's Route (minutes)

Standards for Mathematical Practice

Ben and Julia analyze the data on the calendar by finding the average of their mother's driving times and the average of their father's driving times. They know that the mean (average) is one way to represent a typical travel time.

How would you analyze the data to answer the question?

To find their mother's mean travel time, they find the sum of her travel times and divide by the number of days that their mother drove. They use the same method to find their father's mean travel time.

Mom's Route: Mean Travel Time
12 + 10 + 11 + 13 + 12 + 13 + 12 + 11 + 13 + 12 + 13 + 12 = 144
144 / 12 = 12
The average time to get to school using Mom's route is 12 minutes.

Dad's Route: Mean Travel Time
10 + 9 + 12 + 17 + 13 + 9 + 12 + 11 + 10 + 11 + 18 = 132
132 / 11 = 12
The average time to get to school using Dad's route is 12 minutes.

Julia: I'm surprised! The average times are the same. It *seems* different when we actually drive with them.

Ben: I wonder if analyzing the data a different way will help us see what the difference is. Let's try a dot plot.

GMP5.1 Choose appropriate tools.

Ben and Julia found the mean time (average) of each route to get a sense of the time each route typically takes. When they found that the mean was the same for both sets of data, Ben realized they needed to use a different tool to better understand how the two routes differ. Ben suggested using a dot plot to analyze the data in a different way.

Standards for Mathematical Practice

Ben and Julia make a dot plot for each set of data.

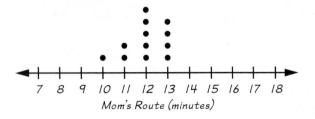
Mom's Route (minutes)

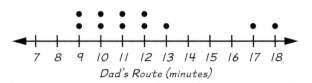
Dad's Route (minutes)

Julia

If you look at both dot plots, Dad's times are more spread out and Mom's times are more clumped in the center. Dad's route usually takes less time, but every once in awhile it is very slow. Look at the two outliers of 17 and 18 minutes. One of those times was when someone had a flat tire and traffic stopped.

Ben

When Dad drove, we were usually a little early, but sometimes we were very late. When Mom drove, we were never late. Mom's route is better if we never want to be late to school. But if we want to get there early most of the time, Dad's route is better. We just might be late once in awhile.

GMP5.2 Use tools effectively and make sense of your results.

Ben and Julia used the mean effectively when they found that the average travel time was the same for both routes. They used dot plots effectively to see how their mother's and father's data sets varied in different ways. They made sense of the difference in variability when they interpreted the dot plots and described the advantages and disadvantages of the two routes.

Mathematical Practice 5: Use appropriate tools strategically.

Standards for Mathematical Practice

Be Precise and Accurate

An important part of doing mathematics is being **accurate** and **precise** when you calculate to solve problems and explain your thinking.

Ms. Kim is buying pencils for the school. Pencils are sold in packs of different sizes at the prices shown in the table below. Ms. Kim asks her class which size pack she should buy so she spends the least amount per pencil.

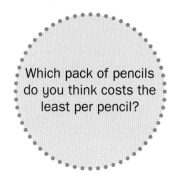

Which pack of pencils do you think costs the least per pencil?

Number of Pencils	2	5	12	48	72	144
Cost per Pack	$0.59	$0.99	$1.39	$4.99	$6.49	$13.59

Amir begins the problem using estimation and mental math.

Look at the pack of 2 pencils. Each pencil costs more than 25 cents, because the total cost is more than 2 * 25 cents, or 50 cents.

Compare that to the cost of each pencil in a pack of 5. Five pencils cost almost a dollar. So that is about 20 cents each, since 100 divided by 5 is 20. 20 cents per pencil is a much better deal.

Amir

twenty-three 23

Standards for Mathematical Practice

Erik says he knows that the pack of 48 pencils is an even better deal and writes on the board:

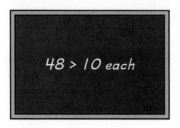

48 > 10 each

Amir asks Erik to explain what he means.

I know that 48 pencils at 10 cents each would be 48 * 10 cents = 480 cents, or $4.80. Since the actual cost of the pack is $4.99, which is only a little more than $4.80, each pencil must cost a little more than 10 cents.

Erik

GMP6.1 Explain your mathematical thinking clearly and precisely.
When Erik's first statement that he wrote on the slate was not clear to Amir, Erik clarified his thinking. He used precise language to explain the steps he took to figure out the approximate cost of each pencil in a 48-pack.

GMP6.3 Use clear labels, units, and mathematical language.
When Erik explained his thinking, he used labels and units to clarify the meaning of the numbers in his first statement. Erik talked about finding the approximate cost of a pack of 48 pencils and the approximate cost of each pencil in the pack, using dollars and cents. Erik used mathematical language to describe the steps he took to figure out that each pencil in a pack of 48 pencils costs about 10 cents.

Standards for Mathematical Practice

Amir wonders if the pencils in the 12-pack might be a better deal. He knows that 10 cents per pencil * 12 pencils = 120 cents = $1.20.

Amir

Ella

Look, $1.20 is less than $1.39, the cost of a pack of 12 pencils. I think the 12-pack is less than 10 cents per pencil.

I disagree. If each pencil costs 10 cents, 12 pencils would cost $1.20. On the price chart, a pack of 12 pencils costs $1.39, which is more than $1.20. So the pencils in a 12-pack cost *more than* 10 cents each.

GMP6.4 Think about accuracy and efficiency when you count, measure, and calculate.

When Amir made a mistake interpreting the answer from his mental math, Ella thought about accuracy as she recognized his mistake and helped him correct it.

Erik and Amir think about the cost of the last two packs: 72 pencils for $6.49 and 144 pencils for $13.59.

Erik reasons, "If those pencils cost 10 cents each, the prices would be $7.20 for the 72-pack and $14.40 for the 144-pack. So both packs cost less than 10 cents per pencil."

Amir realizes that 144 pencils is twice the number of pencils in a pack of 72. He compares the prices by doubling the cost of the 72-pack. "Twice $6.49 is $12.98, and that's less than $13.59. If we buy 2 packs of 72 pencils, it would cost $12.98. That's cheaper than buying one pack of 144 pencils at $13.59."

```
         1
      6 . 4 9
  *         2
  ─────────────
    1 2 . 9 8
```

GMP6.2 Use an appropriate level of precision for your problem.

When Amir, Erik, and Ella estimated and compared the cost of each pencil in the packs of 2, 5, 12, and 48 pencils, they didn't need to figure out the exact cost per pencil to determine which was the best deal. But when Erik and Amir compared the cost of the 72-pack with the 144-pack, Amir knew he needed a more precise answer to show that two packs of 72 pencils was cheaper than one pack of 144 pencils.

Mathematical Practice 6: Attend to precision.

Standards for Mathematical Practice

Use Structures to Solve Problems

An important part of doing mathematics is looking for **mathematical structures** such as properties, categories, and patterns. Finding relationships between numbers, expressions, shapes, and mathematical ideas can help you solve problems and think about math in new ways.

Think about this problem:

There are 100 sixth-grade students at a school and 100 lockers lining the hallway outside their classrooms. The lockers are numbered from 1 to 100.

For Math Day, the students plan to decorate the lockers with different stickers following this pattern:

- The first student will put a sticker on Locker 1 and then all multiples of 1.
- The second student will put a sticker on Locker 2 and then all multiples of 2.
- The third student will put a sticker on Locker 3 and then all multiples of 3.
- The fourth student will put a sticker on Locker 4 and then all multiples of 4.
- The fifth student will put a sticker on Locker 5 and then all multiples of 5.

If all 100 students continue to put stickers on the lockers in this pattern, which lockers will have an odd number of stickers? Explain how you know.

Caleb and Camila work together to find out what happens when the first 16 students decorate the first 16 lockers with stickers. They draw a diagram to represent the lockers and the stickers each student put on the lockers.

How would you begin to solve this problem? What mathematics might help you solve the problem?

For the first student, Caleb and Camila write "1" on each locker that is a multiple of 1. For the second student, they write "2" on each locker that is a multiple of 2. For the third student, they write "3" on each locker that is a multiple of 3, and so on through the first 16 students.

1	2	3	4	5	6	7	8	9	10	11	12	13	14	15	16
1	1	1	1	1	1	1	1	1	1	1	1	1	1	1	1
	2	3	2	5	2	7	2	3	2	11	2	13	2	3	2
			4		3		4	9	5		3		7	5	4
					6		8		10		4		14	15	8
											6				16
											12				

Camila notices that Locker 1 has 1 sticker, Locker 2 has 2 stickers, Locker 3 has 2 stickers, and Locker 4 has 3 stickers. She continues looking for a pattern that she can relate to the number of stickers on each locker.

Camila: Look! The only lockers that have an odd number of stickers are 1, 4, 9, and 16. Those are square numbers, so I think that other lockers with square numbers, like Lockers 25, 36, 49 and on up to 100, will also have an odd number of stickers.

Caleb: Hmm. Why do you think just those lockers have an odd number of stickers?

Caleb and Camila look at Locker 9 and notice that it has three stickers—one from the first student, one from the third student, and one from the ninth student.

Caleb sees that 1, 3, and 9 are the **factors** of 9, and that 9 is a **multiple** of 1, 3, and 9. But he still doesn't see why only the square-numbered lockers have an odd number of stickers.

GMP7.1 Look for mathematical structures such as categories, patterns, and properties.
Caleb and Camila drew a diagram to help them look for patterns and relationships between the locker numbers and the number of stickers. Camila noticed a mathematical structure when she saw that the lockers with an odd number of stickers were the lockers with square numbers. Caleb noticed a mathematical structure when he pointed out that the relationship between the stickers and locker numbers was the same as the relationship between factors and multiples.

Standards for Mathematical Practice

Camila thinks about Caleb's observation and has a new idea. She says, "Square numbers are special. They come from multiplication facts that have a repeated factor, like 4 * 4 = 16. Let's compare the factors of a non-square number, like 15, with a square number, like 16. We can make a factor rainbow for each one."

Factors of 15: 1, 3, 5, 15

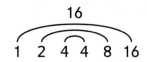

Factors of 16: 1, 2, 4, 8, 16

Camila lists the factors of each number below the rainbows. She reasons, "The factor rainbow for 15 shows all the factor pairs. 1 and 15 make a factor pair, and 3 and 5 make the other factor pair. There are four factors, so that's four stickers. The factor rainbow for 16 shows that 1 and 16, 2 and 8, and 4 are the factors. Even though 4 * 4 = 16, we don't list 4 twice when we list the factors. So, 16 has five factors."

Caleb

> I get it. In the factor rainbow for 16, every factor has a different partner except for 4. Since all factor rainbows of square numbers will have one factor without a different partner, the number of factors has to be odd.

Camila

> That's right. For Locker 16, the fourth student puts just one sticker on the locker, not two. And in factor rainbows for all non-square numbers, each factor has a different partner, so there is an even number of factors. Lockers with non-square numbers always have an even number of stickers.

GMP7.2 Use structures to solve problems and answer questions.
Caleb and Camila used the mathematical structure of factor pairs along with their understanding of even and odd numbers to solve the problem.

Mathematical Practice 7: Look for and make use of structure.

Check Your Understanding
Which lockers will have exactly 2 stickers on them? How do you know?

Standards for Mathematical Practice

Create and Explain Rules and Shortcuts

It is often useful to look for shortcuts, rules, and generalizations to make procedures and operations more efficient. When you see problems that have similar mathematical ideas or structures, you can create a shortcut, rule, or generalization to solve the problems.

How would you answer this question?

Think about this problem: If you continue the pattern of shapes shown below, how many squares will be in Figure 40?

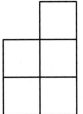 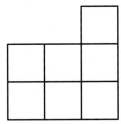 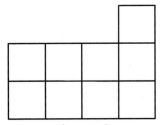 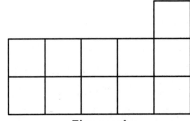

Figure 1   Figure 2   Figure 3   Figure 4

Ashley

I started a table. I noticed that the number of squares increases by 2 every time you go from one figure number to the next. But when I try to add 2 repeatedly to get to Figure 40, the table is too long. I keep losing track of how many 2s I added. We need to figure out a different way.

Figure Number	Number of Squares
1	5
2	7
3	9
4	11

Standards for Mathematical Practice

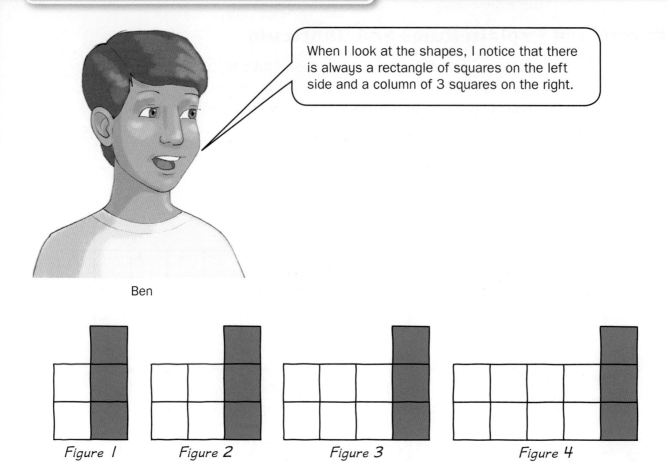

Ben

"When I look at the shapes, I notice that there is always a rectangle of squares on the left side and a column of 3 squares on the right."

Figure 1   Figure 2   Figure 3   Figure 4

Ben and Ashley realize they can sketch the shape of any figure number. They make a list of rules for drawing any figure:

- Every figure has a rectangle on the left side that is made up of 2 rows of squares.
- The number of squares in each row of that rectangle is the same as the figure number.
- Every figure has a column of 3 squares on the right.

Ashley and Ben think about their rules and sketch Figure 40. They write numerical expressions to find the number of squares in the figure.

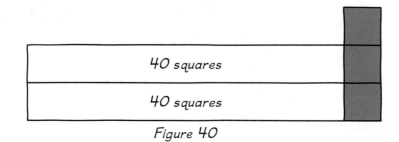

Figure 40

$$40 \text{ squares} + 40 \text{ squares} + 3 \text{ squares} = 83, \text{ or}$$
$$2 * (40 \text{ squares}) + 3 \text{ squares} = 83$$

Figure 40 has 83 squares.

Standards for Mathematical Practice

Ben says, "I bet we can easily figure out the number of squares in any figure." He thinks about how to write a rule for finding the number of squares in any figure using numerical expressions and a variable. He thinks about it as finding the number of squares in Figure n.

For Figure 40: $40 + 40 + 3 = 83$ $\qquad$ $2 * 40 + 3 = 83$

For Figure n: $\; n + n + 3 =$ total number $\quad 2 * n + 3 =$ total number
$\qquad\qquad\qquad$ of squares $\qquad\qquad\qquad\quad$ of squares

That makes sense. Now I can see how to use my table to find the number of squares for any figure.

Ashley

Figure Number	Number of Squares
1	5
2	7
3	9
4	11
...	...
40	$2 * 40 + 3 = 83$
...	...
n	$2 * n + 3$

GMP8.1 Create and justify rules, shortcuts, and generalizations.

Ashley and Ben found a more efficient way to solve the problem when they created rules for drawing any figure number and wrote a numerical expression for Figure 40 based on their diagram. Ben generalized their numerical expression by using a variable, and Ashley applied the generalization to her table.

Mathematical Practice 8: Look for and express regularity in repeated reasoning.

Standards for Mathematical Practice

A Problem-Solving Diagram

When you solve a problem, you work to make sense of the problem. You reflect on your thinking and keep trying when the problem is hard. As a good problem solver, you always check to see whether your answer makes sense. You can check by trying to solve the problem in more than one way and by comparing the strategies you use to what others use.

The diagram below can help you to think about problem solving and to persevere, or keep trying, when the problem is hard. The boxes in the diagram show the types of things you do when you apply mathematical practices to solving problems. The arrows show that you don't always do things in the same order.

Organize the information.
- Study the information in the problem.
- Arrange the information into a list, table, graph, or diagram.
- Look for more information if you need it.
- Get rid of information you don't need.

Understand the problem.
- Retell the problem in your own words.
- Figure out what you want to find.
- Figure out what you know.
- Imagine what the answer will look like.
- Make a guess at the answer.

Play with the information.
- Draw a picture, diagram, or another mathematical representation.
- Write a number model.
- Model the problem using objects such as counters or base-10 blocks.

Check your answer as you work.
- Does your answer make sense?
- Compare your answer with a classmate's.
- Does your answer fit the problem?
- Can you solve the problem another way?

Figure out what math can help.
- Can you use addition? Subtraction? Another operation?
- Can you use geometry? Patterns? Other mathematics?
- Try the math. See what happens.
- What units are you using? Label your numbers with units.

Standards for Mathematical Practice

Problem Solving and the Mathematical Practices

Ms. Kim asks her class to think about how they can use mathematical practices to solve this problem:

On the first day of school, a teacher greets all 31 of her students at the classroom door. She asks the students to introduce themselves to each other by shaking every other student's hand. How many handshakes will take place among the students?

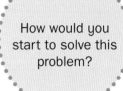

How would you start to solve this problem?

After students have started working, Ms. Kim asks volunteers to report their strategy.

Amir, Camila, and Ben explain and show their group's strategy.

We started by thinking about the 3 people in our group. Since 2 people shake hands at a time and there are 3 of us, we multiplied 2 times 3. That would be 6 handshakes. But Camila doesn't think that is the right answer.

I think that 6 handshakes for 3 people is too many. So I suggested we actually shake hands. Amir and I shook hands, Ben and I shook hands, and then Amir and Ben shook hands. That's only 3 handshakes for the 3 of us.

Amir

Camila

Ben draws a diagram to model their real handshakes. He uses a line to represent each handshake.

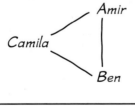

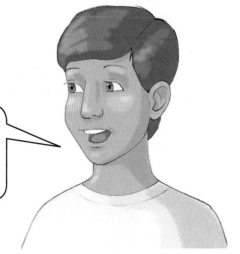

Since we found 3 handshakes for the 3 people in our group, I think there would be 31 handshakes for 31 people.

Ben

thirty-three 33

Standards for Mathematical Practice

Ashley disagrees with Ben and shares her model of the handshakes for a group of 5 students.

1 → 2 1 → 3 1 → 4 1 → 5	
2 → 3 2 → 4 2 → 5	5 kids make 10 handshakes.
3 → 4 3 → 5	
4 → 5	

Ashley: There are 5 students. I numbered the students 1 to 5. You can see that Students 1 and 2 shake hands. I wrote that as 1 *arrow* 2. I listed all the handshakes and counted 10 handshakes among the 5 kids.

Amir: Why are some handshakes missing? It looks like Student 1 shook 4 students' hands, but Student 2 only shook 3 students' hands.

Ashley: Students 1 and 2 already shook hands in the first row, so Student 2 doesn't have to shake Student 1's hand again in the second row. If you count each student's handshakes, they each shake hands 4 times.

Standards for Mathematical Practice

Amir agrees with Ashley's reasoning. Ms. Kim asks students to continue working to find the number of handshakes among 31 students. London and Ashley work together using Ashley's diagram and then share their results in a table.

Number of Students	Number of Handshakes
2 +1	1 +2
3 +1	3 +3
4 +1	6 +4
5 +1	10 +5
6 +1	15 +6
7	21

London

We used Ashley's diagram to model the situation with different numbers of students, then made a table of the results and looked for a pattern. We noticed that the number of students increases by 1 going down the first column. In the second column, the number of handshakes we add to get to the next row is 2, 3, 4, 5, and then 6. The number we add keeps getting larger by one.

That makes sense, because if you add an eighth student to the group, that student would have to shake 7 hands in order to shake everyone's hand. We would add those 7 handshakes to the column on the right, in addition to all of the previous handshakes. So for 8 students there are 28 handshakes.

Erik

Ashley uses a calculator and keeps adding the next larger number to the number of handshakes until she gets to 31 students. She finds a total of 465 handshakes.

Standards for Mathematical Practice

Ms. Kim asks whether other students agree or disagree with the answer of 465 handshakes and to share their reasoning.

Ella

Ben

> I got the same answer. But instead of adding the next bigger number to the number of handshakes, I *started* at 30 and kept adding the next smaller number. I used a calculator to add 30 + 29 + 28 + 27, and so on, until I got to + 1.

> I don't understand why you started with 30, and then kept adding the next smaller number.

> I looked back at Ashley's first diagram. With 5 students, I saw that there were 4 + 3 + 2 + 1 handshakes. I continued the diagram for 6 students and saw that the total was 5 + 4 + 3 + 2 + 1. Then I used the same pattern for 31 students. I added all of the numbers from 30 down to 1. That's 465 handshakes in all.

Ben thinks about Ella's reasoning and summarizes their thinking. "That makes sense because the first student shakes everyone else's hand. That would be 30 handshakes. Since the first and second students already shook hands, the second student would have 29 new handshakes. Each student after that would have one fewer new handshake than the student before."

Check Your Understanding

Use the mathematical practices described on page 3 to answer the questions.

1. What mathematical practices did Amir, Camila, and Ben use? Explain.
2. What mathematical practices did London and Ashley use?
3. What mathematical practices did Ella use?

Ratios and Proportional Relationships

Ratios and Proportional Relationships

Ratios

A **ratio** is a relationship between two quantities that involves multiplication or division.

> **Examples**
> - The ratio of pet cats to pet dogs in a city is 2:3, which means there are 2 pet cats for every 3 pet dogs.
> - If a person's heart beats 11 times every 10 seconds, that is a ratio of 11 heartbeats every 10 seconds (11:10). The person's heart rate is 66 beats per minute (60 seconds).

Ratios give information about the *relationship* between two quantities. Ratios alone do not provide information about the actual number in a part or a whole.

- Think about the ratio of cats to dogs in a city (2:3) in the example above. Neither 2 nor 3 tells you about the actual number of cats or dogs in the city. The actual numbers may be in the hundreds of thousands.

- Think about the ratio of heartbeats to seconds (11:10). Neither 11 nor 10 tells you about the actual number of heartbeats or seconds. The actual numbers depend on the length of time being considered.

Types of Ratios

Part-to-whole ratios compare a part of a whole to the whole. **Part-to-part ratios** compare a part of a whole to another part of the same whole.

> **Example**
>
> 1 out of every 2 students in Caleb's school rides the bus to get to school. The ratio of students who ride the bus to students in the school is 1 out of 2.
>
> This ratio is part-to-whole, because the students who ride the bus are compared to all students in the school. Note that the numbers in this ratio do not provide information about the actual number of students in the school or the actual number of students who ride the bus.

> **Example**
>
> A softball team's record is shown below. The ratio of games won to games lost is 3 to 2.
>
> | Game | 1 | 2 | 3 | 4 | 5 | 6 |
> |--------|-----|------|------|-----|-----|------|
> | Status | Won | Tied | Lost | Won | Won | Lost |
>
> This ratio is part-to-part, because the games won are part of the set of games played and the games lost are part of the same set of games played. Note that the numbers in this ratio do not include all of the games played, since another part, games tied, is not reflected in the ratio.

Ratios and Proportional Relationships

Part-to-whole and part-to-part ratios always compare quantities that have the same unit. For example, the ratio in the example about Caleb's school compares students to students, and the ratio in the example about the softball team compares games to games.

Rates are ratios that compare two quantities with unlike units.

> **Example**
>
>
> 3 apples for $1.89
>
> This ratio is a rate because different units are being compared: a number of apples is being compared to an amount of money.
>
> This is the same rate as 1 apple for $0.63 (since 1.89 / 3 = 0.63).

Note Part-to-whole ratios are like **part-to-whole fractions**. In Caleb's school, the ratio of students who ride the bus to students in the school can be thought of as $\frac{1}{2}$ or 1 ÷ 2. For each set of two students, one rides the bus. So 2 represents the whole and 1 represents the part. One-half, or $\frac{1}{2}$, of the students ride the bus.

Representations of Ratios

Quantities compared in ratios are related through multiplication and division, so you can use words such as *per, out of,* and *for each* to describe ratios. You can also use a colon (:), a slash (/), or fraction notation ($\frac{a}{b}$) to represent ratios.

> **Example**
>
> 2 out of the 12 eggs in the carton are cracked or damaged.
>
> Other ways to describe this part-to-whole ratio include:
>
> 2 eggs out of 12 eggs are damaged.
>
> 2 in 12 eggs are damaged.
>
> Some ways to write the ratio include:
>
> 2 : 12
>
> 2 to 12
>
> $\frac{2}{12}$
>
> You can describe the ratio of damaged eggs to all eggs in the carton as:
>
> 1 out of every 6 eggs is damaged.
>
> Some ways to write this ratio include:
>
> 1 : 6
>
> 1 to 6
>
> $\frac{1}{6}$

Note Since the example at the left is a part-to-whole ratio, you can think of the ratio as a fraction of the whole that is damaged: $\frac{2}{12}$ can be read two-twelfths of the eggs, and $\frac{1}{6}$ can be read one-sixth of the eggs.

thirty-nine 39

Ratios and Proportional Relationships

Example

Ashley's training for the track team requires her to run 3 minutes for every 2 minutes she walks.

This is a part-to-part ratio that compares minutes to minutes.

Some ways to write the ratio include:

3 to 2, or 3:2, or $\frac{3}{2}$.

To read $\frac{3}{2}$ as a ratio, you can say "three to two" or "three for every two."

Note that this ratio does not include information about the actual numbers of minutes run and walked.

Note Part-to-part ratios are sometimes written using fraction notation, but a part-to-part ratio cannot be read like a fraction because the bottom number does not describe the whole. For example, the ratio of minutes run to minutes walked cannot be read as "three-halves." It must be read using the language of ratios.

Example

This sign means that the speed limit is 55 mph, or 55 miles per hour. That means a vehicle should travel no faster than 55 miles in 1 hour. Because this ratio uses two different units, miles and hours, the relationship is called a rate.

Some ways to write this rate include:

55 miles to 1 hour

55 miles : 1 hour

$\frac{55 \text{ miles}}{1 \text{ hour}}$

You can express rates using the word *per*, as in 55 miles per hour. Note that when *per* is followed by a unit with no quantity, you can assume the number of units in the second quantity is 1.

Sometimes a slash (/) is used instead of the word *per*. For example, 55 miles per hour can be written as 55 miles/hour.

Note Part-to-whole and part-to-part ratios can be expressed as numerical relationships without stating the units. But rates should be expressed using units along with the quantities.

Did You Know?

Some rates have special abbreviations. For example, *miles per hour* can be abbreviated by writing "mph." A car may be able to travel 35 miles on one gallon of gas, or 35 mpg.

Ratios and Proportional Relationships

Equivalent Ratios

When two different ratios make the same multiplicative comparison, they are called **equivalent ratios.** An equation that states that two ratios are equivalent is sometimes called a *proportion*. To decide whether two or more ratios are equivalent, you can write them using fraction notation and try to rename them as equivalent fractions.

> **Examples**
>
> 92 of the 219 students at Lincoln School live within walking distance of their school. 104 of the 197 students at Central School live within walking distance of their school. Is the ratio of students living within walking distance to the whole student population the same for both schools?
>
> | | |
> |---|---|
> | Write the ratios using fraction notation: | Write the Lincoln School ratio, 92 : 219, as a fraction: $\frac{92}{219}$.
 Write the Central School ratio, 104 : 197, as a fraction: $\frac{104}{197}$.
 Is $\frac{92}{219}$ equivalent to $\frac{104}{197}$? |
> | Decide whether the fractions are equivalent: | Use logical reasoning and what you know about fraction comparisons.

 $\frac{92}{219}$ is less than half of the students at Lincoln School.

 $\frac{104}{197}$ is more than half of the students at Central School. |
> | Answer the question: | Since one fraction is more than $\frac{1}{2}$ and the other fraction is less than $\frac{1}{2}$, the fractions are not equivalent. So the two schools have different ratios of students living within walking distance. |
>
> A printer prints 4 pages per minute. Is that equivalent to a rate of 20 pages in 5 minutes?
>
> | | |
> |---|---|
> | Write the ratios using fraction notation: | Write 4 pages per minute using fraction notation: $\frac{4 \text{ pages}}{1 \text{ minute}}$.
 Write 20 pages per 5 minutes as a fraction: $\frac{20 \text{ pages}}{5 \text{ minutes}}$.
 Is $\frac{4 \text{ pages}}{1 \text{ minute}}$ equivalent to $\frac{20 \text{ pages}}{5 \text{ minutes}}$? |
> | Decide whether the fractions are equivalent: | Use what you know about equivalent fractions.
 Think about $\frac{4}{1}$. You can multiply both the numerator and denominator by 5 to get $\frac{20}{5}$.

 $\frac{4}{1} = \frac{4 * 5}{1 * 5} = \frac{20}{5}$ |
> | Answer the question: | Since the two fractions are equivalent, the ratios are equivalent. Printing at a rate of 4 pages per minute is the same as printing at a rate of 20 pages in 5 minutes. |

forty-one   SRB   41

Ratios and Proportional Relationships

Comparing Ratios

Sometimes you need to compare two ratios to solve a problem.

For some problems you can use mathematical reasoning to compare ratios.

Example

Ella is reading a book that has 12 chapters and 325 pages. Amir's book is 127 pages long and has 18 chapters. On average, which book has more pages in each chapter?

You can summarize the problem using rates.

Ella's book: 325 pages per 12 chapters   Amir's book: 127 pages per 18 chapters

You can use mathematical reasoning.

Ella's book has fewer chapters but more pages than Amir's book, so on average Ella's book must have longer chapters. Another way to think about the comparison is to consider that Amir's book has more chapters but fewer pages, so on average Amir's book must have shorter chapters.

For some problems you can use equivalent fractions to compare ratios.

Example

At which grocery store are avocados less expensive?

You can write the ratios using fraction notation:

Grocery Store A: $2 for 3 avocados can be written as $\frac{\$2}{3 \text{ avocados}}$.

Grocery Store B: $3 for 8 avocados can be written as $\frac{\$3}{8 \text{ avocados}}$.

Use what you know about equivalent fractions to compare $\frac{2}{3}$ and $\frac{3}{8}$.

Fractions equivalent to $\frac{2}{3}$:

$\frac{4}{6}, \frac{6}{9}, \frac{8}{12}, \frac{10}{15}, \frac{12}{18}, \frac{14}{21}, \boxed{\frac{16}{24}}$

Fractions equivalent to $\frac{3}{8}$:

$\frac{6}{16}, \boxed{\frac{9}{24}}$

$\frac{16}{24}$ is the same as $16 for 24 avocados.

$\frac{9}{24}$ is the same as $9 for 24 avocados.

Since $\frac{9}{24} < \frac{16}{24}$, Grocery Store B has less expensive avocados.

$2 for 3 Avocados
Grocery Store A

$3 for 8 Avocados
Grocery Store B

Ratios and Proportional Relationships

Using Ratio/Rate Tables to Solve Problems

You can display equivalent ratios, including rates, in a ratio/rate table.

Examples

A ratio/rate table for the exchange rate between euros and U.S. dollars can look like the one below.

Euros	U.S. Dollars
1.29	1
2.58	2
3.87	3
5.16	4

Amir trains for his track team almost every day by running and walking. He makes sure that for every mile he walks, he runs 3 miles. He has kept a record of how many miles he has run and walked each week in a table.

	Week 1	Week 2	Week 3	Week 4
Miles run	6	15	9	18
Miles walked	2	5	3	6

These two ratio/rate tables are alike because each shows equivalent ratios.
- In the table comparing euros to U.S. dollars, equivalent ratios are found in each row.
- In the table comparing miles run to miles walked, equivalent ratios are found in each column.

These two ratio/rate tables are different because:
- The table for equivalent euros and U.S. dollars is formatted vertically, while Amir's table for miles run and miles walked is horizontal.
- The table for euros and U.S. dollars includes a **unit ratio,** in which one of the quantities being compared is 1. In the exchange rate table, the first ratio, 1.29 euros to 1 U.S. dollar, is a unit ratio. Amir's table does not include a unit ratio because none of the quantities being compared is 1.
- The table for euros and U.S. dollars is arranged in order of increasing euros and U.S. dollars, but Amir's table is not in numerical order. Instead, it shows the ratios of miles run to miles walked from week to week.

forty-three 43

Ratios and Proportional Relationships

You can use ratio/rate tables to solve problems.

One way to solve problems is to use the tables to find combinations of equivalent ratios.

Example

The table below compares the number of miles that Amir has run to the number of miles that Amir has walked for the past four weeks. He plans to run 21 miles this coming week using the same ratio of 3 miles run for every 1 mile walked.

	Week 1	Week 2	Week 3	Week 4
Miles run	6	15	9	18
Miles walked	2	5	3	6

To figure out how many miles he will walk when he runs 21 miles, look for a combination of miles run equivalent to 21 miles. Notice that 6 miles run and 15 miles run are both in the table, and 6 + 15 = 21 miles run.

- 6 miles of running is associated with 2 miles of walking.
- 15 miles of running is associated with 5 miles of walking.

Using the first two columns, a total of 6 miles run + 15 miles run = 21 miles run is associated with a total of 2 miles walked + 5 miles walked = 7 miles walked.

So when Amir runs 21 miles, he walks 7 miles.

Another way to solve problems is to find equivalent ratios by multiplying or dividing both numbers in a ratio by the same number.

Example

Use the exchange rate in the table to find how many euros are equivalent to 20 U.S. dollars.

Complete the entry for 20 U.S. dollars in the ratio/rate table.

Since you can multiply 1 U.S. dollar by 20 to get 20 U.S. dollars, you can also multiply 1.29 euros by 20 to find the equivalent number of euros.

1.29 euros * 20 = 25.80 euros

So 25.80 euros are equivalent to 20 dollars.

Euros	U.S. Dollars
1.29	1
2.58	2
3.87	3
5.16	4
25.80	20

* 20 (applied to both columns)

Ratios and Proportional Relationships

Using Arrays and Tape Diagrams to Solve Ratio Problems

You can use arrays and tape diagrams to represent many ratio problems. An **array** uses equal rows and columns to represent the quantities in a ratio. A **tape diagram** uses a series of boxes to represent the quantities in a ratio.

Examples

Grandma Sue receives a 24-piece package of her favorite almond clusters. She gives away 1 almond cluster for every 7 almond clusters she keeps for herself. Represent three different ratios using this information.

Ratio 1
Represent the ratio of the number of almond clusters she keeps to the number she gives away.

Draw the 7 clusters she keeps and the 1 she gives away in one row. Then mark the cluster given away with an X. This is a 1-by-8 array.

Use a tape diagram. Draw a row of 7 boxes labeled "keeps" and a row of 1 box labeled "gives."

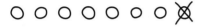

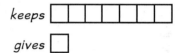

Both the array and tape diagram above show that the ratio of the number of almond clusters she keeps to the number she gives away is 7 to 1, or 7:1.

Ratio 2
Represent the ratio of the number of clusters given away to the total number of clusters in a package. You can revise the array and tape diagram to represent this ratio.

For the array, draw identical rows until you have the total of 24 almond clusters in a package.

For the tape diagram, think: The total number of almond clusters is 24. You need to evenly distribute the clusters among the parts she keeps (7) and the part she gives away (1). 24 clusters divided by 8 parts is 3 clusters. Write a 3 in each box.

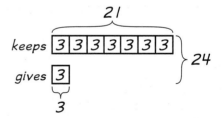

Both representations show that she gave away 3 of the 24 almond clusters. So the ratio of almond clusters given away to the total number of almond clusters is 3:24.

Ratio 3
Represent the ratio of clusters kept to the total number of clusters in a package.
You can use the array and the tape diagram above. Both show that she kept 21 of the 24 almond clusters. So the ratio of almond clusters kept to the total number of almond clusters is 21:24.

These representations provide another way to represent Ratio 1. The ratio of almond clusters kept to almond clusters given away is 21:3.

Ratios and Proportional Relationships

You can solve many ratio problems using arrays and tape diagrams.

Example

Ella's mother realized that her best-selling quilts are those made up of large colored triangles. The ratio of the number of blue triangle pieces to the number of orange triangle pieces is 5 to 3. Ella has prepared 30 blue pieces for the new quilt her mother is making. How many orange pieces does Ella need to prepare?

Draw an array.
Show 5 blue triangles and 3 orange triangles in one row to represent the ratio of the number of blue pieces to the number of orange pieces.

Draw a tape diagram.
Draw and label a row of 5 boxes to represent the blue pieces and a row of 3 boxes to represent the orange pieces.

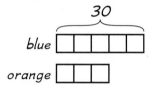

Think: Ella has prepared 30 blue pieces. Since each row in the array has 5 blue pieces, how many rows are needed to have 30 blue pieces?

5 * 6 = 30, so draw an array with 6 identical rows.

Think: The number of boxes representing blue pieces is 5, and those 5 boxes need to represent 30 blue pieces in all. So each box represents 30 / 5, or 6 pieces. Since each box in the tape diagram represents the same number of pieces in all parts of the diagram, write 6 in each of the boxes.

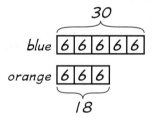

Both the array and tape diagram show that the number of blue pieces is 5 * 6 = 30 and the number of orange pieces is 3 * 6 = 18.

Ella needs to prepare 18 orange triangle pieces for her mother's quilt.

Ratios and Proportional Relationships

Arrays represent every object in a problem, so drawing arrays to solve problems can be inefficient when the quantities in a problem are large. Tape diagrams, on the other hand, represent objects grouped together in boxes. Since these boxes are labeled with the number of objects in each group, they are a more efficient tool for solving problems with large quantities.

Example

For every 3 steps London takes, her dog Sherbie takes 5. If London takes 75 steps, how many steps does Sherbie take?

Draw an array.

In the first row, write 3 Ls for London's 3 steps and 5 Ss for Sherbie's 5 steps:

L L L S S S S S

An array is inefficient for this problem because you would need to draw enough identical rows to get 75 Ls, and it would be difficult to keep track of the counts.

Draw a tape diagram instead.

In the first row, draw 3 boxes to represent London's 3 steps.

In the second row, draw 5 boxes to represent Sherbie's 5 steps.

Since London takes 75 steps in all, each of the three boxes in the first row represents 25 steps (75 / 3 = 25). Write 25 in each box in both rows.

London | 25 | 25 | 25 |
Sherbie | 25 | 25 | 25 | 25 | 25 |

To figure out the number of steps Sherbie takes, look at the numbers in the second row. Sherbie takes 5 * 25 steps, or 125 steps.

Sherbie takes 125 steps for London's 75 steps.

Ratios and Proportional Relationships

You can use tape diagrams to identify **equivalent ratios.** Two ratios are equivalent if the tape diagram for one ratio can be used to represent the other ratio.

Example

Puppy Palace charges $4 for 1 hour of doggy daycare. Canine Care says its rates are comparable at $32 for a full day (8 hours) of doggy daycare. Use tape diagrams to show whether the two ratios comparing dollars to hours are the same.

Try to draw a tape diagram that will work for both ratios, even though each ratio may have different numbers in the boxes.

The ratio of dollars compared to hours for Puppy Palace is $4 : 1 hour. Show the 4 dollars in the top row of boxes and the 1 hour in the bottom row of boxes.

The tape diagram looks like this:

dollars | 1 | 1 | 1 | 1 |
hours | 1 |

The ratio of dollars to hours for Canine Care is $32 : 8 hours.

Try using the same tape diagram as above, but change the amount in each box.

For Canine Care, the 4 boxes at the top represent $32, so each box represents $32 / 4, or $8. Write 8 in each box in the top and bottom rows. This tape diagram works because the number of hours in the second row represents the number of hours at Canine Care.

dollars | 8 | 8 | 8 | 8 |
hours | 8 |

Even though the numbers in the boxes are different, the tape diagrams for both ratios can be drawn with the same number of boxes in the top and bottom rows. This means the ratios are equivalent.

So the doggy daycare rates for Puppy Palace and Canine Care are the same.

Ratios and Proportional Relationships

Unit Ratios

A **unit ratio** is a ratio in which one of the quantities being compared is 1.
A **unit rate** is a rate in which one of the quantities being compared is 1.

You can use ratio/rate tables to find unit ratios or unit rates.

> **Example**
>
> Ava walked 9 miles in 3 hours. Find a unit rate.
>
> **One way:** Make a ratio/rate table. Then notice that you can divide both numbers in the ratio by 3 so that one of the quantities is 1.
>
>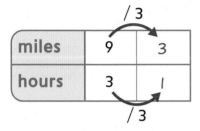
>
> So a unit rate is 3 miles in 1 hour, or 3 mph.
>
> **Another way:** Make a ratio/rate table. Then notice you can divide both numbers in the ratio by 9 so that one of the quantities is 1.
>
>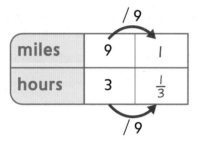
>
> This unit rate means 1 mile in $\frac{1}{3}$ hour, or 1 mile in 20 minutes (because 60 minutes is 1 hour, and 60 minutes / 3 = 20 minutes).
>
> So a unit rate is 1 mile in $\frac{1}{3}$ hour, or 1 mile : 20 minutes.

forty-nine 49

Ratios and Proportional Relationships

You can change any ratio to a unit ratio by using what you know about equivalent fractions. Write the ratio in fraction notation and decide what to multiply or divide both numbers in the ratio by to get a numerator or denominator of 1. Then interpret the result as a unit ratio.

Example

The ratio of pet dogs to pet cats in Erik's class is 5:2. What is an equivalent unit ratio?

Write the ratio using fraction notation:

5:2 can be written as $\frac{5}{2}$.

Find an equivalent fraction in which the numerator or denominator is 1.

$\frac{5}{2} = \frac{?}{1}$

If you divide the numerator and denominator by 2, the equivalent fraction will have 1 in the denominator.

$\frac{5 \div 2}{2 \div 2} = \frac{2.5}{1}$

Identify the unit ratio and what it means. Answer the question.

So an equivalent unit ratio is 2.5:1. This means there are 2.5 pet dogs for every 1 pet cat in Erik's class.

Note You can use fraction notation to help you solve part-to-part ratio problems, but do not read part-to-part ratios as fractions. The second quantity does *not* show the amount in the whole. In this example, 5:2 can be written as $\frac{5}{2}$, but should be read as "5 to 2" or "5 for each 2."

You can use **unit ratios** to solve most problems involving ratios.

Example

Ella walked 5 miles in 2 hours on Monday and 3 miles in 1 hour on Tuesday. On which day did she walk faster?

To compare the situations, represent each rate as a unit rate. Be sure that the quantity 1 is in the same position in each unit rate.

5 miles in 2 hours:

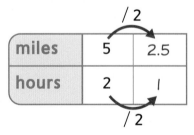

3 miles in 1 hour:

Tuesday's rate of 3 miles in 1 hour is already a unit rate, since one of the quantities is 1.

So the unit rate is 2.5 miles:1 hour.

The unit rate for Monday was 2.5 mph.   The unit rate for Tuesday was 3 mph.

Since Ella walked 2.5 mph on Monday and 3 mph on Tuesday, she walked faster on Tuesday.

Ratios and Proportional Relationships

Solve Problems Using Ratio/Rate Graphs

When equivalent ratios or rates are written as ordered pairs and displayed on a graph, they always form a straight line. To make a ratio/rate graph:

- Make a table of 3 or 4 equivalent ratios.
- Write the equivalent ratios as ordered pairs.
- Plot the ordered pairs on a labeled graph. Connect the points.
- Solve problems by tracing between a point on the line to or from the corresponding point or points on one or both of the axes.

You can use a graph of equivalent ratios to solve problems.

Note While you can use exactly two equivalent ratios to draw a useful ratio/rate graph, using 3 or 4 values allows you to check to make sure you didn't make any mistakes. All of the points should fall on one line.

Example

Ben has a photo of his first pet dog that is 2 inches wide and 3 inches tall. He wants to order an enlarged photo that will fit exactly into a 12-inch wide space above his desk. He knows that the enlargement must have the same ratio of width to height as the original for the enlargement to show everything that is in the original picture. What size enlargement should he order?

Ben creates a graph to solve the problem. Once he has a graph, he can use it to figure out other sizes of the photo to order for items such as book covers or calendars.

Make a table of 3 or 4 equivalent ratios.

To make a ratio/rate table, Ben begins with the ratio he knows and then doubles each quantity repeatedly to get additional equivalent ratios.

width (in.)	2	4	8	16
height (in.)	3	6	12	24

Write the equivalent ratios as ordered pairs.

Ben writes (2, 3), (4, 6), (8, 12), and (16, 24).

fifty-one 51

Ratios and Proportional Relationships

Example (continued)

Plot the ordered pairs on a labeled graph.

Ben draws a graph with the horizontal axis labeled Width (inches) and the vertical axis labeled Height (inches). To graph the ordered pair (16, 24), he knows that the axis labeled Width should show inches ranging from 0 to at least 16 and the axis labeled Height should show inches ranging from 0 to at least 24.

Ben plots the points using the ordered pairs from the ratio/rate table. He sees the points form a straight line and connects them.

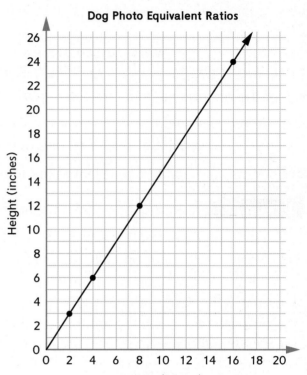

Solve the problem by tracing between a point on the line to or from the corresponding point or points on one or both of the axes.

Ben wants an enlargement that is 12 inches wide. So he starts at the width, the 12-inch mark on the horizontal axis. He traces a dashed line up to the correct point on the equivalent ratios line. The red dashed line shows the path he follows. Note that this trace must be perpendicular to the horizontal axis.

At the equivalent ratios line, Ben traces a dashed line over to the vertical axis to find the measurement needed for the height of the picture. The green dashed line shows the path he follows. Note that this trace must be perpendicular to the vertical axis.

Ben needs to order a picture that is 12 inches wide by 18 inches tall.

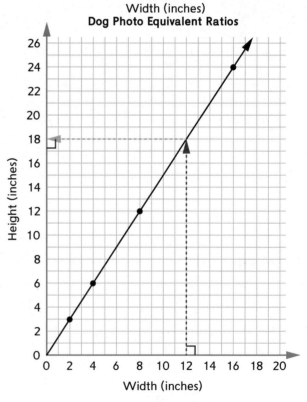

Ratios and Proportional Relationships

Estimating Solutions Using Ratio/Rate Graphs

Sometimes you can use a ratio/rate graph to estimate solutions for problems.

Examples

Use the Dog Photo Equivalent Ratios graph from pages 51–52 to solve this problem.

Ben's math notebook is $8\frac{1}{2}$ inches wide by 11 inches long. If Ben orders a picture that is 11 inches long, will the picture be the correct width to cover his notebook exactly?

Tracing over from 11 inches on the vertical axis to the equivalent ratios line, and then tracing down to the horizontal axis, shows that the width of the picture will be less than $8\frac{1}{2}$ inches. Ben can see that the enlarged picture will not be wide enough to exactly cover the notebook.

If Ben orders a picture that is 11 inches long, and the ratio of the lengths of the sides is the same as the original picture, about how wide will the picture be?

Tracing down to the horizontal axis shows that the width of the picture will be more than 7 inches, but less than 8 inches. The width will be a little less than $7\frac{1}{2}$ inches.

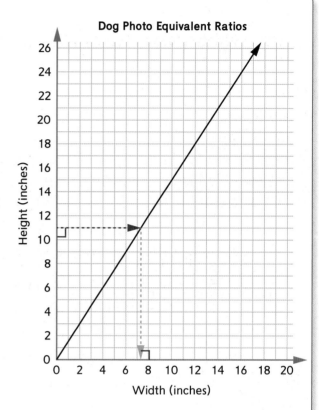

Check Your Understanding

Use the ratio/rate graph above to answer the following questions.

1. Describe the sizes of 3 more enlargements that Ben could order that have the same ratio of side lengths as the original photo.
2. Which of the following photo sizes have the same ratio as the original photo?

    a. 4 inches by 6 inches   **b.** 3 inches by 5 inches   **c.** 6 inches by 9 inches

    Check your answers in the Answer Key.

Ratios and Proportional Relationships

Percents

A percent is a special part-to-whole ratio that compares some amount to 100.

A **percent** means a part out of 100, a quantity out of 100, or a quantity per 100.

The symbol for percent, %, is read as *percent*, or *per cent*.

Percents are frequently used in everyday life as part-to-whole ratios in which the whole is 100.

> **Did You Know?**
>
> The word *percent* comes from the Latin *per centum*. *Per* means *for* and *centum* means *one hundred*.

Examples

Business: "75% off" means that the price of an item will be reduced by 75 cents for every 100 cents that the item usually costs.

**SALE – 75% OFF
Everything Must Go**

Statistics: "45% voter turnout" means that 45 out of every 100 registered voters actually voted. That's fewer than half of all the registered voters.

Voter Turnout Pegged at 45% of Registered Voters

School: A 90% score on a spelling test means that a student scored 90 out of 100 possible points on that test. One way to score 90% is to spell 90 out of 100 words correctly. Another way to score 90% is to spell 9 out of 10 words correctly.

Ratios and Proportional Relationships

In everyday situations, a part-to-whole ratio can be renamed as a percent even when the whole is not 100. All part-to-whole ratios have an equivalent ratio that can compare a part to 100 as the whole.

Note A part-to-part ratio cannot be expressed as a percent.

Example

Crandon Middle School was built for 200 students, but only 150 students attend the school.

The school enrollment is 150 out of the possible 200. The part-to-whole ratio is 150:200.

You can divide both of the numbers in the ratio by 2 to find an equivalent ratio of 75:100 with 100 as the whole.

The ratio means 75 out of every hundred, or 75 percent. This is written as 75%.

So you can say that Crandon Middle School is operating at 75% of its capacity.

When a situation is described using a percent, the *actual numbers* from the situation are not necessarily known. So the statement, "40% of students in a school are absent," does not necessarily mean that there are exactly 100 students in the school and that 40 of those students are absent. It does mean that for every 100 students, 40 of them are absent.

To find the actual numbers, use the percent and the known part or whole.

Example

There are 350 students enrolled at Clifford School. One day, 40% of the students were absent. How many students were absent that day?

Think: 350 = 100 + 100 + 100 + 50

For every 100 students, 40 were absent.

So for every 50 students ($\frac{1}{2}$ of 100), 20 were absent ($\frac{1}{2}$ of 40).

$$350 = 100 + 100 + 100 + 50$$
$$\downarrow \quad \downarrow \quad \downarrow \quad \downarrow$$
$$40 \quad 40 \quad 40 \quad 20$$
$$\text{absent absent absent absent}$$

40 + 40 + 40 + 20 = 140 students were absent that day.

Ratios and Proportional Relationships

Percents, Fractions, and Decimals

Equivalent percents, fractions, and decimals each describe the same part of a whole compared to 100.

One way to identify equivalents among percents, fractions, and decimals is to visualize shaded amounts in a 10-by-10 grid that represents one whole. Think about the ratio of squares shaded compared to the whole set of 100 squares in the grid.

Example

2% of the 10-by-10 grid is shaded.

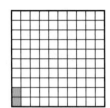

2% shaded means the same as 2 per hundred, or 2 out of 100, shaded.

So, 2% means the same as the fraction $\frac{2}{100}$ and is read as two hundredths, or 2 hundredths.

2% means the same as the decimal 0.02 and is also read as two hundredths, or 2 hundredths.

40% of the 10-by-10 grid is shaded.

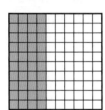

40% shaded means the same as 40 per hundred, or 40 out of 100 shaded.

So, 40% means the same as the fraction $\frac{40}{100}$.

40% also means the same as $\frac{4}{10}$, because $\frac{40}{100} = \frac{4}{10}$. Each column of the grid is $\frac{1}{10}$ of the whole.

40% means the same as the decimal 0.40.

40% means the same as 0.4, because 0.40 = 0.4. Each column of the grid is 0.1 of the whole.

100% of the 10-by-10 grid is shaded.

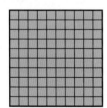

100% shaded means the same as 100 per hundred, or 100 out of 100 shaded.

So, 100% means the same as $\frac{100}{100}$.

100% means the same as 1, because $\frac{100}{100} = 1$.

100% has the same meaning as 1.00, or 1 whole and no hundredths.

200% of a 10-by-10 grid is shaded.

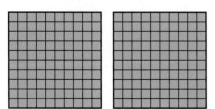

200% shaded means the same as 200 per hundred shaded, or 200 out of 100 shaded.

So, 200% means the same as $\frac{200}{100}$.

200% means the same as 2, because $\frac{200}{100} = 2$.

200% has the same meaning as 2.00, or 2 wholes and no hundredths.

Ratios and Proportional Relationships

A fraction can always be renamed as a percent when the denominator is hundredths.

Examples

Ella is required to put $\frac{25}{100}$ of what she earns into her savings account. What percent of Ella's earned money does she put into her savings account?

The denominator of the fraction is 100, so a percent can be written. The ratio of money saved to the total money earned is 25 per 100, or 25%.

So, Ella puts 25% of her earnings into her savings account.

What percent of the students in Ben's school is $\frac{9}{10}$ of the entire student population?

Since the denominator is not 100, think of an equivalent fraction to $\frac{9}{10}$ in hundredths: $\frac{90}{100}$.

$\frac{90}{100}$ is the same as 90 students per 100, or 90 out of 100 students, or 90%.

So, 90% is equivalent to $\frac{9}{10}$ of the school population.

To identify a percent that is equivalent to a fraction, you can visualize shading a 10-by-10 grid.

Example

Erik has earned $\frac{3}{4}$ of the money he needs to buy a bicycle. What percent of the amount of money he needs has Erik earned?

Erik sketches a 10-by-10 grid to represent the whole cost of the bicycle. He then divides it into fourths and figures out how many small squares are in 3 of the 4 equal parts.

There are 100 small squares in all. Dividing the grid into 4 equal parts results in 25 little squares in each of the four parts. Each part is $\frac{25}{100}$. Three shaded parts is $3 * \frac{25}{100}$, or $\frac{75}{100}$. So, $\frac{3}{4}$, or $\frac{75}{100}$, is the same as 75%.

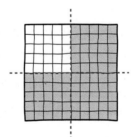

Erik has saved 75% of what he needs to buy the bicycle.

Note You can rename part-to-whole ratios as percents. However, percents do not make sense for part-to-part ratios, even when they are written using fraction notation, because the second quantity (the denominator) does not reflect the whole. For example, if $\frac{3}{5}$ is used to represent 3 pet cats for every 5 pet dogs, the denominator of 5 does not represent the whole set of pets. So renaming this part-to-part ratio as a percent does not make sense.

Fraction and percent equivalents commonly used in everyday life:

fraction	$\frac{1}{2}$	$\frac{1}{4}$	$\frac{3}{4}$	$\frac{1}{5}$	$\frac{2}{5}$	$\frac{3}{5}$	$\frac{4}{5}$	$\frac{1}{10}$	$\frac{3}{10}$	$\frac{7}{10}$	$\frac{9}{10}$
fraction in hundredths	$\frac{50}{100}$	$\frac{25}{100}$	$\frac{75}{100}$	$\frac{20}{100}$	$\frac{40}{100}$	$\frac{60}{100}$	$\frac{80}{100}$	$\frac{10}{100}$	$\frac{30}{100}$	$\frac{70}{100}$	$\frac{90}{100}$
percent	50%	25%	75%	20%	40%	60%	80%	10%	30%	70%	90%

Ratios and Proportional Relationships

When a decimal represents part of a whole, it can be expressed as a percent once it is named or renamed in terms of hundredths.

Examples

Amir attempts to complete a fitness test that involves running 1 kilometer. He sprains his ankle when he is 0.62 of the required distance and stops running. What percent of the required distance does Amir complete?

Change 0.62 to its fraction equivalent in hundredths: $\frac{62}{100}$.

$\frac{62}{100}$ is the same as 62 out of 100, or 62%.

So Amir completes 62% of the required distance.

Erik looks online for his spelling test grade and finds that he answered 0.8 of the test questions correctly. What is his score as a percent?

Erik answered 0.8 of the test items correctly.

Think of 0.8 as an equivalent decimal to the hundredths place, or 0.80.

0.80 is the same as $\frac{80}{100}$, and that is the same as 80 out of 100, or 80%.

0.8 is equivalent to 80%.

So Erik's score is 80%.

A typical baby weighs 4 times his or her birth weight after one year. What percent of the baby's original birth weight can be expected on the baby's first birthday?

Think of 4 as an equivalent decimal written to the hundredths place, or 4.00.

4.00 is the same as $\frac{400}{100}$. That is the same as 400 parts per 100, or 400 for every 100.

4 wholes is equivalent to 400%.

So on his or her first birthday, a typical baby weighs 400% of the original birth weight.

Check Your Understanding

1. Write each of the following percents as a fraction and as a decimal.
   a. 25%   b. 40%   c. 70%   d. 500%   e. 300%

2. Each of the following expressions represents part of a whole. Write an equivalent percent for each.
   a. 0.65   b. 0.9   c. $\frac{300}{100}$

Check your answers in the Answer Key.

Ratios and Proportional Relationships

Using Unit Percents to Solve Problems

Unit percent is another name for 1%. Other names for 1% are $\frac{1}{100}$ and 0.01. Finding the unit percent, or 1%, of the whole can be useful in solving problems.

Example

The sales tax on a $400 item is 7%. How much will be paid in sales tax?

Think about the unit percent: 1% of $400.

$1\% = \frac{1}{100}$, so 1% of $400 is the same as $\frac{1}{100}$ of $400.

If you divide $400 into 100 equal groups, there are $4 in each group.

So, 1% of $400 is equal to $4, and 7% of $400 is 7 * $4, or $28.

7% sales tax on $400 is $28.

Did You Know?

The United States Congress passed the first income tax law in 1862. If a person's income was between $800 and $10,000, the tax was 3% of the income. Those with incomes greater than $10,000 paid taxes at a higher rate.

Example

London's mother spends 30% of her monthly income on rent. Her monthly income is $3,200. How much does London's mother spend on rent each month?

Think about 1% of $3,200.

$1\% = \frac{1}{100}$, so 1% of $3,200 is the same as $\frac{1}{100}$ of $3,200.

If you divide $3,200 into 100 equal groups, there are $32 in each group.

So, 1% of $3,200 is equal to $32, and 30% of $3,200 is 30 * $32, or $960.

London's mother spends $960 on rent each month.

Check Your Understanding

Use a unit percent to solve each problem. Explain your thinking.

1. Ben's goal is to walk 200 miles during the school year. He has completed 20% of his goal. How many miles has Ben walked?
2. 22% of the sixth-grade students at Ashton School are in the school band. There are 300 sixth-grade students in the school. How many of the sixth-grade students are in the school band?

Check your answers in the Answer Key.

Ratios and Proportional Relationships

Using Percents to Find a Part

In some everyday situations, you know the whole and the percent and you have to figure out what the part is. While you can always use a unit percent to solve these problems, you can also use other strategies. One way to solve this type of problem is to change the percent to an equivalent fraction that you can use to calculate mentally.

Example

Ben wants to buy a backpack that sells for $70. He told his brother that he has already saved 30% of what he needs. How much money has Ben saved so far?

One way to solve the problem mentally is to think about 30% as $\frac{3}{10}$.
30% is the same as $\frac{3}{10}$, because 30% = $\frac{30}{100}$ = $\frac{3}{10}$.
Figure out $\frac{3}{10}$ of 70.
Imagine dividing 70 dots into 10 equal groups. Each group has 7 dots.
Imagine circling 3 of the groups.
There are 3 * 7 = 21 dots circled. $\frac{3}{10}$ of 70 is 21.

Ben has saved $21 for the backpack he wants to buy.

Another way to solve this type of problem is to use a ratio/rate table.

Example

Tax on food is 4%. If you buy $50 worth of groceries, how much tax will you pay?

First identify the ratios, and then make a ratio/rate table.
4% means that there would be a $4 tax on $100 worth of groceries.
The ratio is 4 out of 100, or 4:100.
Identify a number you can multiply or divide each quantity in the ratio by to find an equivalent ratio with the original cost of $50.
Since 100 / 2 = 50, divide 4 by 2.
The resulting ratio is 2:50

So the tax on $50 of groceries is $2.

amount of tax ($)	4	2
original cost ($)	100	50

/ 2 (arrows indicating division by 2)

Check Your Understanding

1. London always deposits 60% of her birthday money into her savings account. She received $120 for her birthday. How much will she put into savings?
2. A pair of shoes that regularly sells for $48 is on sale for 25% off. How much will the shoes be discounted? How much will the shoes cost?

Check your answers in the Answer Key.

Ratios and Proportional Relationships

Finding a Percent

In many everyday situations, you need to find the percent when you know the part and the whole.

Since any part-to-whole ratio can be renamed as some number out of 100, you can use what you know about finding equivalent ratios to solve these problems.

You can use a ratio/rate table. Look for combinations or extend the table to make a whole of 100 by multiplying or dividing both of the quantities in the ratio by the same number.

Example

Erik's parents ask him to put a certain percent of the money he earns into a savings account. The ratio/rate table below shows the amount he saved compared to the amount he earned based on his parents' request. The first three columns show his savings and earnings for three weeks. What percent of his earnings is he asked to save?

amount saved ($)	8	6	34	?
amount earned ($)	20	15	85	100

One way:
Use combinations of quantities in the columns in the table above. In this case, you can see that one combination of amounts earned has a sum of $100: $15 earned + $85 earned = $100 earned. So if you add the amounts saved for each of those amounts earned, you can find the amount saved out of every $100. When $15 is earned, $6 is saved. When $85 is earned, $34 is saved. So the amount saved is $6 + $34 = $40 when the amount earned is $100.
The ratio of money saved to money earned is 40 to 100. That is the same as 40%.

Another way:
Multiply or divide each quantity in the ratio by the same number to obtain $100 as the amount earned. Notice that in the ratio 8:20, the $20 that is earned can be multiplied by 5 to obtain $100 earned. Then multiply $8 by 5 to obtain the amount saved. $8 * 5 = $40
The ratio of amount saved to amount earned is 40:100, or 40%.

Erik is required to save 40% of the money he earns.

Ratios and Proportional Relationships

You can use a tape diagram to solve many percent problems.

Example

At Allen Middle School, 3 of every 5 students wore a necktie on Tie Day. What percent of the students wore a tie? What percent of the students did not wear a tie?

First show a row of 3 squares to represent the students who wore ties, and then show a row of 2 squares to represent the students who did not wear ties. Write "out of 5" to show the whole in the diagram.

wore a tie ▢▢▢
did not wear a tie ▢▢
} out of 5

Next think about how you would distribute a whole amount of 100 evenly to the 5 boxes in the tape diagram.
100 / 5 = 20
Write 20 in each box, show the amount of 100 in all, and find the solution to the problem.

wore a tie |20|20|20|
did not wear a tie |20|20|
} 100

The part of students out of 100 who wore a tie is 3 * 20, or 60.
That is 60 out of 100, or 60%.
The part of students out of 100 who did not wear a tie is 2 * 20, or 40. That is 40 out of 100, or 40%.

So, 60% of the students wore a tie, and 40% of the students did not wear a tie.

Check Your Understanding

Solve.

Under normal circumstances $\frac{1}{5}$ of a telephone pole is underground. What percent of the telephone pole do you see above ground?

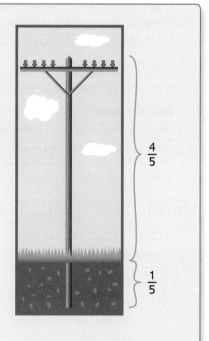

Check your answer in the Answer Key.

Ratios and Proportional Relationships

For some problems, you can think about shading a grid that represents the whole.

Example

The large rectangle below represents the whole, and its area is divided into 20 equal parts (4 rows of 5 boxes). Identify the percent of the area that is shaded.

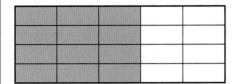

The ratio of the number of boxes shaded to the total number of boxes is 12:20.

The percent will be an equivalent ratio that is some number of boxes to 100 boxes.

Think: If you need to have 100 equal parts as the whole, that amount is 5 times as many as in the grid above (since 5 ∗ 20 = 100). So think about dividing each box in the grid into 5 equal parts to make 100 boxes in all.

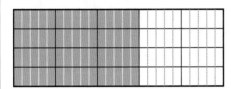

The number of shaded parts in the equivalent ratio is 5 ∗ 12 shaded boxes, or 60 shaded boxes.

Since the ratio of the number of parts shaded to all the parts is 60 to 100, the equivalent percent is 60%.

So, 60% of the boxes are shaded.

Check Your Understanding

1. Write a percent for each of the following part-to-whole ratios.
    a. 15 to 20   b. 17:50   c. 80 to 200   d. 250:100

2. Any class that returns at least 80% of their parent-signed report cards will be given extra recess time. What percent of each class turned in their signed report cards? Will the class get extra recess time?
    a. Amir's class: 15 out of 25 students turned in their signed report cards.
    b. Ella's class: 24 out of 30 students turned in their signed report cards.
    c. Ashley's class: 18 out of 24 students turned in their signed report cards.

Check your answers in the Answer Key.

Ratios and Proportional Relationships

Using Percents to Find the Whole

In many everyday situations, the whole can be found when you know the part and the percent. One way to solve this type of problem is to use a ratio/rate table.

Example

London made 6 successful free-throw shots during basketball season. Her percent of successful free-throw shots for the season is 30%. How many free-throw shots did she attempt during the season?

First identify the ratios to make a ratio/rate table.
30% success on all free-throw shots is the same as making 30 out of 100 shots.
Exactly 6 free-throw shots were successful out of an unknown number of attempts.
Decide what number to multiply or divide both numbers in the ratio by to find an equivalent ratio in the table.
Since 30 / 5 = 6, you can divide both numbers in the ratio by 5 to get an equivalent ratio.
30 / 5 = 6, and 100 / 5 = 20.
The resulting ratio is 6 : 20 in which 6 is the number of successful shots and 20 is the number of attempted shots.

| Successful free-throw shots | 30 | 6 |
| Total attempted free-throw shots | 100 | 20 |

/ 5

London attempted 20 free-throw shots.

Sometimes you can sketch a 10-by-10 grid to help you solve a problem.

Example

Ella bought a shirt that was discounted by $9. This was 25% off the regular price. What was the regular price of the shirt? What was the sale price?

Think: Using the table on page 57, you know that 25% means the same as $\frac{1}{4}$.
That means that $9 is 25%, or $\frac{1}{4}$, of the whole cost of the shirt.
Sketch a diagram that divides a whole 10-by-10 grid into fourths.
Each part is 25 out of 100 total small squares, which is an equivalent ratio to 1 out of 4.
Since $9 is $\frac{1}{4}$ of the price, you can show $9 in each of the four regions.
The value of the whole grid can be represented as 4 * $9 = $36.

The regular cost of the shirt was $36. The sale price of the shirt was $36 − $9, or $27.

Check Your Understanding

Ashley made 12 of her free-throw shots this month. She exactly met her goal to make 75% of her attempts. How many free-throw shots did she attempt?

Check your answer in the Answer Key.

Ratios and Proportional Relationships

Ratios in Systems of Measurement

Systems of weights and measures have been used in many parts of the world since ancient times. People measured and compared lengths and weights for centuries before there were rulers and scales. Most of the units we use in the U.S. customary system of measurement have roots in historical units of measure, but many ancient units of measure are no longer used. For example, a unit of weight that is no longer used is the Chinese *zhu*. One *zhu* was defined as the weight of 100 millet seeds. The unit rate was 1 *zhu* : 100 millet seeds.

People throughout history have used natural measures based on the human body to measure length and distance. Some of these units are shown below.

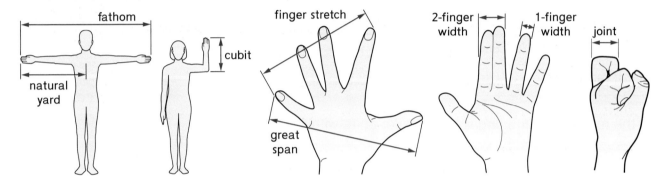

Some natural measures have multiplicative relationships that can be written as unit rates. For example, 1 yard : 2 cubits and 2 spans : 1 cubit.

Using grains or body lengths to measure often resulted in inconsistent measurements, because the size of the units, such as arm lengths, varied greatly from person to person. Devising a set of **standard units** solved the problem of inconsistent units. For example, the unit of a foot became standardized when the length of the foot of an important person, such as a king, was declared to be the official length and a tool of that length was distributed. As trade increased, countries had to agree on common units of measure. Today all foot-long rulers are the same length and are marked in smaller standard units of inches. Since standard units never change and are the same for everyone, two people who measure the same object will obtain the same, or almost the same, measure depending on the degree of accuracy.

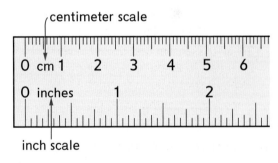

sixty-five 65

Ratios and Proportional Relationships

The Metric System

The metric system was developed about 200 years ago as a standard system of weights and measures. The metric system is based on scientifically defined standard units for length, mass and weight, capacity and liquid volume, and temperature. The metric system is used internationally by scientists and by people in most countries around the world.

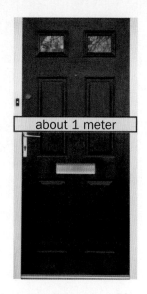
about 1 meter

The base units and personal references for each unit are shown in the table below. (For more personal references, see page 375.)

Type of Measure	Base Unit	Personal Reference
length	1 meter (m)	width of a front door
mass	1 gram (g)	mass of 1 large paper clip
liquid volume	1 liter (L)	volume of 2 large glasses of milk

Note The metric unit used to measure temperature is degrees Celsius (°C). Water freezes at 0°C and boils at 100°C.

In the metric system, larger and smaller units within a type of measure are related by powers of 10, so the metric system is especially easy to use to convert units. You can convert units by expressing the relationship between units as a **unit rate**, and then using **equivalent ratios** to make the **unit conversions** (see pages 68–69).

You can find a full reference table of metric units on page 374. The examples below show relationships between commonly used metric units along with a unit rate.

Examples

- 1 meter is the same length as 100 centimeters.
  1 meter : 100 centimeters
- 1,000 grams is the same mass as 1 kilogram.
  1,000 grams : 1 kilogram
- 1 liter is the same volume as 1,000 milliliters.
  1 liter : 1,000 milliliters

Note You can use the prefixes of metric measures to interpret relationships within units of length, volume, mass, and weight.

- *kilo-* means 1,000
- *deci-* means $\frac{1}{10}$
- *centi-* means $\frac{1}{100}$
- *milli-* means $\frac{1}{1,000}$

The U.S. Customary System

In the United States, the U.S. customary system is commonly used in everyday life. The U.S. customary system is based on historically defined standard units of measure. There are a variety of standard units for each type of measure.

The types of standard units and personal references for some units are shown in the table below. (For more personal references, see page 375.)

Type of Measure	Units of Measure	Personal Reference
length	inch, foot, yard, mile	A yard is about the width of a door.
weight	ounce, pound, ton	A pound is about the weight of a baseball.
liquid volume	fluid ounce, cup, pint, quart, gallon	A gallon is the amount of milk in a large container.

Note The U.S. customary unit used to measure temperature is degrees Fahrenheit (°F). Water freezes at 32°F and boils at 212°F.

A gallon container

In the U.S. customary system, larger and smaller units within each type of measure are not defined by powers of 10. However, you can convert between two units by first expressing the relationship between the units as a **unit rate,** and then using what you know about **equivalent ratios** to make the **unit conversion** (see pages 68–69).

You can find a full reference table of U.S. customary units on page 374. The examples below show relationships between some commonly used U.S. customary units along with a unit rate.

Examples

- 1 foot is the same length as 12 inches.
  1 foot : 12 inches
- 16 ounces is the same weight as 1 pound.
  16 ounces : 1 pound
- 1 quart is the same liquid volume as 4 cups.
  1 quart : 4 cups

Ratios and Proportional Relationships

Unit Conversions

When you rename a measure in one unit to an equivalent measure in another unit, you are making a **unit conversion**. You can make a unit conversion by identifying a **unit rate** that is based on the relationship between the two units and then using the unit rate and the known measure to find an equivalent rate with the desired measure. You can use ratio/rate tables to make unit conversions between U.S. customary units of measure.

Note *Unit conversion* refers to renaming a measure as an equivalent measure in a different *unit*. *Unit rate*, on the other hand, refers to comparing exactly one of one unit to some number of another unit.

Example

The Golden Gate Bridge is just over 2,993 yards long. What is its length in feet?

1 yard is the same length as 3 feet. So a unit rate is 1 yard : 3 feet.

Make a ratio/rate table. Write the known quantities in the table.

Find an equivalent rate for 2,993 yards.

Since 1 * 2,993 = 2,993 and 3 * 2,993 = 8,979, the equivalent rate is 2,993 yards : 8,979 feet.

So the Golden Gate Bridge is about 8,979 feet long.

yards (yd)	1	2,993
feet (ft)	3	8,979

* 2,993

You can use ratio/rate tables to make unit conversions between metric units.

Example

An oarfish about 5 meters long was found off the coast of California. About how many millimeters long is the fish?

There are 1,000 mm in 1 m. So a unit rate is 1 meter : 1,000 mm.

Make a ratio/rate table. Write the known quantities in the table.

Find an equivalent rate for 5 meters.

Since 1 * 5 = 5 and 1,000 * 5 = 5,000, the equivalent rate is 5 meters : 5,000 millimeters.
There are 5,000 millimeters in 5 meters.

So the oarfish is about 5,000 millimeters long.

meters (m)	millimeters (mm)
1	1,000
5	5,000

* 5

Ratios and Proportional Relationships

To make unit conversions, you can write and use a rule.

Example

A recipe for making a batch of homemade soap requires 3 pounds of lard, $\frac{3}{4}$ pound of lye, and 3 cups of water. How many ounces of lard and lye are needed for one recipe?

Find a unit rate to show the relationship between ounces and pounds: 1 pound : 16 ounces.

Write a rule to convert from pounds to ounces.
Think: If 1 pound is 16 ounces, then 3 pounds is 3 groups of 16 ounces. Multiplication makes sense for multiple groups.

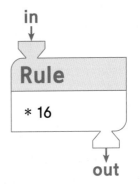

in	out
1 lb	16 oz
3 lb	48 oz
$\frac{3}{4}$ lb	12 oz

Rule: number of pounds * 16 ounces per pound = number of ounces
Use the rule above to convert the units.
3 pounds of lard * 16 ounces per pound = 48 ounces of lard
$\frac{3}{4}$ pound of lye * 16 ounces per pound = 12 ounces of lye

So, 48 ounces of lard and 12 ounces of lye are needed for the soap recipe.

Example

Ella's mom bought 1,500 grams of wheat flour and 2,000 grams of oat flour. How many kilograms of flour did she buy in all?

Find a unit rate to show the relationship between grams and kilograms: A unit ratio is 1,000 grams to 1 kilogram.

Write a rule to convert from grams to kilograms.
Think: If 1,000 grams make 1 kilogram, then how many groups of 1,000 grams are in 1,500 grams?
Division makes sense for splitting an amount into equal groups.
Rule: number of grams / 1,000 grams per 1 kilogram = number of kilograms

Use the rule above to convert the units.
1,500 grams / 1,000 grams per kilogram = 1.5 kilograms
2,000 grams / 1,000 grams per kilogram = 2.0 kilograms
1.5 kilograms + 2.0 kilograms = 3.5 kilograms

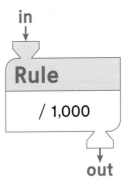

in	out
1,000 g	1 kg
1,500 g	1.5 kg
2,000 g	2 kg

The bag of wheat flour is 1.5 kilograms, and the bag of oat flour is 2.0 kilograms. Ella's mom bought 3.5 kilograms in all.

sixty-nine

Ratios and Proportional Relationships

Multi-Step Unit Conversions

Unit conversions require more than one step when a reference table does not directly state the relationship between the units in question. For example, the number of cups in 1 quart may not be listed in a reference table, but the table may list that 1 pint = 2 cups and 2 pints = 1 quart.

You can use a ratio/rate table for each of the steps of the multi-step unit conversion.

Example

How many cups are in 2 quarts?

Identify the unit rates that refer to cups and quarts with a common third unit.
1 quart : 2 pints
1 pint : 2 cups
You can identify 2 unit rates, so you can do this problem in 2 steps.

Step 1 You can use a ratio/rate table to convert 2 quarts to some number of pints. The result is an *intermediate rate*.

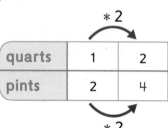

2 quarts : ? cups

You can multiply both quantities in the unit rate by 2 to find an equivalent rate. This intermediate rate is 2 quarts : 4 pints. So, 2 quarts is the same as 4 pints.

Step 2 You can use another ratio/rate table to convert the resulting number of pints (in the intermediate rate) to cups.

pints	1	4
cups	2	8

*4

To find an equivalent rate with 4 pints, you can multiply both quantities in the unit rate by 4.
4 pints : 8 cups

Since 2 quarts is the same as 4 pints and 4 pints is the same as 8 cups, there are 8 cups in 2 quarts.

Ratios and Proportional Relationships

You can use mathematical reasoning, estimation, and mental math in each of the steps in a multi-step unit conversion.

Example

60,000 minutes equals about how many days?

Identify useful unit relationships that relate to minutes and days.

1 hour = 60 minutes

1 day = 24 hours

Since there are 2 unit rates, you can complete this conversion in two steps.

Step 1 Convert 60,000 minutes to hours to obtain an intermediate rate.

60 minutes make up 1 hour. How many groups of 60 are in 60,000?

60,000 minutes / 60 minutes in an hour = 1,000 hours

Step 2 Convert the number of hours in the intermediate rate to days.

24 hours make 1 day. How many groups of 24 are in 1,000?

Think: 25 is close to 24 and more easily divides into 1,000.
100 / 25 = 4, so 1,000 / 25 = 40.

Estimate: 1,000 hours / about 25 hours in a day = 40 days

So, 60,000 minutes is about 40 days.

Check Your Understanding

Use the measurement conversion tables on page 374 to solve the following problems.

1. How many inches are in 3 yards?
2. 64 fluid ounces are equal to how many pints?
3. How many millimeters are equal to 5 decimeters?

Check your answers in the Answer Key.

Ratios and Proportional Relationships

Using Ratios to Describe Size Changes

There are various ways to display an image or object as a different size. For example, a magnifying glass, a microscope, and a document camera all produce size changes that enlarge the original images. Most copy machines can create a variety of size changes—both enlargements and reductions of the original document

Enlargements and reductions have the same shape as the original object, but not necessarily the same size. They are described using size-change factors. The size-change factor is a number that tells the amount of enlargement or reduction that takes place.

original

size-change factor 2

size-change factor 0.5

Did You Know?

Binoculars that are labeled as "8X" or "8 power" magnify all the lengths you see without the binoculars to 8 times their actual size.

- When a copy machine is used to make a 2X change in size, every length in the copy is twice the size of the original. The size-change factor is 2, or 200%.

- When a copy machine is used to make a 0.5X change in size, every length in the copy is half the size of the original. The size-change factor is $\frac{1}{2}$, 0.5, or 50%.

Note The word *length* refers to any linear measurement, including the length, width, or height of an object.

You can think of the size-change factor as a ratio. For a 2X size change, the unit ratio of a length in the copy to the corresponding length in the original is 2 to 1. You can use a ratio/rate table to generate many equivalent ratios.

copy length	2	8 in.	4 cm
original length	1	4 in.	2 cm

If the original image is 4 inches wide, the copy image is twice as wide, or 8 inches. If the copy image is 4 centimeters long, the original image is half as long, or 2 centimeters.

For a 0.5X size change, the ratio of a length in the copy to the corresponding length in the original is 0.5 to 1. You can use a ratio/rate table to generate many equivalent ratios.

copy length	0.5	0.25 cm	3 in.
original length	1	0.5 cm	6 in.

Note If the size-change factor is greater than one, it is an *enlargement* of the original. If the size-change factor is less than one, it is a *reduction* of the original.

seventy-two

Ratios and Proportional Relationships

You can find the size-change factor between two images that are the same shape by comparing the lengths in the copy to the lengths in the original.

One way to find a size-change factor is to use a ratio/rate table. Use the known ratio to identify a unit ratio. When the second quantity in the unit ratio is 1, the first quantity in the unit ratio is the size-change factor. For example, if the unit ratio is 8 : 1, the size-change factor is 8.

Note Sometimes the symbol " is used to represent inches. The image on the original negative in the example below is 2" by 2", or 2 inches by 2 inches. The symbol ' is often used to represent feet.

Example

A photographer uses an enlarger to make prints from negatives. The size of the image on the original negative is 2" by 2". The size of the image on the enlargement is 6" by 6".

The ratio of a length in the enlargement, or copy, to a corresponding length in the original is 6 to 2. To find the size-change factor, determine the unit ratio from the known ratio.

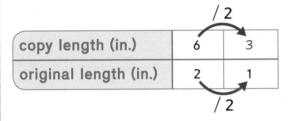

Write the known ratio in the first column after the labels, and then write an original length of 1 in the next column to help find a unit ratio.
Notice that the original length of 2 inches can be divided by 2 to get an original length of 1 in the unit ratio. To find this equivalent unit ratio, divide both quantities in the known ratio by 2.
2 / 2 = 1, and 6 / 2 = 3.
The unit ratio of a length in the copy to a length in the original is 3 to 1. For every 3 inches of length in the copy, the corresponding length in the original is 1 inch.
So the size-change factor is a 3X enlargement.

Ratios and Proportional Relationships

You can use size-change factors to find an unknown linear measure in a copy when you know the corresponding linear measure in the original. One way is to use a ratio/rate table. Use the size-change factor as a unit ratio in the table. Then use the known linear measure in the original to find the unknown linear measure in the copy.

> **Example**
>
> Amir drew a logo for his bowling team. He is reducing the logo to put on T-shirt pockets using a size-change factor of 0.2. The height of the logo in the original is 12 centimeters, and the width is 8 centimeters. What will be the height and width in the copy on the T-shirts?
>
> Write a unit ratio that corresponds to the size-change factor: the length of a measure in the copy to the length of a measure in the original is 0.2 to 1.
>
> Write the unit ratio in the first column after the labels in a rate/ratio table. Write the known measures, 12 centimeters and 8 centimeters, in the second and third columns as the original height and width of the logo.
>
>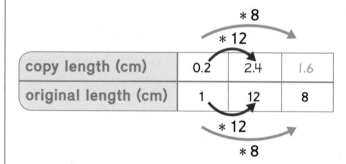
>
> To find the corresponding measures in the copy, find the equivalent ratios.
>
> Notice that the original length of 1 in the unit ratio can be multiplied by 12 to obtain the original length of 12 in the second column. So to find the equivalent ratio, multiply both quantities in the unit ratio by 12.
>
> Notice that the original length of 1 in the unit ratio can be multiplied by 8 to obtain the original length of 8 in the third column. So to find the equivalent ratio, multiply both quantities in the unit ratio by 8.
>
> $1 * 12 = 12$, and $0.2 * 12 = 2.4$.
> So, 12 centimeters in the original corresponds to 2.4 centimeters in the copy.
> $1 * 8 = 8$, and $0.2 * 8 = 1.6$.
> So, 8 centimeters in the original corresponds to 1.6 centimeters in the copy.
>
> So, the height of the logo on the T-shirt pocket will be 2.4 centimeters, and the width will be 1.6 centimeters.

Ratios and Proportional Relationships

Scale Models

A model that is a careful, reduced copy of an actual object is called a **scale model.** You have probably seen scale models of cars, trains, and airplanes. A **scale factor** tells the relationship between the scale model and the actual object. Sometimes the scale factor is given as a size-change factor, and sometimes it is given as a rate.

You can use scale factors just like you use ratios and rates.

Examples

Dollhouses often have a scale factor and a size-change factor of $\frac{1}{12}$. This relationship between the length measurements of the doll house and a real house can be described as "$\frac{1}{12}$ of actual size," "scale 1:12," or

$$\frac{\text{dollhouse length}}{\text{real house length}} = \frac{1 \text{ inch}}{12 \text{ inches}}.$$

An O-scale model railroad has a scale factor of $\frac{1}{48}$. This is a reduction with a size-change factor of $\frac{1}{48}$. This relationship can be described as "$\frac{1}{48}$ of the actual size," "scale 1:48," or

$$\frac{\text{model railroad length}}{\text{real railroad length}} = \frac{1 \text{ inch}}{48 \text{ inches}}.$$

Ratios and Proportional Relationships

When the difference in size between a model and the actual object is very large, the size-change factor can be difficult to interpret because it requires the measures in the model and the actual object to use the same unit.

Example

Think about the $\frac{1}{48}$ size-change factor of an O-scale model railroad.

The scale factor, $\frac{1}{48}$, can be written as a ratio, 1:48.

- Why might it be difficult to think about this ratio using the common unit of inches?

  Consider a building in the model that is 8 inches tall. The actual height of the building is 8 * 48 inches, or 384 inches. A height of 384 inches is difficult to visualize. It would be easier to visualize the actual height by converting the inches to feet: 384 inches / 12 feet per inch = 32 feet. It may be easier to imagine a height of 32 feet than 384 inches.

- Why might it be difficult to think about this ratio using the common unit of feet?

  Consider constructing a model of 16 feet of railroad track. How long will that section of the model railroad track be?

  Use a ratio/rate table to find the length of the model in feet.

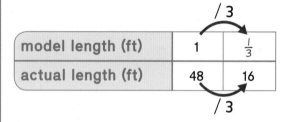

Measuring $\frac{1}{3}$ foot of model railroad track is inconvenient and can be inaccurate. It would be easier to measure and build the model using inches. $\frac{1}{3}$ of a foot is 4 inches (because 12 / 3 = 4). It is easier to measure and construct a track using the measurement of 4 inches than using a measurement of $\frac{1}{3}$ foot.

Ratios and Proportional Relationships

When the difference in size between a model and the actual object is very large and makes measuring and interpreting measures inconvenient, a rate that compares quantities with different units can be used as the scale factor.

Example

A standard scale for model train scales is to compare a length in the model in inches to 1 foot in the object's actual measure. Convert the O-scale model railroad scale factor of $\frac{1}{48}$ to a rate that compares a length in the model to 1 foot in the object's actual measure.

First you can convert 48 inches to feet: 48 inches / 12 inches per foot = 4 feet.

So a scale that is written as 1 inch : 48 inches can also be written as the rate 1 inch : 4 feet.

That rate compares 1 inch in the model to 4 feet in the actual object. But to find the standard scale, you need to compare a length in inches in the model to 1 foot in the *actual* railroad.

A ratio/rate table can be used to find the equivalent unit rate. Write the unit rate you know in the first column after the labels. In the second column, write 1 foot for the actual length. Find the equivalent rate.

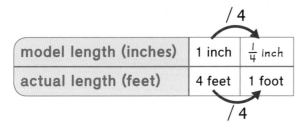

4 / 4 = 1, and 1 / 4 = $\frac{1}{4}$.

So, 1 foot in the actual object is $\frac{1}{4}$ inch in the model.

A scale factor for the O-scale model railroad that compares the number of inches in the model to 1 foot in the actual object is $\frac{1}{4}$ inch : 1 foot.

The scale factor in the example is the same as 0.25 inch : 1 foot, because $\frac{1}{4}$ inch = 0.25 inch. This is a **rate** that compares a length in inches in the model to 1 foot in the actual railroad.

Ratios and Proportional Relationships

Scale Drawings

The size-change factor for scale drawings is usually referred to simply as the *scale*. While the scale on a drawing can be described in terms of a size-change factor, the scale is more commonly displayed as a unit rate or unit ratio.

> **Example**
>
> An architect's scale drawing of a newly designed home may show the scale as a unit rate:
>
> The scale is $\frac{1}{4}$ inch : 1 foot.
>
> This means that $\frac{1}{4}$ inch on the drawing represents 1 foot in actual length.
>
> This scale can be expressed as a ratio by using the same unit for both quantities.
>
> *Think:* 1 foot = 12 inches, so the ratio is $\frac{1}{4}$ inch : 12 inches.

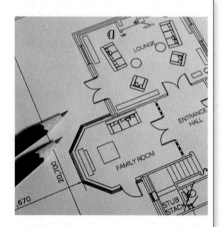

You can use ratio/rate tables to find different, but equivalent, scales.

> **Example**
>
> Use the architect's scale above of $\frac{1}{4}$ inch : 1 foot to identify equivalent scales.
>
> Write the scale you know as a ratio in the first column. Be sure your ratio compares quantities using the same units, as shown in the example above ($\frac{1}{4}$ inch : 12 inches, or $\frac{1}{4}$: 12)
>
>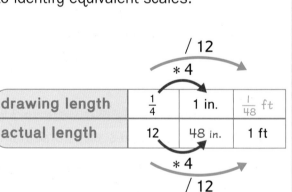
>
> You can find an equivalent ratio, or scale, in which 1 inch in the drawing corresponds to some number of inches in the actual house.
>
> A ratio comparing the drawing to the actual house can be written as 1 to 48. That is, 1 inch in the drawing represents 48 inches in actual house.
>
> Another equivalent ratio, or scale, can be found in which 1 foot in the actual house is represented by some number of feet in the drawing.
>
> A ratio comparing a length in the actual house to a length in the drawing can be written as 1 to $\frac{1}{48}$. That is, $\frac{1}{48}$ foot in the drawing represents 1 foot in the actual house. Note that this is not an easy scale to use to make a scale drawing.

Ratios and Proportional Relationships

Map Scales

Cartographers (mapmakers) show large areas of land and water in small areas on paper or electronically on computer monitors. Map scales are displayed on maps to compare a distance on the map to an actual distance. Typically, a map scale shows the relationship between 1 inch or 1 centimeter on the map and some number of actual miles or kilometers. However, other units or quantities may be used in map scales. You can use map scales to solve problems in the same way that you use ratios and rates.

Did You Know?

The U.S. Geological Survey (USGS) has made a detailed set of maps that covers the entire area of the United States. Their best known maps have a scale of 1:24,000 (1 inch represents 24,000 inches or 2,000 feet).

The map scale of 1 : 24,000 used by the U.S. Geological Survey can be thought of as a reduction of the actual distance with a size change of $\frac{1}{24,000}$X to find the map distance. Or, the same map scale can be thought of as an enlargement of the map distance with a size change of 24,000X to find the actual distance. A unit ratio comparing the map distance to the real distance can be written as 1 to 24,000. This size-change factor relationship can be represented as the unit ratio 1 : 24,000 or in a ratio/rate table:

map distance (units)	$\frac{1}{24,000}$	1
actual distance (units)	1	24,000

This relationship can be expressed as:

$$\frac{\text{map distance}}{\text{real distance}} = \frac{1}{24,000}$$

Map scales sometimes show a line segment on a map that is partitioned and labeled. The labels show the length of the drawn line segment on one side and the real-world distance being represented on the other side.

Example

The line segment in this map scale is 2 inches long, and the full length represents 10 actual miles.

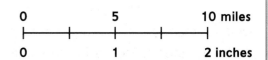

This map scale can also be represented as a rate:

$$\frac{\text{map distance}}{\text{actual distance}} = \frac{2 \text{ inches}}{10 \text{ miles}}$$

Another way to represent this map scale is with a ratio/rate table.

map distance (inches)	2 in.	1 in.
actual distance (miles)	10 mi	5 mi

seventy-nine 79

Ratios and Proportional Relationships

Finding Distances Using a Map Scale

Using a map and a map scale, there are many ways you can find actual distances. One method requires that you use a ruler and a string.

Step 1 Measure the map distance.

If the distance is along a straight path, such as the distance from Home to School, use the ruler to measure the map distance directly.

If the distance is along a curved path, such as the distance from Home to the Park:

- Lay the string along the path. Mark the beginning and ending points on the string.
- Straighten out the string. Use a ruler to measure the beginning and ending points.

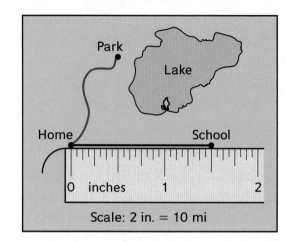

Step 2 Use the map scale to find the actual distance that corresponds to the map distance you measured.

For example, the map distance from Home to School is 1.5 inches and the scale is 2 inches to 10 miles. To find the actual distance from Home to School, you can use a ratio/rate table.

map distance (inches)	2 inches	1.5 inches	
actual distance (miles)	10 miles		

One way to find the actual distance that corresponds to 1.5 inches on the map is to first find a unit rate.

2 inches for every 10 miles is the same as 1 inch for each 5 miles.

Enter this unit rate into the ratio/rate table in an empty column.

map distance (inches)	2 inches	1.5 inches	1 inch
actual distance (miles)	10 miles		5 miles

Use the unit rate to find the number of actual miles that corresponds to 1.5 inches.

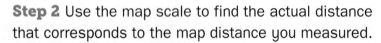

map distance (inches)	2 inches	1.5 inches	1 inch
actual distance (miles)	10 miles	7.5 miles	5 miles

So the actual distance between Home and School is about 7.5 miles.

Scale Models

A scale model is a three-dimensional representation of an object that is larger or smaller than the original thing. All of the parts in a model are proportional to the size of the parts in the original object. Workers use scale models in many fields, such as architecture, design, science, engineering, education, and entertainment.

Many scale models are smaller versions of the object they represent. A globe is a scale model of Earth. It shows features that can only be seen from locations hundreds of miles above the planet's surface.

This is a model of a microscopic DNA molecule. The actual molecule is so tiny it cannot be seen with the naked eye. In science and industry, a large-scale model can help people study microscopic parts.

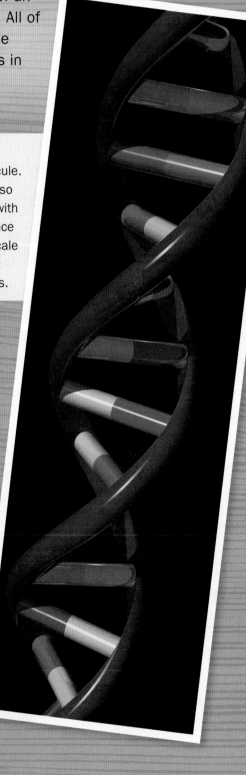

Understanding Scale

Models are built according to different scales. The scale of a model is the ratio of the size of the model to the size of the actual object.

This space shuttle model is part of a NASA display. Its scale is 1:15.

The scale 1:15 means the dimensions of the actual space shuttle, shown here at liftoff, are 15 times as large as the dimensions of the NASA display model.

The scale of 1:87 means, for example, that the length of the real train engine is 87 times the length of the model train engine.

Model railways are very common throughout the world. A popular scale is known as HO. The HO scale is 1:87.

Models in Architecture and Design

Architects and designers use scale models when planning structures and developing products. This enables them to make modifications to their plans before the actual structures or products are created.

These architects are using a scale model of a building design to solve problems before construction begins. It is easier and cheaper to work with a model to make sure the plan for a structure is workable than to fix the structure after it has been built.

This scale model is being used to plan the layout of streets, buildings, and other structures in a downtown area.

This scale model shows apartment buildings with landscaping for a new housing development. It offers city planners and potential buyers a view of the completed project before it is finished.

eighty-three

Models in Science and Engineering

Scientists and engineers create scale models to design new equipment and machines and to test existing ones. They test the function and safety of machines by simulating the environment in which the machines will operate.

In the early 1900s, the Wright brothers built and tested hundreds of scale models before finding a successful airplane design.

This is a replica of a wind tunnel that the Wright brothers used to test their designs. To simulate flight conditions, a fan blew air over models placed in the tunnel. From these tests, they could evaluate the performance and safety of their airplane designs.

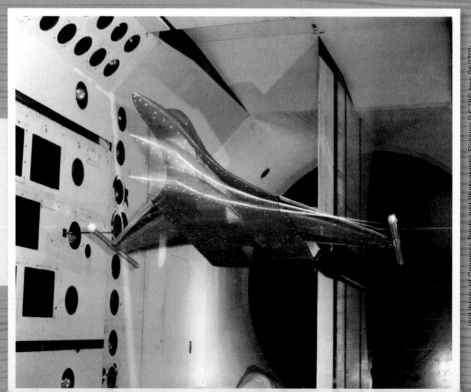

Aircraft designers still use scale models and wind tunnels to collect data about their designs. They test the safety and reliability of their aircraft during takeoffs and landings at a variety of speeds.

Engineers use scale models like these to test the performance and improve the efficiency of wind turbines. According to the Department of Energy, wind power generated more than 4% the electricity used in the United States in 2013.

This man is holding a scale model of his home. How could the model be helpful to him in choosing solar panels for his roof?

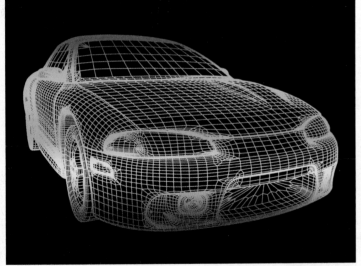

Scientists and engineers also create scale models using computers. This is a computer model of an automobile.

This scientist is using a handheld molecular model and a computer-generated model to carry out his investigation.

eighty-five

Models in Education and Entertainment

With the help of this oversized model, a young girl learns how to correctly brush her teeth.

Scale models are now being made in science labs and classrooms with 3D printers like this one.

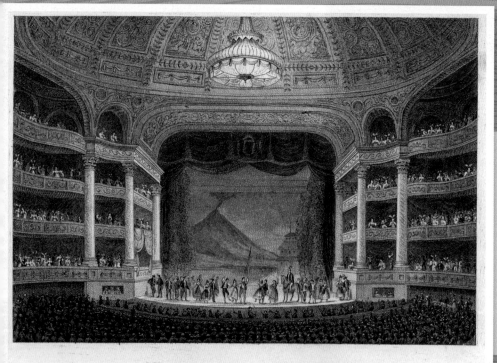

ACADEMIE ROYALE DE MUSIQUE.

Models are also created for the sets of operas, plays, ballets, and musicals. This painting shows a model set and backdrop used in the stage design for the opera *The Mute Girl of Portici* by Daniel Auber. The final act of the opera occurs with Mount Vesuvius erupting in the background.

The Number System

The Number System

Uses of Numbers

It is hard to live even one day without using or thinking about numbers. Numbers are used on clocks, calendars, license plates, rulers, scales, and so on.

- Numbers are used for counting.

Examples

Students sold 129 tickets to the school play.
The first U.S. Census counted 3,929,326 people.
The population of Malibu, California is 15,272.

- Numbers are used for measuring.

Examples

Ivan swam the length of the pool in 34.5 seconds.
The package is 31 inches long and weighs $4\frac{3}{8}$ pounds.

- Numbers are used to show where something is in a reference system.

Examples

Situation	Type of Reference System
Normal room temperature is 21°C.	Celsius temperature scale
David was born on June 22, 2006.	Calendar
The time was 2:39 P.M.	Clock time
Detroit is located at 42°N and 83°W.	Earth's latitude and longitude system

- Numbers are used to compare measures or counts.

Examples

The cat weighs $\frac{1}{2}$ as much as the dog.
There were 4 times as many boys as girls at the game.

- Numbers can be used for identification and as codes.

Examples

phone number: (709) 555-1212
ZIP code: 60637
driver's license number: M286-423-2061
bar code (used to identify product and manufacturer):
9 780021 308088
car license plate: HTX 585

The Number System

Kinds of Numbers

The **counting numbers** are the numbers used to count things: 1, 2, 3, 4, and so on. The **whole numbers** include all of the counting numbers and zero: 0, 1, 2, 3, 4, and so on.

Long ago, people found that these numbers did not meet all of their needs.

- Counting numbers cannot be used to express measures between two consecutive whole numbers, such as $2\frac{1}{2}$ inches or 1.6 kilometers.
- If you use only the counting numbers, you will not be able to give a precise answer to division problems such as $8 \div 5$ and $3 \div 7$.

Fractions and **decimals** were invented to meet those needs. Fractions are all numbers written as $\frac{a}{b}$, where a and b can be any **integers** ($b \neq 0$). Any fraction can be renamed as a decimal. Fractions are often used in the United States for cooking recipes and for measurements made in carpentry and other building trades. Decimals are used for almost all measurements made in science and industry. Money amounts are usually written as decimals.

Examples

The recipe calls for $2\frac{1}{2}$ cups of water.

The width of a quarter is 2.4 centimeters.

Most of the numbers you use, such as 8, $\frac{1}{2}$, $5\frac{1}{6}$, $1.75, and 1.23, are either fractions or can be renamed as fractions. With the invention of fractions, it became possible to name many more points on the number line, to solve any division problem involving whole numbers (except division by 0), and to express rates and ratios efficiently.

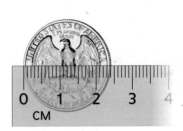

Note Every whole number can be renamed as a fraction. For example, 0 can be written as $\frac{0}{1}$ and 8 can be written as $\frac{8}{1}$.

The Number System

Numbers less than zero are needed to solve problems such as $5 - 7$ and $2\frac{3}{4} - 5\frac{1}{4}$, since they have answers that are less than 0. This led to the invention of **negative numbers.** Negative numbers are numbers that are less than 0. The numbers greater than 0 are called **positive numbers.** Negative numbers are used to describe some locations when there is a zero point or to indicate changes in quantities.

Examples

A temperature of 40 degrees below zero is written as $-40°F$ or $-40°C$.

A depth of 179 feet below sea level is written as -179 feet.

A weight loss of $6\frac{1}{2}$ pounds can be recorded as $-6\frac{1}{2}$ pounds.

Note A positive number may be written using a positive sign (+), but is usually written without one. For example, $+10 = 10$ and $\frac{1}{2} = +\frac{1}{2}$. A negative number is written using a negative sign (−). For example, -5 is read as "negative 5" or "the opposite of 5."

Situation	Negative (−)	Zero (0)	Positive (+)
bank account	withdrawal	no transaction	deposit
weight	loss	no change	gain
games	behind	even	ahead
elevation	below sea level	at sea level	above sea level

Two numbers are **opposite** numbers if they are the same distance from 0 on the number line, but on opposite sides of 0. The opposite of every positive number is a negative number. And the opposite of every negative number is a positive number. For example, the opposite of 3 is −3 and the opposite of −3 is 3. The diagram shows this relationship. The number 0 is neither positive nor negative. Zero is its own opposite.

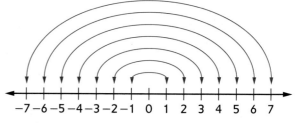

All of the whole numbers, together with their opposites, are called **integers.** The integers are 0, 1, −1, 2, −2, 3, −3, and so on.

Rational and Irrational Numbers

The **rational numbers** are all numbers that can be renamed as fractions. Counting numbers, whole numbers, integers, and fractions are all rational numbers. There are other numbers called **irrational numbers.** For example, to reverse squaring numbers, you take the square root. The number 36 is the **square** of 6 because $6 * 6$ or $6^2 = 36$. If you "unsquare" 36, the result is 6. The number 6 is called the **square root** of 36. Your calculator has a $\sqrt{}$ key to calculate square roots. If you take the square root of 2, the calculator reports a decimal with digits that do not repeat. The square root of 2 is an example of an irrational number. An irrational number cannot be renamed as a fraction. See the number line on the next page for examples of rational and irrational numbers.

The Number System

The Real Number Line

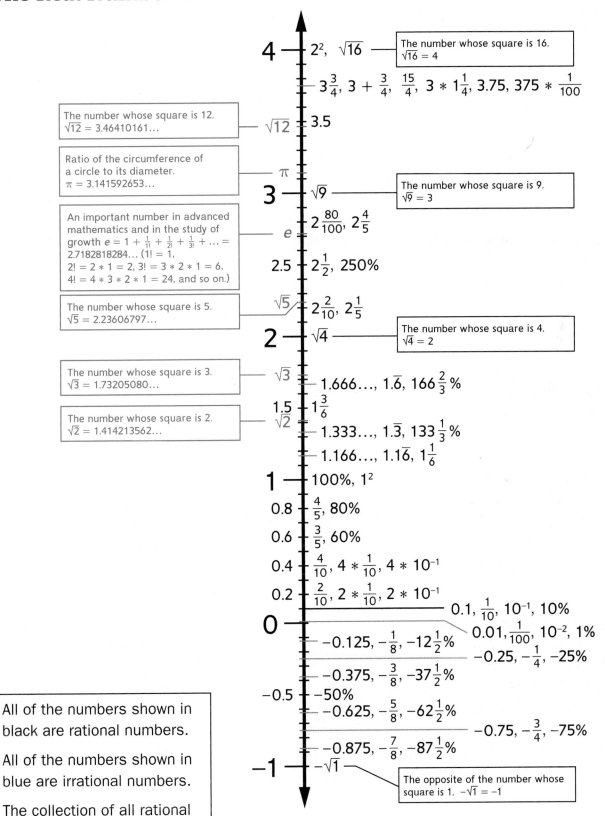

All of the numbers shown in black are rational numbers.

All of the numbers shown in blue are irrational numbers.

The collection of all rational and irrational numbers is called the **real numbers**.

ninety-one

The Number System

Absolute Value

The **absolute value** of a number is the distance between the number and 0 on the number line. The absolute value of any number is either positive or 0. The absolute value of a positive number is the number itself. For example, the absolute value of 4.5 is 4.5. The absolute value of a negative number is the opposite of the number. For example, the absolute value of -6 is 6. The absolute value of 0 is 0. Absolute value is shown by vertical lines before and after a number. Read $|-2|$ as "the absolute value of negative 2."

Examples
$|-7.25| = 7.25$ $|39| = 39$ $|0| = 0$

The absolute value of a number can tell you about the number's **magnitude,** or size. For example, Death Valley in California has the lowest elevation in the United States at -282 feet. This means that the elevation there is $|-282| = 282$ feet below sea level.

Absolute value can also be useful in finding the distance between two numbers. When both numbers are positive or both numbers are negative, you can find the distance between them by finding their absolute values and subtracting the smaller absolute value from the larger.

Example
Find the distance between -12 and -5.

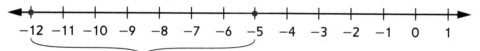

The distance is 7.

Both numbers are negative, so find their absolute values and subtract.
$|-12| = 12$ $|-5| = 5$
$12 - 5 = 7$, so the distance between -12 and -5 is 7.
This makes sense because you can count 7 unit intervals between -12 and -5.

The Number System

When one number is positive and the other is negative, you can add their absolute values to find the distance between them.

Example

Find the distance between $1\frac{3}{4}$ and $-2\frac{1}{4}$.

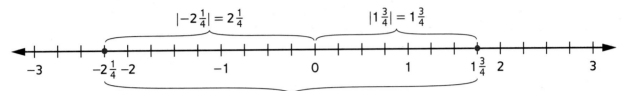

The two points are on opposite sides of 0, so the distance between the two points is the sum of their distances from 0.

$1\frac{3}{4}$ is $|1\frac{3}{4}|$ from 0. $-2\frac{1}{4}$ is $|-2\frac{1}{4}|$ from 0.

$|1\frac{3}{4}| + |-2\frac{1}{4}| = 1\frac{3}{4} + 2\frac{1}{4} = 4$

The distance between $1\frac{3}{4}$ and $-2\frac{1}{4}$ is 4.

This makes sense because you can count 16 fourths between $1\frac{3}{4}$ and $-2\frac{1}{4}$. $\frac{16}{4} = 4$

Check Your Understanding

Find the opposite and absolute value of each number.

1. -4.5
2. $7\frac{2}{3}$
3. 74
4. $-1,452$
5. Find the distance between 5 and -7.
6. Find the distance between -3 and -18.

Check your answers in the Answer Key.

ninety-three

The Number System

Number Lines

You can locate and plot positive and negative numbers on a number line. Number lines have arrows on each end to indicate that they extend forever in both directions. The number 0 is called the zero point or **origin.** By convention, on a horizontal number line, positive numbers are to the right of zero and negative numbers are to the left. On a vertical number line, positive numbers are above zero and negative numbers are below.

Each number and its opposite are the same distance from zero in opposite directions.

Example

Name the location of points A, B, and C.

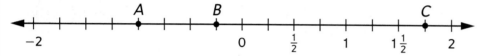

Point A and 1 are the same distance from 0, but in opposite directions. Point A must be the opposite of 1, or −1.

There are 4 equal spaces between whole numbers, so each space represents $\frac{1}{4}$.
Point B is $\frac{1}{4}$ to the left of 0, or $-\frac{1}{4}$.
Point C is $\frac{1}{4}$ more than $1\frac{1}{2}$ and $\frac{1}{4}$ less than 2. Point C is located at $1\frac{3}{4}$.

Comparing on a Number Line

When two numbers are shown on a horizontal number line, the number that is farther to the right is the larger number. For any pair of numbers on a vertical number line, the number that is higher is larger and the number that is lower is smaller.

Examples

Compare 3 and −1.

3 > −1 because 3 is higher than −1 on the vertical number line.

Compare −2.435 and −3.627.

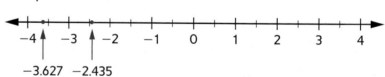

−2.435 > −3.627 because −2.435 is to the right of −3.627.

The Number System

Coordinate Grids

A rectangular **coordinate grid** is used to name points in a plane. It is made up of two number lines called **axes** that meet at right angles at their zero points. The horizontal number line is called the **x-axis,** and the vertical number line is called the **y-axis.** The point where the two lines meet is called the **origin.**

Every point on a rectangular coordinate grid can be named with an **ordered pair.** The two numbers that make up an ordered pair are called the **coordinates** of the point. The first coordinate, called the **x-coordinate,** always tells the horizontal distance and direction from the vertical axis. The second coordinate, called the **y-coordinate,** always tells the vertical distance and direction of the point from the horizontal axis. For example, the ordered pair (2, 4) names point A on the grid at the right. The numbers 2 and 4 are the coordinates of point A.

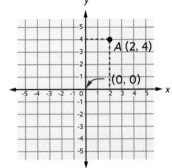

The ordered pair (0, 0) names the origin.

The four sections of the coordinate grid determined by the axes are called **quadrants.** The quadrants are typically numbered with Roman numerals in order to identify them. The point at (2, 4) has 2 positive coordinates, so it is in the top right quadrant, called Quadrant I (one). The other quadrants, II (two), III (three), and IV (four), are numbered in a counterclockwise direction around the origin.

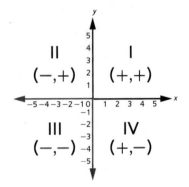

> **Example**
>
> Plot the ordered pair (4, 2).
>
> Locate 4 on the horizontal axis and draw a vertical line.
>
> Locate 2 on the vertical axis and draw a horizontal line.
>
> The point (4, 2) is located at the intersection of the two lines.
>
> The order of the numbers in an ordered pair is important. The ordered pair (4, 2) does not name the same point as the ordered pair (2, 4).
>
>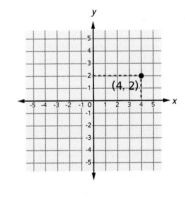

ninety-five **SRB 95**

The Number System

Example

Locate $(-4, 3)$, $(-4, -3)$, and $(2, 0)$.

For each ordered pair:
Locate the first coordinate on the horizontal axis and draw a vertical line.
Locate the second coordinate on the vertical axis and draw a horizontal line.
The two lines intersect at the point named by the ordered pair.

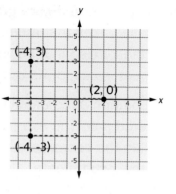

Finding Distance on a Coordinate Grid

You can find the distance between two points on a coordinate grid. To find the distance between two points on the same horizontal line (same y-coordinate) or on the same vertical line (same x-coordinate):

- Find the absolute values of the unlike coordinates (the coordinates in the ordered pairs that are different).
- If the points are in the same quadrant, subtract the absolute values of the unlike coordinates.
- If the points are in different quadrants, add the absolute values of the unlike coordinates.

Examples

Find the distance between point A and point B.

Point A is located at $(4, 3)$, and point B is located at $(1, 3)$. They are in the same quadrant and are on the same horizontal line. They have the same y-coordinate. Find the absolute values of the x-coordinates of point A and point B, then subtract.
$|4| - |1| = 3$

The distance between point A and point B is 3 units.

Find the distance between point A and point C.

Point A is located at $(4, 3)$, and point C is located at $(4, -2)$. They are in different quadrants, but are on the same vertical line. They have the same x-coordinate. Point A is $|3|$ units above the x-axis, and point C is $|-2|$ units below the x-axis.
The distance between point A and point C is the sum of their distances from the x-axis:
$|3| + |-2| = 3 + 2 = 5.$

The distance between point A and point C is 5 units.

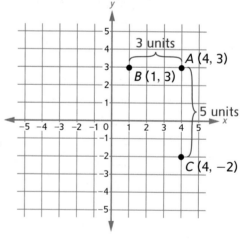

The Number System

Place Value for Whole Numbers

Any number, no matter how large or small, can be written using one or more of the **digits** 0, 1, 2, 3, 4, 5, 6, 7, 8, and 9. A place-value chart shows how much each digit in a number is worth. The **place** of a digit is its position in the number. The **value** of a digit is how much it is worth according to its place in the number.

Study the place-value chart below. Look at the numbers that name the places in the top row. Each number in the top row is **10 times as large as the number to its right** and $\frac{1}{10}$ **the size of the number to its left.** Finding $\frac{1}{10}$ of a number is the same as dividing that number by 10.

10,000s ten thousands	1,000s thousands	100s hundreds	10s tens	1s ones
9	4	8	0	5

Example

The number 94,805 is shown in the place-value chart above.

The value of 9 is 90,000 (9 ∗ 10,000).
The value of 4 is 4,000 (4 ∗ 1,000).
The value of 8 is 800 (8 ∗ 100).
The value of 0 is 0 (0 ∗ 10).
The value of 5 is 5 (5 ∗ 1).

94,805 is read as "ninety-four thousand, eight hundred five."

In larger numbers, groups of 3 digits are separated by commas. Commas help identify the thousands, millions, billions, trillions, quadrillions, quintillions, and so on.

Example

The number 246,357,026,909,389 is shown in the place-value chart.

trillions			billions			millions			thousands			ones		
100	10	1	100	10	1	100	10	1	100	10	1	100	10	1
2	4	6	3	5	7	0	2	6	9	0	9	3	8	9

This number is read 246 **trillion,** 357 **billion,** 26 **million,** 909 **thousand,** 389.

Check Your Understanding

Read each number to yourself. What is the value of the 6 in each number?

1. 26,482   **2.** 45,678,910   **3.** 207,464   **4.** 8,765,432

Check your answers in the Answer Key.

The Number System

Exponential Notation

A square array is an arrangement of objects into rows and columns that form a square. All rows and columns must be filled, and the number of rows must equal the number of columns. A counting number that can be represented by a square array is called a **square number.** Any square number can be written as the product of a counting number with itself.

two square arrays

> **Example**
>
> 16 is a square number. It can be represented by an array consisting of 4 rows and 4 columns. $16 = 4 * 4$

square array for 16

Here is a shorthand way to write the square number 16: $16 = 4 * 4 = 4^2$. 4^2 is read as "4 times 4," "4 squared," or "4 to the second power." The raised 2 is called an **exponent.** It tells that 4 is used as a factor two times (two 4s are multiplied). The 4 is called the base. Numbers written with an exponent are in **exponential notation.**

4^2 — exponent / base

Exponents are also used to show that a factor is used more than twice.

> **Examples**
>
> $2^3 = 2 * 2 * 2$
> 2 is used as a factor 3 times.
> 2^3 is read "2 cubed" or "2 to the third power."
>
> $9^5 = 9 * 9 * 9 * 9 * 9$
> 9 is used as a factor 5 times.
> 9^5 is read "9 to the fifth power."
>
> Any number raised to the first power is equal to itself. For example $5^1 = 5$.

Some calculators have special keys for changing numbers written in exponential notation to standard notation. Look at your calculator to see which keys it has for exponents.

> **Example**
>
> Use a calculator. Find the value of 2^6.
> On some calculators, key in 2 [^] 6 [Enter]. Answer 64.
> On other calculators, you may key in 2 [x^y] 6 [=]. Answer 64.
> $2^6 = 64$ You can verify by 2 [×] 2 [×] 2 [×] 2 [×] 2 [×] 2 [=].

> **Check Your Understanding**
>
> Write each number in standard notation.
>
> 1. 6^2 2. 4^3 3. 10^5 4. 9^1 5. 225^2 6. 11^5
>
> Check your answers in the Answer Key.

The Number System

Powers of 10

Numbers like 10, 100, and 1,000 are called **powers of 10**. They are numbers that can be written as products of 10s. For example, 100 can be written as 10 * 10 or 10^2. 1,000 can be written as 10 * 10 * 10 or 10^3.

A number written using place value, like 1,000, is in **standard form,** or **standard notation.** A number written with an exponent, like 10^3, is in **exponential notation.** The chart below shows powers of 10 from 10 through 1 billion.

Powers of Ten		
Standard Notation	Product of 10s	Exponential Notation
10	10	10^1
100	10*10	10^2
1,000 (1 thousand)	10*10*10	10^3
10,000	10*10*10*10	10^4
100,000	10*10*10*10*10	10^5
1,000,000 (1 million)	10*10*10*10*10*10	10^6
10,000,000	10*10*10*10*10*10*10	10^7
100,000,000	10*10*10*10*10*10*10*10	10^8
1,000,000,000 (1 billion)	10*10*10*10*10*10*10*10*10	10^9

Note 10^2 is read as "10 to the second power" or "10 squared." 10^3 is read as "10 to the third power" or "10 cubed." 10^4 is read as "10 to the fourth power."

Powers of 10 can be used to write other numbers using exponential notation. For example, the population of the world is about 7 billion people. The number 7 billion can be written in standard notation as 7,000,000,000 or using exponential notation as $7 * 10^9$.

Look at 10^9. 10^9 is the product of 10 used as a factor 9 times.

10^9 = 10 * 10 * 10 * 10 * 10 * 10 * 10 * 10 * 10
     = 1,000,000,000 = 1 billion

So $7 * 10^9$ = 7 * 1,000,000,000
            = 7,000,000,000
            = 7 billion

The world's population is about $7 * 10^9$.

Examples

Write the number using exponential notation.

5,000,000 = **?**
5,000,000 = 5 * 1,000,000
1,000,000 = 10 * 10 * 10 * 10 * 10 * 10
         = 10^6
So, 5,000,000 = **$5 * 10^6$**.

Write in standard notation.

$5.6 * 10^7$ = **?**
10^7 = 10 * 10 * 10 * 10 * 10 * 10 * 10
     = 10,000,000
So, $5.6 * 10^7$ = 5.6 * 10,000,000
               = **56,000,000.**

ninety-nine

The Number System

Expanded Form

The number 481,926 is written in **standard form,** or **standard notation,** the most common way of writing a number. When a number is written as the sum of the values of each digit, it is written in **expanded form.** There are several ways to represent a number in expanded form.

> **Note** Based on patterns with powers of 10, mathematicians agree that $10^0 = 1$. Each power of 10 is $\frac{1}{10}$ of the next power:
>
> $\frac{1}{10}$ of $10^1 = 10$
> $\frac{1}{10}$ of $10^2 = 100$
> $10^3 = 1,000$
>
> 10^0 must be $\frac{1}{10}$ of 10^1. $\frac{1}{10}$ of 10 is 1, so $10^0 = 1$.
> See page 110 for more information on 10^0 and other powers of 10.

Example

100,000s hundred thousands	10,000s ten thousands	1,000s thousands	100s hundreds	10s tens	1s ones
4	8	1	9	2	6

Write the number using words to show the place names:
4 hundred thousands + 8 ten thousands + 1 thousand + 9 hundreds + 2 tens + 6 ones

Write the number using the value of each digit in standard notation:
400,000 + 80,000 + 1,000 + 900 + 20 + 6

Write the number using equal-groups notation:
4 [100,000s] + 8 [10,000s] + 1 [1,000s] + 9 [100s] + 2 [10s] + 6 [1s]

Write the number as a sum of products of powers of 10:
(4 * 100,000) + (8 * 10,000) + (1 * 1,000) + (9 * 100) + (2 * 10) + (6 * 1)

Write the number as a sum of products using exponential notation:
$(4 * 10^5) + (8 * 10^4) + (1 * 10^3) + (9 * 10^2) + (2 * 10^1) + (6 * 10^0)$

Check Your Understanding

Write each number in standard notation.

1. (8 * 10,000) + (3 * 1,000) + (4 * 100) + (5 * 1)
2. $(2 * 10^6) + (6 * 10^5) + (7 * 10^4) + (3 * 10^3) + (9 * 10^2) + (5 * 10^1) + (2 * 10^0)$

Write each number in expanded form using exponential notation.

3. 8,744   4. 1,456,900

Check your answers in the Answer Key.

The Number System

Comparing Numbers and Amounts

When two numbers or amounts are compared, there are two possible results: They are equal, or they are not equal because one is larger than the other.

Different symbols are used to show that numbers and amounts are equal or not equal.

- Use an *equal sign* (=) to show that the numbers or amounts *are equal*.
- Use a *not-equal sign* (≠) to show that they are *not equal*.
- Use a *greater-than symbol* (>) or a *less-than symbol* (<) to show that they are *not equal* and to show which is larger.

Symbol	=	≠	>	<
Meaning	"equals" or "is the same as"	"is not equal to"	"is greater than"	"is less than"
Examples	$\frac{1}{2} = 0.5$ $3^3 = 27$ $2 * 5 = 9 + 1$	$2 \neq 3$ $3^2 \neq 6$ $1 \text{ m} \neq 100 \text{ mm}$	$1.42 > 1.4$ 16 ft 9 in. > 15 ft 11 in. $10^3 > 100$	$3 < 5$ $100 - 2 < 99 + 2$ $\frac{1}{10^3} < 1$

When you compare amounts that include units, use the *same unit* for both amounts.

Example

Compare 30 yards and 60 feet.

The units are different—yards and feet. Change yards to feet, then compare.
1 yd = 3 ft, so 30 yd = 30 * 3 ft, or 90 ft. Now compare feet. 90 ft > 60 ft

Therefore, 30 yd > 60 ft.

1 yard

1 foot

Or change feet to yards, and then compare. 1 ft = $\frac{1}{3}$ yd, so 60 ft = 60 * $\frac{1}{3}$ yd, or 20 yd.
Now compare yards. 30 yd > 20 yd

So, 30 yd > 60 ft.

Check Your Understanding

True or false?

1. $8^2 < 16$
2. 37 in. > 3 ft
3. $6 * 5 \neq 90 / 3$
4. $20 - 1 > 20 - 1$

Check your answers in the Answer Key.

The Number System

Factors of a Counting Number

One way to identify **factors** of a counting number is to make a rectangular **array,** an arrangement of objects in rows and columns that forms a rectangle. Each row has the same number of objects. Each column has the same number of objects. A multiplication **number model** can represent a rectangular array.

> **Note** When you find factors of a counting number, the factors must also be counting numbers. The counting numbers are 1, 2, 3, and so on.

Example

Find factors of 15.

Begin by making a rectangular array with 15 red dots.
This array has 3 rows with 5 dots in each row.
3 * 5 = 15 is a number model for this array.
3 and 5 are **factors** of 15.
15 is the **product** of 3 and 5.

3 and 5 are a **factor pair** for 15.

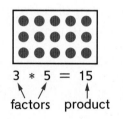

3 * 5 = 15
↑   ↑   ↑
factors product

Counting numbers can have more than one factor pair. 1 and 15 are another factor pair for 15 because 1 * 15 = 15.

To test whether a counting number m is a factor of another counting number n, divide n by m. If the result is a counting number and the remainder is 0, then m is a factor of n.

Examples

4 is a factor of 12 because 12 / 4 gives 3 with a remainder of 0.
6 is *not* a factor of 15 because 15 / 6 gives 2 with a remainder of 3.

You can find all the factors of a counting number by finding all its factor pairs.

Example

Find all the factors of the number 24.

Write all multiplication number sentences that equal 24. Drawing a factor rainbow with arcs to connect factor pairs can help you check to make sure you have found all of the factors.

Number Sentences	Factor Pairs	Factor Rainbow
24 = 1 * 24	1, 24	24
24 = 2 * 12	2, 12	
24 = 3 * 8	3, 8	
24 = 4 * 6	4, 6	1 2 3 4 6 8 12 24

The factors of 24 are 1, 2, 3, 4, 6, 8, 12, and 24.

Check Your Understanding

List all the factors of each number.

1. 51 **2.** 81 **3.** 44 **4.** 72 **5.** 128 **6.** 50

Check your answers in the Answer Key.

The Number System

Divisibility

When one counting number is divided by another counting number and the quotient is a counting number with a remainder of 0, then the first number is **divisible by** the second number. If the quotient has a non-zero remainder, then the first number is *not divisible by* the second number. A counting number is divisible by all of its **factors.**

Example

128 / 4 → 32 R0 The remainder is 0, so 128 is divisible by 4.

92 / 5 → 18 R2 The remainder is not 0, so 92 is not divisible by 5.

It is possible to test for divisibility without dividing.

Here are divisibility tests that you can use instead of dividing:

- All counting numbers are **divisible by 1.**
- Counting numbers with a 0, 2, 4, 6, or 8 in the ones place are **divisible by 2.** They are the **even numbers.**
- Counting numbers with 0 in the ones place are **divisible by 10.**
- Counting numbers with 0 or 5 in the ones place are **divisible by 5.**
- If the sum of the digits in a counting number is divisible by 3, then the number is **divisible by 3.**
- If the sum of the digits in a counting number is divisible by 9, then the number is **divisible by 9.**
- A counting number divisible by both 2 and 3 is **divisible by 6.**

Did You Know?

In books, magazines, and newspapers, the pages on the left side are almost always even-numbered.

Examples

Find some numbers 324 is divisible by. 324 is divisible by:
- 2 because 4 in the ones place is an even number.
- 3 because the sum of its digits is 3 + 2 + 4 = 9, and 9 is divisible by 3.
- 9 because the sum of its digits is divisible by 9.
- 6 because it is divisible by both 2 and 3.
- 324 is not divisible by 10 or 5 because it does not have a 0 or a 5 in the ones place.

Check Your Understanding

Which numbers are divisible by 2? By 3? By 5? By 6? By 9? By 10?

105   4,470   526   621   13,680

Check your answers in the Answer Key.

The Number System

Prime and Composite Numbers

A **prime number** is a counting number greater than 1 that has exactly two *different* factors: 1 and the number itself. A prime number is divisible only by 1 and itself. A **composite number** is a counting number that has more than two different factors. A composite number is divisible by at least three different counting numbers.

Examples

13 is a prime number because its only factors are 1 and 13. 13 has exactly two different factors.
18 is a composite number because it has more than two different factors. Its factors are 1, 2, 3, 6, 9, and 18.
1 is neither prime nor composite because the only factor of 1 is 1 itself.

Every composite number can be renamed as a product of prime numbers. This is called the **prime factorization** of that number.

Example

Find the prime factorization of 80.

The number 80 can be renamed as the product $2 * 2 * 2 * 2 * 5$.
The prime factorization of 80 can be written as $2^4 * 5$.

Note The prime factorization of a prime number is that number. For example, the prime factorization of 7 is 7.

You can make a **factor tree** to find the prime factorization of a composite number. First write the number. Then below the number, write any two factors whose product is that number. Repeat the process for these two factors. Continue until all the factors are prime numbers.

Example

Find the prime factorization of 36.

No matter which two factors are used to start the tree, the tree will always end with the same prime factors.

$36 = 2 * 2 * 3 * 3$

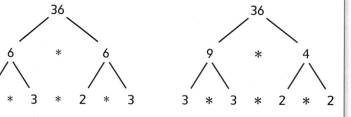

The prime factorization of 36 is $2 * 2 * 3 * 3$, or $36 = 2^2 * 3^2$.

Check Your Understanding

Make a factor tree to find the prime factorization of each number.
1. 12
2. 32
3. 42
4. 24
5. 50
6. 100

Check your answers in the Answer Key.

The Number System

Greatest Common Factor

The **greatest common factor (GCF)** of two counting numbers is the largest counting number that is a factor of both numbers.

> **Example**
> Find the greatest common factor of 20 and 24.
> **Step 1** List all the factors of 20: **1, 2, 4,** 5, 10, and 20.
> **Step 2** List all the factors of 24: **1, 2,** 3, **4,** 6, 8, 12, and 24.
> 1, 2, and 4 are common factors of 20 and 24.
> 4 is the greatest common factor of 20 and 24.

Note Problems like these are sometimes written using *greatest common factor notation*. This problem and solution could be written as GCF (20, 24) = 4.

Two numbers are called **relatively prime** if the only factor they have in common is 1. Two numbers can be relatively prime even when neither number is a prime number.

> **Example**
> Find the greatest common factor of 15 and 16.
> Factors of 15: **1,** 3, 5, 15   Factors of 16: **1,** 2, 4, 8, 16
> GCF (15, 16) = 1
> 15 and 16 don't have any common factors except for 1. They are relatively prime.

Another way to find the greatest common factor of two counting numbers is to use the grid method. This method is especially useful with larger numbers.

> **Example**
> Find the greatest common factor of 48 and 64.
>
> **Step 1**
> Make 3 columns. Write the numbers at the top of the middle and right columns.
>
> | | 48 | 64 |
> |---|---|---|
> | | | |
>
> **Step 2**
> Find a factor of both numbers, and record this number at the top of the left column.
> Both numbers are even, so 2 is a factor of both numbers.
>
> | 2 | 48 | 64 |
> |---|---|---|
> | | | |
>
> **Step 3**
> Divide each number by the factor and record the quotient under each number.
> 48 ÷ 2 = 24, and 64 ÷ 2 = 32.
>
> | 2 | 48 | 64 |
> |---|---|---|
> | | 24 | 32 |
>
> **Step 4**
> Repeat Steps 1–3 until the numbers in the last row are relatively prime (they have no common factors). 3 and 4 are relatively prime.
>
> | 2 | 48 | 64 |
> |---|---|---|
> | 2 | 24 | 32 |
> | 2 | 12 | 16 |
> | 2 | 6 | 8 |
> | | 3 | 4 |
>
> **Step 5** The product of all the common factors in the left column of the grid is the greatest common factor.
> The greatest common factor of 48 and 64 is 2 * 2 * 2 * 2 = 16. GCF (48, 64) = 16

The Number System

Least Common Multiples

A **multiple of a number n** is the product of any counting number and the number n. A multiple of a counting number n is always divisible by n. For example, 6 is a multiple of 3 because 3 * 2 = 6 and 6 is divisible by 3.

The **least common multiple (LCM)** of two numbers is the smallest number that is a multiple of both numbers.

> **Example**
> Find the least common multiple of 8 and 12.
> **Step 1** List multiples of 8: 8, 16, **24**, 32, 40, **48**, ...
> **Step 2** List multiples of 12: 12, **24**, 36, **48**, 60, ...
> 24 and 48 are common multiples of 8 and 12.
> 24 is the smallest possible common multiple, so it is the least common multiple of 8 and 12.

Note Problems like these are sometimes written using least common multiple notation. This problem and solution could be written as LCM (8, 12) = 24.

Another way to find the least common multiple of two counting numbers is to use the grid method. The grid is completed the same way as it is when finding the greatest common factor (see page 105). To find the least common multiple, multiply the common factors in the left column by the relatively prime factors of each number in the bottom row.

> **Example**
> Find the least common multiple of 48 and 64.
> Use the grid method to find greatest common factors from page 105.
> The least common multiple of 48 and 64 is 2 * 2 * 2 * 2 * 3 * 4 = 192.
> Check: Multiples of 48: 48, 96, 144, **192**, 240, ...
> Multiples of 64: 64, 128, **192**, 256, ...
>
> | 2 | 48 | 64 |
> |---|----|----|
> | 2 | 24 | 32 |
> | 2 | 12 | 16 |
> | 2 | 6 | 8 |
> | | 3 | 4 |

It is an interesting fact that the product of the least common multiple and the greatest common factor of two counting numbers is the same as the product of the two numbers themselves.

> **Example**
> The least common multiple of 4 and 6 is 12. The greatest common factor of 4 and 6 is 2.
> The product of the LCM and the GCF of 4 and 6 is 12 * 2, or 24, and the product of 4 and 6 is also 24.

The Number System

Decimals

Mathematics in everyday life uses more than **whole numbers.** Other numbers, called **decimals** and **fractions,** name numbers that are between whole numbers. Decimals and fractions are used to name a part of a whole thing or a part of a collection. We use decimals and fractions to make more precise measurements than can be made using only whole numbers.

Note Some fractions between 1 and 2: $\frac{3}{2}, \frac{7}{4}, \frac{11}{8}, \frac{19}{16}$
Some decimals between 1 and 2:
1.5, 1.75, 1.875, 1.9999

You probably see many uses of decimals every day.

- Fractional parts of a dollar are almost always written as decimals.
- Weather reports give rainfall amounts in decimals.
- Digital scales in supermarkets show the weight of fruits, vegetables, and meat with decimals.
- Many sports events are timed to a tenth or hundredth of a second, and the times are reported as decimals.
- Sports statistics often use decimals. For example, batting averages and average points scored per game are usually reported as decimals.

Did You Know?

Jim Hines was the first person to complete the 100-meter dash in less than 10 seconds. His time in the 1968 Olympics was 9.95 seconds and his record stood for 15 years.

Decimals are another way to write fractions. Many fractions have denominators of 10, 100, 1,000, and so on. It is easy to write the decimal names for fractions like these.

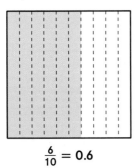

$\frac{6}{10} = 0.6$

This square is divided into 10 equal parts. Each part is $\frac{1}{10}$ of the square. The decimal name for $\frac{1}{10}$ is 0.1.
$\frac{6}{10}$ of the square above is shaded. The decimal name for $\frac{6}{10}$ is 0.6.

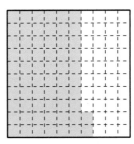

$\frac{62}{100} = 0.62$

This square is divided into 100 equal parts. Each part is $\frac{1}{100}$ of the square. The decimal name for $\frac{1}{100}$ is 0.01.
$\frac{62}{100}$ of the square above is shaded. The decimal name for $\frac{62}{100}$ is 0.62.

Like mixed numbers, decimals can be used to name numbers greater than one.

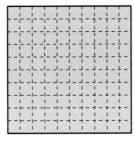 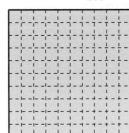 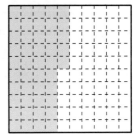

$2\frac{45}{100} = 2.45$

one hundred seven SRB 107

The Number System

Writing Decimals

In a decimal, the dot is called the **decimal point.** It separates the whole-number part from the decimal part. A decimal with one digit after the decimal point names *tenths*. A decimal with two digits after the decimal point names *hundredths*. A decimal with three digits after the decimal point names *thousandths*.

$$12\underset{\text{whole-number part}}{.}\underset{\text{decimal part}}{105}$$

Examples

tenths	hundredths	thousandths
$0.5 = \frac{5}{10}$	$0.43 = \frac{43}{100}$	$0.291 = \frac{291}{1,000}$
$0.7 = \frac{7}{10}$	$0.75 = \frac{75}{100}$	$0.003 = \frac{3}{1,000}$
$0.9 = \frac{9}{10}$	$0.08 = \frac{8}{100}$	$0.072 = \frac{72}{1,000}$

Did You Know?

Decimals were invented by the Dutch scientist Simon Stevin in 1585. But there is no single, worldwide form for writing decimals. People in the United States, Australia, and most Asian countries write the decimal 3.25. In the United Kingdom, the decimal is written 3·25, and in some parts of Europe and South America, the decimal is written 3,25.

Reading Decimals

One way to read a decimal is to say it as you would a fraction or mixed number. For example, $0.001 = \frac{1}{1,000}$ and can be read as "one thousandth." $7.9 = 7\frac{9}{10}$, so 7.9 can be read as "seven and nine tenths."

You can read decimals by first saying the whole-number part, then saying "point," and finally saying the digits in the decimal part. For example, 6.8 can be read as "six point eight"; 0.15 can be read as "zero point one five." This way of reading decimals is often useful when there are many digits in the decimal.

Examples

0.26 is read as "26 hundredths" or "zero point two six."

34.5 is read as "34 and 5 tenths" or "34 point 5."

0.004 is read as "4 thousandths" or "0 point zero zero four."

Check Your Understanding

Read each decimal to yourself in two ways.

1. 0.4   **2.** 1.65   **3.** 0.872   **4.** 16.04   **5.** 0.003   **6.** 59.061

Check your answers in the Answer Key.

The Number System

Extending Place Value to Decimals

The base-10 system works the same way for decimals as it does for whole numbers.

Example

1,000s thousands	100s hundreds	10s tens	1s ones	.	0.1s tenths	0.01s hundredths	0.001s thousandths
		4	7	.	8	0	5

In the number 47.805,
    8 is in the **tenths** place; its value is 8 tenths, or $\frac{8}{10}$, or 0.8.
    0 is in the **hundredths** place; its value is 0.
    5 is in the **thousandths** place; its value is 5 thousandths, or $\frac{5}{1,000}$, or 0.005.

Decimals can be written in **expanded form.** For the example above,

$$47.805 = 40 + 7 + 0.8 + 0.00 + 0.005$$
$$= (4 * 10) + (7 * 1) + (8 * 0.1) + (0 * 0.01) + (5 * 0.001)$$
$$= (4 * 10) + (7 * 1) + \left(8 * \frac{1}{10}\right) + \left(0 * \frac{1}{100}\right) + \left(5 * \frac{1}{1,000}\right)$$
$$= (4 * 10^1) + (7 * 10^0) + \left(8 * \frac{1}{10^1}\right) + \left(0 * \frac{1}{10^2}\right) + \left(5 * \frac{1}{10^3}\right)$$

Study the place-value chart below. Look at the numbers that name the places. A digit in one place represents **10 times as much as it would represent in the place to its right.**

*10	*10	*10	*10	*10	*10

1,000s thousands	100s hundreds	10s tens	1s ones	.	0.1s tenths	0.01s hundredths	0.001s thousandths

1 [1,000] = 10 [100s]   1 [1] = 10 [$\frac{1}{10}$s]

1 [100] = 10 [10s]   1 [$\frac{1}{10}$] = 10 [$\frac{1}{100}$s]

1 [10] = 10 [1s]   1 [$\frac{1}{100}$] = 10 [$\frac{1}{1,000}$s]

one hundred nine

The Number System

Study the place-value chart below. Look at the numbers that name the places. A digit in one place represents $\frac{1}{10}$ **of what it would represent in the place to its left.** Finding $\frac{1}{10}$ of a number is the same as dividing that number by 10.

1,000s thousands	100s hundreds	10s tens	1s ones	.	0.1s tenths	0.01s hundredths	0.001s thousandths

Each arrow above shows $\frac{1}{10}$ of the place to its left.

$1\ [100] = \frac{1}{10}$ of 1,000

$1\ [10] = \frac{1}{10}$ of 100

$1\ [1] = \frac{1}{10}$ of 10

$1\ [\frac{1}{10}] = \frac{1}{10}$ of 1

$1\ [\frac{1}{100}] = \frac{1}{10}$ of $\frac{1}{10}$

$1\ [\frac{1}{1,000}] = \frac{1}{10}$ of $\frac{1}{100}$

Powers of 10 for Decimals

All of the numbers across the top of a place-value chart are **powers of 10.** A power of 10 is a whole number that can be written using only 10s as factors. Powers of 10 can be written in exponential notation.

Powers of 10 (Greater than 1)

Standard Notation	Product of 10s	Exponential Notation
10	10	10^1
100	10*10	10^2
1,000	10*10*10	10^3
10,000	10*10*10*10	10^4

Note A number written in the usual place-value way, like 100, is in **standard notation**. Another name for standard notation is **standard form**. A number written with an exponent, like 10^2, is in **exponential notation**.

Decimals that can be written using only 0.1s as factors are also powers of 10. They can be written in exponential notation with negative exponents.

Powers of 10 (Less than 1)

Standard Notation	Product of 0.1s	Exponential Notation
0.1	0.1	10^{-1}
0.01	0.1*0.1	10^{-2}
0.001	0.1*0.1*0.1	10^{-3}
0.0001	0.1*0.1*0.1*0.1	10^{-4}

Note A number raised to a negative exponent power is equal to the fraction that is written with one over the number raised to the positive exponent power. For example,

$10^{-2} = \frac{1}{10^2}$

$= \frac{1}{10 * 10}$

$= \frac{1}{100} = 0.01.$

The number 1 is a power of 10 because $1 = 10^0$. The pattern in the table below shows why mathematicians define 10^0 as equal to 1. Each exponent is one less than the exponent in the place to its left.

Note the pattern in the exponents:

10,000s	1,000s	100s	10s	1s		0.1s	0.01s	0.001s	0.0001s
10^4	10^3	10^2	10^1	10^0	.	10^{-1}	10^{-2}	10^{-3}	10^{-4}

one hundred ten

The Number System

Comparing Decimals

You may use place value to compare decimals in the same way you compare whole numbers. One way is to model the decimals with base-10 blocks or shorthand pictures of the blocks.

For the examples on this page:

A flat ☐ is worth 1.

A long | is worth 0.1.

A cube ■ is worth 0.01

Example

Compare 2.3 and 2.16.

2.3   2.16

2 flats and 3 longs are more than 2 flats, 1 long, and 6 cubes.

So, 2.3 is greater than 2.16.

2.3 > 2.16

Another way to compare decimals is to attach one or more zeros to the end of the decimal part of a number without changing the value of the number. This can make comparing easier.

Examples

Compare 0.3 and 0.06.
You can attach a 0 to 0.3 without changing its value by writing 0.3<u>0</u>.
Now compare 0.30 and 0.06.
30 hundredths is more than 6 hundredths.
0.30 > 0.06, so 0.3 > 0.06.

Compare 1.172 and 1.4.
You can attach 0s to 1.4 without changing its value by writing 1.4<u>00</u>.
Now compare 1.172 and 1.400.
Both show one whole and a part of a whole.
172 thousandths is less than 400 thousandths.
1.172 < 1.400, so 1.172 < 1.4.

Here is another way to compare decimals. Start by comparing digits in the leftmost place. Continue to the right until the digits in a place do not match.

Examples

Compare 1.35 and 1.288.

1.35
1.288
  ↑ different
  3 tenths > 2 tenths
So, 1.35 > 1.288.

Compare 0.5 and 0.105.

0.5
0.105
 ↑ different
 5 tenths > 1 tenth
So, 0.5 > 0.105.

one hundred eleven

The Number System

Comparing Decimals on a Number Line

Decimals are often shown on a number line. You can use a number line to compare decimals or to find decimals between decimals. For any pair of numbers on the number line, the number to the left is less than the number to the right.

Example

Compare 1.25 and 1.4.

Sketch a number line.

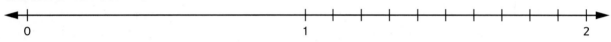

Plot the points on the number line.

1.4 is 4 tenths more than 1.

25 is halfway between 20 and 30, so 1.25 is halfway between 1.20 and 1.30.

1.25 < 1.4

To find decimals between decimals, you can imagine "zooming in" on the number line.

Example

Find a decimal number between 1.45 and 1.46.

Step 1 Begin with a number line with whole numbers.
Step 2 Zoom in to see the tenths between 1 and 2. Locate 1.4 and 1.5.
Step 3 Zoom in to see the hundredths between 1.4 and 1.5. Locate 1.45 and 1.46 between 1.4 and 1.5.
Step 4 Zoom in again between 1.45 and 1.46. You have found some of the decimal numbers between 1.45 and 1.46.

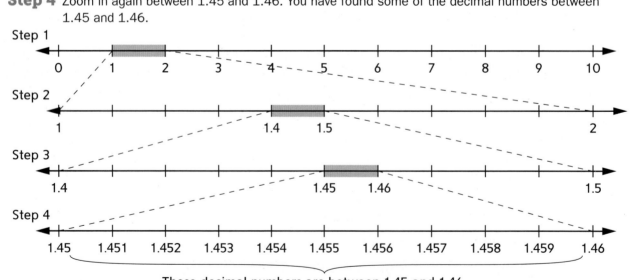

These decimal numbers are between 1.45 and 1.46.

The Number System

Estimation

An **estimate** is a number that is close to an exact answer. You **estimate** when you use a nearby number to help you make sense of a situation or make an approximate calculation. Making estimates requires mathematical reasoning.

Example

The school play will be performed in 1 month, 3 weeks, and 4 days.

When a girl says, "The play will be performed in about 2 months," she's using an estimate based on knowing that there are about 4 weeks in a month.

Estimation is also useful when you need to find an exact answer to a calculation. You can make an estimate when you start working on a problem to help you understand the problem better. You can also use your estimate after you calculate to check whether your answer makes sense. If your answer is not close to your estimate, then you need to use another method of estimation or check your work, correct it, and check whether the new answer makes sense.

One way to produce an estimate is to keep the digit in the highest place value and replace the rest of the digits with zeros. This is called **front-end estimation.**

Examples

How much will 6 pens cost if the price is 74¢ per pen?
  The digit in the highest place value in **74¢** is the **7** in the tens place.
  Use **70¢**.
  Calculate: 6 * 70¢ = 420¢, or $4.20
Estimate: The 6 pens will cost a little more than $4.20.

A boy calculates 4.2 + 21.35. His answer is 21.77. Is he correct?
  Find front-end estimates for each number in the problem:
    4.2 → 4.0, or 4
    21.35 → 20.00, or 20
  Calculate: 4 + 20 = 24

Estimate: The exact answer is more than 24. The boy's answer is incorrect. He needs to check his work.

Did You Know?

Sometimes estimates are used to make sense of a situation when the exact number cannot be known. For example, scientists can estimate the number of fish in a lake by counting a sample and using that information to estimate the total number of fish in the lake. But it is impossible to know the exact number because the scientists cannot see all the fish and the fish move about, entering and leaving the lake at different times.

Note Sometimes *front-end estimation* is called *leading-digit estimation.*

Note Whenever you use front-end estimation for addition and multiplication with numbers greater than one, your resulting estimate is always less than the exact answer. This is because all of the digits other than the front end are replaced with zeros. So the numbers you use to calculate the estimate are always less than those in the original problem.

one hundred thirteen

The Number System

When calculating, some numbers are easier to mentally compute than others. Selecting **close-but-easier numbers** can help you estimate the answers for calculations.

Example

249 + 345 + 185 = **?**

Estimate: 249 is close to 250, 345 is close to 350, and 185 is close to 200.

250 + 350 + 200 = 800

Calculate:

```
    1
    2 4 9
    3 4 5
  + 1 8 5
  ───────
    6̶ 7̶ 9̶
```

Check: The numbers used to make the estimate were very close to the original numbers. 679 seems too far away from the estimate of 800. Check the work again.

Correct the work:

```
    1 1
    2 4 9
    3 4 5
  + 1 8 5
  ───────
    7 7 9
```

Check again: There was an error before. 779 seems more reasonable because it is closer to the estimate of 800.

Example

Estimate: 23.2 ∗ 0.43 = ?

One way:

0.43 is close to a half, or 0.5, and 23.2 is close to 24.
24 ∗ 0.5 → half of 24 is 12

So, 23.2 ∗ 0.43 is about 12.

Another way:

0.43 is close to a half, or 0.5, and 23.2 is close to 20.
20 ∗ 0.5 → half of 20 is 10

So, 23.2 ∗ 0.43 is about 10.

Another way:

Use an easy mental math problem.

25 ∗ 4 = 100 is easy.
Since 23.2 is close to 25 and 0.43 is close to 0.4, you can think about 25 ∗ 0.4 = ?
Since 25 ∗ 4 = 100, 25 ∗ 0.4 (or 25 ∗ 4 tenths) would be 100 tenths, or 10.

So, 23.2 ∗ 0.43 is about 10.

10 and 12 are both reasonable estimates for 23.2 ∗ 0.43.

Check Your Understanding

Estimate. Show how you estimated for Problems 3 and 6.

1. 16.24 − 8.95 = ?
2. 9.87 ∗ 3.104 = ?
3. 35.534 / 11.8 = ?
4. 152 + 387 + 212 = ?
5. 679 ∗ 43 = ?
6. 758 / 31 = ?

Check your answers in the Answer Key.

The Number System

Rounding

Rounding is one way to make sense of numbers in real-world situations or to estimate an answer for a calculation. You can round by using or thinking about number lines, or you can use a shortcut.

Rounding Using Number Lines

When using a number line to round a given number:

- Decide what place you are rounding to. Identify the two numbers you can round to. These numbers will be multiples of 10, 100, or 1,000 (and so on) on both sides of the number.
- Sketch a number line that shows the distance between the two numbers you can round to, mark and label the location of the halfway point, and mark and label the approximate location of the number to be rounded.
- Decide whether to round up to the higher number or down to the lower number.

Example

Round 4,611 to the nearest 1,000.

Find the two multiples of 1,000 that 4,611 is between.
4,000 and 5,000 are multiples of 1,000. These are the two numbers you can round to.
4,611 is more than 4,000 and less than 5,000.
Next sketch (or visualize) a number line showing 4,000 and 5,000 and the point halfway between.

Then estimate where 4,611 is on the number line. Label it.

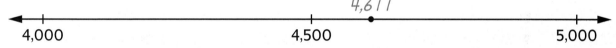

Think: Is 4,611 closer to 4,000 or 5,000?

4,611 is closer to 5,000. 4,611 rounded to the nearest thousand is 5,000.

Example

Round 4,611 to the nearest 100.

Find the two multiples of 100 that 4,611 is between.
4,600 and 4,700 are multiples of 100. These are the two numbers you can round to.
4,611 is more than 4,600 and less than 4,700.
Next sketch (or visualize) a number line showing 4,600 and 4,700. Mark the halfway point and 4,611.

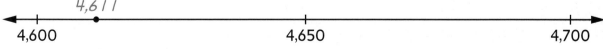

Think: Is 4,611 closer to 4,600 or 4,700?

4,611 is closer to 4,600 than 4,700. 4,611 rounded to the nearest hundred is 4,600.

one hundred fifteen **SRB 115**

The Number System

Rounding Numbers Using a Shortcut

Here is another way to round numbers.

Step 1 Find the digit in the place you are rounding to.

Step 2 Rewrite the number, replacing all digits to the right of this digit with zeros. This is the *lower number*.

Step 3 Add 1 to the digit in the place you are rounding to. If the sum is 10, write 0 and add 1 to the digit to its left. This is the *higher number*.

Step 4 Ask, "Is the number I am rounding closer to the lower number or the higher number?"

Step 5 Round to the closer of the two numbers.

Note: If the number being rounded is exactly halfway between the higher and lower numbers, use real-world information or the problem situation to decide whether to round up or down. For example, to estimate the sum of 3.67 + 2.5, round the addends to the nearest whole number. Round 3.67 to 4, because 3.67 is closer to 4 than 3. Since 2.5 is exactly halfway between 2 and 3, you can round up or down. It makes sense to round 2.5 down to 2, since you rounded up 3.67. A reasonable estimate for 3.67 + 2.5 is 4 + 2 = 6.

When there is not any information in the real-world or problem situation to help you decide whether to round up or down, the "rounding up" rule usually applies. That means you round up to the higher multiple or number.

Examples

Round each decimal to the place shown in the table.

	2.851 (nearest tenth)	8.35 (nearest tenth)	2.891 (nearest hundredth)
Step 1 Find the place you are rounding to.	2.8̲51	8.3̲5	2.89̲1
Step 2 Find the lower number.	2.800	8.30	2.890
Step 3 Find the higher number.	2.900	8.40	2.900
Step 4 Is it closer to the lower or higher number?	higher	halfway	lower
Step 5 Round to the closer number.	2.900 = 2.9	8.40 = 8.4	2.890 = 2.89

The Number System

Addition and Subtraction of Decimals

There are many ways to add and subtract decimals. One way is to use base-10 blocks. When working with decimals, we usually say that the flat is worth 1.

To add with base-10 blocks, count out blocks for each number and put all the blocks together. Make any trades for larger blocks that you can. Then count the blocks for the sum.

To subtract with base-10 blocks, count out blocks for the larger number. Take away blocks for the smaller number, making trades as needed. Then count the remaining blocks.

Using base-10 blocks is a good idea, especially at first. However, drawing shorthand pictures is usually easier and quicker.

For the examples on this page:
A flat ☐ is worth 1.
A long | is worth 0.1.
A cube ■ is worth 0.01

Example

1.61 + 4.7 = ?

First draw pictures for each number.

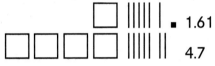

Next draw a ring around 10 longs and trade them for 1 flat.

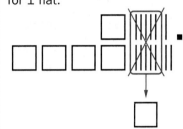

After the trade, there are 6 flats, 3 longs, and 1 cube.
This means that 1.61 + 4.7 = 6.31. This makes sense because 1.61 is near $1\frac{1}{2}$ and 4.7 is near $4\frac{1}{2}$. So the answer should be near $1\frac{1}{2} + 4\frac{1}{2}$, or 6, which it is.
1.61 + 4.7 = **6.31**

Example

4.07 − 2.7 = ?

The drawing for 4.07 shows 4 flats, 0 longs, and 7 cubes.

You want to take away 2.7 (2 flats and 7 longs). To do this, trade 1 flat for 10 longs.

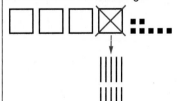

Now remove 2 flats and 7 longs (2.7).

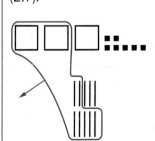

1 flat, 3 longs, and 7 cubes are left. These blocks show 1.37.
4.07 − 2.7 = **1.37**

The Number System

Addition Methods

Partial-Sums Addition

Partial-sums addition is used to find sums mentally or with paper and pencil. For both whole numbers and decimals, you must line up the place values of the numbers correctly. To use partial-sums addition, first add the value of the digits in each place separately. Then add the partial sums. Use an estimate to check whether the answer is reasonable.

Note To make your written work more efficient, you don't need to write the steps shown in green because steps such as estimating or adding the 100s, 10s, and 1s can be done using mental math.

Example

348 + 177 = ?

Estimate: 348 is close to 350, and you can round 177 to 200. 350 + 200 = 550

		3	4	8
	+	1	7	7
Add the 100s. 300 + 100 →		4	0	0
Add the 10s. 40 + 70 →		1	1	0
Add the 1s. 8 + 7 →			1	5
Add the partial sums.		5	2	5

348 + 177 = **525**

The answer is reasonable because it is close to the estimate of 550.

Example

4.56 + 7.9 = ?

Estimate: You can round 4.56 to 4.6 and 7.9 to 8. 4.6 + 8 = 12.6

	1s	0.1s	0.01s
	4 .	5	6
+	7 .	9	0
Add the ones. 4 + 7 →	11 .	0	0
Add the tenths. 0.5 + 0.9 →	1 .	4	0
Add the hundredths. 0.06 + 0.00 →	0 .	0	6
Add the partial sums.	12 .	4	6

4.56 + 7.9 = **12.46**

The answer of 12.46 makes sense because it is close to the estimate of 12.6.

The Number System

Column Addition

Column addition can be used to find sums with paper and pencil. To add numbers using column addition:

- Add the numbers in each place-value column. Write each sum in its column.
- If the sum of any column is a 2-digit number, make a trade with the column to the left.

Use an estimate to check whether the answer is reasonable.

Example

$359 + 289 = $ **?**

Estimate: 359 is close to 350, and 289 is close to 300.
An estimated sum is $350 + 300 = 650$.

	100s	10s	1s
	3	5	9
+	2	8	9
Add the numbers in each column.	5	13	18
Trade 10 ones for 1 ten. Move 1 ten to the tens column.	5	14	8
Trade 10 tens for 1 hundred. Move the 1 hundred to the hundreds column.	6	4	8

$359 + 289 = $ **648**

The answer is reasonable because it is close to the estimate of 650.

Example

$7.3 + 3.98 = $ **?**

Estimate: You can round 3.98 to 4. An estimated sum is $7.3 + 4 = 11.3$.

	1s	.	0.1s	0.01s
	7	.	3	0
+	3	.	9	8
Add the numbers in each column.	10	.	12	8
Trade 12 tenths for 1 and 2 tenths. Move the 1 into the ones column.	11	.	2	8

$7.3 + 3.98 = $ **11.28**

The answer makes sense because it is close to the estimate of 11.3.

one hundred nineteen   SRB 119

The Number System

U.S. Traditional Addition

In **U.S. traditional addition** you add from right to left one column at a time, making trades mentally as you go. Partial sums are not recorded.

> **Did You Know?**
>
> Writing numbers above the addends in U.S. traditional addition is sometimes called "carrying." Digits that are carried, such as the blue 1 in the tens column in Step 1 below, are sometimes called "carry marks" or "carries."

Example

248 + 187 = ?

Estimate: 248 is close to 250, and you can round 187 to 200. An estimated sum is 250 + 200 = 450.

Step 1 Add the ones.

```
    1
  2 4 8
+ 1 8 7
───────
        5
```

8 ones + 7 ones = 15 ones = 1 ten and 5 ones

Step 2 Add the tens.

```
  1 1
  2 4 8
+ 1 8 7
───────
      3 5
```

1 ten + 4 tens + 8 tens = 13 tens = 1 hundred and 3 tens

Step 3 Add the hundreds.

```
  1 1
  2 4 8
+ 1 8 7
───────
    4 3 5
```

1 hundred + 2 hundreds + 1 hundred = 4 hundreds

248 + 187 = **435**

The answer is reasonable because it is close to 450.

> **Note** Addends are numbers that are added. In 3 + 4 = 7, the numbers 3 and 4 are addends.

Example

9.23 + 4.29 = ?

Estimate: 9.23 rounds to 9, and 4.29 rounds to 4. An estimated sum is 9 + 4 = 13.

Step 1 Start with the 0.01s:

```
      1
  9 . 2 3
+ 4 . 2 9
─────────
        2
```

3 hundredths + 9 hundredths = 12 hundredths
12 hundredths = 1 tenth + 2 hundredths

Step 2 Add the 0.1s:

```
      1
  9 . 2 3
+ 4 . 2 9
─────────
      5 2
```

1 tenth + 2 tenths + 2 tenths = 5 tenths

Step 3 Add the 1s: 9 + 4 = 13.

```
      1
  9 . 2 3
+ 4 . 2 9
─────────
 1 3 . 5 2
```

13 ones = 1 ten and 3 ones
Remember to include the decimal point in the answer.

9.23 + 4.29 = **13.52**

The answer of 13.52 makes sense because it is close to the estimate of 13.

The Number System

Subtraction Methods

Most paper-and-pencil strategies for subtracting whole numbers also work for decimals. For both whole numbers and decimals, you must line up the place values of the numbers correctly when adding or subtracting.

Trade-First Subtraction

To use **trade-first subtraction,** compare each digit in the top number with each digit below it and make any needed trades before subtracting.

- If each digit in the top number is greater than or equal to the digit below it, subtract separately in each column.
- If any digit in the top number is less than the digit below it, make a trade with the digit to the left before doing any subtraction. Mark the problem to show each trade.

Use an estimate to check whether the answer is reasonable.

> **Example**
>
> Subtract 275 from 463 using trade-first subtraction.
>
> Estimate: You can round 463 to 500 and 275 to 300.
> An estimated difference is 500 − 300 = 200.
>
> | | | | | | | | | | | | |
> |---|---|---|---|---|---|---|---|---|---|---|---|
> | | 4 | 6 | 3 | | | 4 | 5̶ | 13 | | 3 | 15 | 13 |
> | − | 2 | 7 | 5 | | − | 2 | 7 | 5 | | − | 2 | 7 | 5 |
> | | | | | | | | | | | 1 | 8 | 8 |
>
> Look at the 1s place.
>
> 3 < 5, so you need to make a trade.
>
> Trade 1 ten for 10 ones.
>
> Mark the problem to show the trade. Now look at the 10s place.
>
> 5 < 7, so you need to make a trade.
>
> Trade 1 hundred for 10 tens.
>
> Now subtract in each column.
>
> 463 − 275 = **188**
>
> The answer is reasonable because it is close to the estimate of 200.

The Number System

The example below shows how to use trade-first subtraction with decimals.

Note If you can keep track of the places in your head, you don't need to draw lines between the columns or label the columns.

Example

9.4 − 4.85 = ?

Estimate: 9.4 rounds to 9. You can round 4.85 to 5.
An estimated difference is 9 − 5 = 4.

Remember to line up the digits in the correct place-value columns. For this problem, attach a zero in the hundredths place to 9.4 (9.4 = 9.40) to help you line up the digits correctly.

1s	0.1s	0.01s
9.	4	0
− 4.	8	5

Look at the 0.01s place.

0 < 5, so you need to make a trade.

1s	0.1s	0.01s
	3	10
9.	$\cancel{4}$	$\cancel{0}$
− 4.	8	5

Trade 1 tenth for 10 hundredths.

Mark the problem to show the trade. Now look at the 0.1s place.

3 < 8, so you need to make a trade.

1s	0.1s	0.01s
	13	
8	$\cancel{3}$	10
$\cancel{9}.$	$\cancel{4}$	$\cancel{0}$
− 4.	8	5
4.	5	5

Trade 1 one for 10 tenths. Now subtract in each column.

9.4 − 4.85 = **4.55**

This answer makes sense because it is close to the estimate of 4.

The Number System

Counting-Up Subtraction

You can use **counting-up subtraction** to find the difference between two numbers by counting up from the smaller number to the larger number. There are many ways to count up. One way is to start by counting up to the nearest multiple of 10, then continue counting by 10s and 100s.

> **Example**
>
> 425 − 48 = **?**
>
> Estimate: 48 is close to 50.
>
> An estimated difference is 425 − 50 = 375.
>
> Write the smaller number, 48.
>
> ```
>        4 8
>   +     (2)    Count up to the nearest 10.
>   ─────────
>        5 0
>   +   (5 0)   Count up to the nearest 100.
>   ─────────
>      1 0 0
>   + (3 0 0)   Count up to the largest possible
>   ─────────   hundred.
>      4 0 0
>   +   (2 5)   Count up to the larger number
>   ─────────
>      4 2 5
> ```
>
> As you count from 48 to 425, circle each number that you count up.
>
> Add the numbers you circled: 2 + 50 + 300 + 25 = 377.
> You counted up 377.
>
> 425 − 48 = **377**
>
> The answer is reasonable because it is close to the estimate of 375.

one hundred twenty-three

The Number System

You can use **counting-up subtraction** to find the difference between two decimals.

Example

9.40 − 4.85 = ?

Estimate: 4.85 is close to 5.
An estimated difference is 9.40 − 5 = 4.40.

Write the smaller number, 4.85.

As you count from 4.85 to 9.40, circle each number that you count up.

```
    4.85
+ (0.15)   Count up to the nearest whole.
    5.00
+ (4.00)   Count up to the largest possible
    9.00   whole.
+ (0.40)   Count up to the larger number.
    9.40
```

Add the numbers you circled: 0.15 + 4.00 + 0.40 = 4.55.
You counted up 4.55.

9.40 − 4.85 = **4.55**

The answer 4.55 makes sense because it is close to the estimate of 4.40.

Check Your Understanding

Solve using counting-up subtraction. Estimate to check whether your answers are reasonable.

1. 84 − 38
2. 653 − 362
3. 535 − 293
4. 818 − 746
5. 21.5 − 7.35
6. 10 − 1.79
7. 7.4 − 3.082
8. 1.19 − 0.45

Check your answers in the Answer Key.

U.S. Traditional Subtraction

You can subtract numbers using **U.S. traditional subtraction.** Start at the right. Subtract column by column. Make any necessary trades as you go. Use an estimate to check whether the answer is reasonable.

Sometimes you need to make a trade in only one step.

Example

579 − 385 = ?

Estimate: You can round 579 to 600 and 385 to 400.
An estimated difference is 600 − 400 = 200.

Step 1 Start with the 1s. Since 9 > 5, you do not need to regroup. Subtract the ones.

```
  5 7 9
−  3 8 5
─────────
        4
```

9 ones − 5 ones = 4 ones

Step 2 Go to the 10s. Since 7 < 8, you need to regroup. Trade 1 hundred for 10 tens. Subtract the tens.

```
    4 17
  5̷ 7̷ 9
− 3 8 5
─────────
    9 4
```

17 tens − 8 tens = 9 tens

Step 3 Subtract the hundreds.

```
    4 17
  5̷ 7̷ 9
− 3 8 5
─────────
  1 9 4
```

4 hundreds − 3 hundreds = 1 hundred

579 − 385 = **194**

This answer is reasonable because it is close to the estimate of 200.

The Number System

You can use **U.S. traditional subtraction** to subtract decimals.

Example

9.4 − 4.85 = ?

Estimate: 9.4 rounds to 9. You can round 4.85 to 5.

An estimated difference is 9 − 5 = 4.

Step 1 Write the problem. Write a 0 in the hundredths place of 9.4. This evens out the places without changing the value of the starting number.

```
   9 . 4 0
 − 4 . 8 5
```

Step 2 Start with the hundredths place. Since 0 < 5, you need to regroup. Trade 1 tenth for 10 hundredths. Subtract the hundredths.

```
         3  10
    9 . 4̸  0̸
  − 4 . 8  5
  ─────────
            5
```

10 hundredths − 5 hundredths = 5 hundredths

Step 3 Go to the tenths place. Since 3 < 8, you need to regroup. Trade 1 whole for 10 tenths. Subtract the tenths.

```
           13
      8   ̸3  10
    9̸ . 4̸   0̸
  − 4 . 8   5
  ──────────
          5  5
```

13 tenths − 8 tenths = 5 tenths

Step 4 Go to the ones place. Subtract 8 − 4 = 4.

```
           13
      8   ̸3  10
    9̸ . 4̸   0̸
  − 4 . 8   5
  ──────────
    4 . 5   5
```

8 ones − 4 ones = 4 ones

9.4 − 4.85 = **4.55**

The answer 4.55 makes sense because it is close to the estimate of 4.

The Number System

When the minuend has a zero as one of its digits, sometimes two trades are made in one step. Often after making the double trade, no more trades are needed.

Note Each number in a subtraction problem has a special name. Consider 904 − 385 = 519.

minuend   subtrahend   difference
904 − 385 = 519

Example

904 − 385 = **?**

Estimate: You can round 904 to 900 and 385 to 400.
An estimated difference is 900 − 400 = 500.

Step 1 Compare the digits in the ones place. You need to trade 1 ten for 10 ones, but there are no tens. So two trades are needed.
First trade 1 hundred for 10 tens. Then trade 1 ten for 10 ones. Subtract the ones.

```
    9
  8 10 14
  9  0  4
-    3  8  5
─────────────
             9
```

14 ones − 5 ones = 9 ones

Step 2 Compare the digits in the tens place. No trade is needed. Subtract the tens.

```
    9
  8 10 14
  9  0  4
-  3  8  5
─────────────
       1  9
```

9 tens − 8 tens = 1 ten

Step 3 Subtract the hundreds.

```
    9
  8 10 14
  9  0  4
-  3  8  5
─────────────
   5  1  9
```

8 hundreds − 3 hundreds = 5 hundreds

904 − 385 = **519**

This answer is reasonable because it is close to the estimate of 500.

Note Another way to trade is to think of 904 as 90 tens and 4 ones. By trading 1 ten for 10 ones, the result is 89 tens and 14 ones, which is marked in the problem this way:

```
  8  9  14
  9̶  0̶  4̶
-  3  8  5
─────────────
   5  1  9
```

Check Your Understanding

Subtract using any method. Estimate to check whether your answers are reasonable.

1. 21.5 − 8.8
2. 10 − 1.79
3. 7.8 − 7.763
4. 0.72 − 0.48
5. 647 − 54
6. 751 − 347
7. 449 − 275
8. 5,216 − 1,418

Check your answers in the Answer Key.

one hundred twenty-seven

The Number System

Units and Precision in Decimal Addition and Subtraction

Counts and measures always have units. For addition or subtraction, all the numbers must have the same unit. If they do not, before solving the problem you will have to convert at least one of the numbers so that all units are the same.

Example

Find the perimeter of the triangle.

Method 1
Convert the centimeter measures to millimeters, then add:
6 cm = 60 mm   12 cm = 120 mm
Perimeter = 60 mm + 120 mm + 85 mm = 265 mm

Method 2
Convert the millimeter measures to centimeters, then add:
85 mm = 8.5 cm
Perimeter = 6 cm + 12 cm + 8.5 cm = 26.5 cm

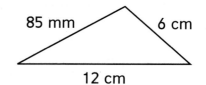

Measures may have different degrees of **precision.** For example, measuring to the nearest tenth of a meter is more precise than measuring to the nearsest meter. If some measures are more precise than others, you can convert them so that they have the same precision as the *least precise* measure.

Example

The winning times in the men's 100-meter dash in the 1936 and 1988 Olympic Games are shown at the right.

Year	Winner	Time
1936	Jesse Owens, U.S.A.	10.3 seconds
1988	Carl Lewis, U.S.A.	9.92 seconds

How much faster did Carl Lewis run than Jesse Owens?

Carl Lewis was timed to the nearest hundredth of a second.
Jesse Owens was timed to the nearest tenth of a second.
Round the more precise measure, 9.92 seconds, to match the less precise measure, 10.3 seconds. 9.92 seconds rounded to the nearest tenth of a second is 9.9 seconds.
Since 10.3 − 9.9 = 0.4, Carl Lewis ran the 100-meter dash about 0.4 second faster then Jesse Owens.

Check Your Understanding

Solve 4.7 m − 3.62 m. Use the degree of precision of the less precise measure.

Check your answers in the Answer Key.

The Number System

Multiplying by Powers of 10

The numbers in the chart are powers of 10. The powers of 10 are represented as whole numbers, as repeated factors, and with exponential notation.

10,000	1,000	100	10	1	.	0.1	0.01	0.001	0.0001
10 * 10 * 10 * 10	10 * 10 * 10	10 * 10	10	1	.	$\frac{1}{10}$	$\frac{1}{10} * \frac{1}{10}$	$\frac{1}{10} * \frac{1}{10} * \frac{1}{10}$	$\frac{1}{10} * \frac{1}{10} * \frac{1}{10} * \frac{1}{10}$
10^4	10^3	10^2	10^1	10^0	.	10^{-1}	10^{-2}	10^{-3}	10^{-4}

You can use what you know about place value and exponential notation to multiply by powers of 10.

Multiplying Whole Numbers by Powers of 10

In the base-10 place-value system, each digit in a place is worth 10 times as much as the place to its right. This place-value structure helps you multiply by powers of 10.

Example

1,000 * 100 = ?

1,000s thousands	100s hundreds	10s tens	1s ones
1	0	0	0

Identify the place value of one of the factors. 1 is in the thousands place

Write the second factor as a product of 10s. 100 = 10 * 10

Write a new number sentence for 1,000 * 100 = ? 1,000 * <u>10 * 10</u> = ?

100

Each time a number is multiplied by 10, the digits shift one place to the left. So the 1 in the first factor (1,000) is written two places to the left in the 100 thousands place, and 2 zeros are attached to show the place shift.

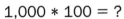

100,000

2 zeros are attached to show that the 1 was moved two places to the left.

So, 1,000 * 100 = **100,000**.

one hundred twenty-nine **SRB 129**

The Number System

Because of the base-10 place-value system, you can use patterns with zeros to multiply whole numbers by powers of 10.

Example

What patterns do you notice?

8 * 10 = 80	8 is multiplied by 10 one time. There is one zero in the product.
8 * 100 = 800	100 = 10 * 10 = 10^2, so 8 is multiplied by 10 two times. There are two zeros in the product.
8 * 1,000 = 8,000	1,000 = 10 * 10 * 10 = 10^3, so 8 is multiplied by 10 three times. There are three zeros in the product.

Each time 8 is multiplied by 10, the 8 shifts one place to the left and one zero is attached to the product.

When multiplying a whole number by a power of 10, the exponent tells how many places the digits of the number are shifted to the left and how many zeros to attach to show that the place value has shifted.

Multiplying Decimals by Powers of 10

One way to multiply decimals by powers of 10 greater than 1 is to use **partial-products multiplication**. You can use patterns you already know.

Example

Solve 1,000 * 45.6 using partial-products multiplication.

Step 1 Solve the problem as if there were no decimal point.

Step 2 Estimate the answer to 1,000 * 45.6 and place the decimal point where it belongs.

1,000 * 45 = 45,000, so 1,000 * 45.6 must be near 45,000.

```
              1 0 0 0
          *     4 5 6
          ───────────
400 * 1,000 → 4 0 0 0 0 0
 50 * 1,000 →   5 0 0 0 0
  6 * 1,000 →     6 0 0 0
          ───────────
              4 5 6 0 0 0
```

So the answer to 1,000 * 45.6 is **45,600**.

Another way to multiply a number by a power of 10 is to move the decimal point. Think of this as an efficient shortcut.

When multiplying by a power of 10 greater than 1, the original digits shift to the left and the decimal point shifts to the right because the product is larger than both factors. In the example above, the original digits 4, 5, and 6 shifted to the left and the decimal point moved three places to the right.

The Number System

When multiplying a number by a power of 10 that is less than one, the original digits shift to the right and the decimal point moves to the left as the product gets smaller.

To decide whether to move the decimal point to the right or left, think: "Should the answer be *greater than* or *less than* the number I started with?" If the answer should be *greater*, move the decimal point to the right. If the answer should be *less*, move the decimal point to the left.

Examples

1,000 * 45.6 = ?   0.001 * 45.6 = ?

Step 1 Locate the decimal point in the power of 10.

1,000 = 1000.   0.001

Step 2 Move the decimal point LEFT or RIGHT until it is right of the number 1.

1.0 0 0.   0.0 0 1.

Step 3 Count the number of decimal places you moved the decimal point.

3 places to the left   3 places to the right

Step 4 Move the decimal point in the other factor the same number of places but in the OPPOSITE direction. Insert 0s as needed.

4 5.6 0 0.   0.0 4 5.6

1,000 * 45.6 = **45,600**   0.001 * 45.6 = **0.0456**

one hundred thirty-one   **131**

The Number System

Extended Multiplication Facts

You are using **extended multiplication facts** when you combine basic multiplication facts with multiplying by powers of 10.

One way to find answers to extended multiplication facts is to use basic facts with multiples of 10.

Examples

8 * 70 = **?**

Think: 8 [7s] = 56

Then 8 [70s] is 10 times as much.

8 * 70 = 10 * 56 = 560

So, 8 * 70 = **560**.

6,000 * 3 = **?**

Think: 6 [3s] = 18

Then 6,000 [3s] is 1,000 times as much.

6,000 * 3 = 1,000 * 18 = 18,000

So, 6,000 * 3 = **18,000**.

You can use exponential notation to find the products of extended multiplication facts, especially when the numbers are large. First rename the numbers using exponential notation. Then use the Associative and Commutative Properties to regroup and reorder the numbers before multiplying.

Examples

3,000 * 200 = **?**

$$\begin{aligned}3{,}000 * 200 &= (3 * 10^3) * (2 * 10^2) \\ &= (3 * 2) * (10^3 * 10^2) \\ &= 6 * (10^5) \\ &= 600000, \text{ or } 600{,}000\end{aligned}$$

3,000 * 200 = **600,000**

60,000 * 5,000 = **?**

$$\begin{aligned}60{,}000 * 5{,}000 &= (6 * 10^4) * (5 * 10^3) = ? \\ &= (6 * 5) * (10^4 * 10^3) = ? \\ &= (30) * (10^7) \\ &= 300000000, \text{ or } 300{,}000{,}000\end{aligned}$$

60,000 * 5,000 = **300,000,000**

Check Your Understanding

Solve these problems using mental math.

1. 6 * 400 = ?   2. 9,000 * 20 = ?   3. 160 * 20,000 = ?   4. 3,000 * 700 = ?

Check your answers in the Answer Key.

The Number System

Multiplication Methods

You can use the same methods for multiplying whole numbers and decimals. The main difference is that with decimals you have to decide where to place the decimal point in the product. To multiply decimals:

- Estimate the product.
- Multiply as if the factors were whole numbers, ignoring the decimal points.
- Use your estimate to place the decimal point in the product.

Partial-Products Multiplication

When you multiply using **partial-products multiplication,** the value of each digit in one factor is multiplied by the value of each digit in the other factor. The final product is the sum of these **partial products.**

When you use partial-products multiplication, factors are broken up using place value. This results in partial products that are either basic multiplication facts or extended multiplication facts you can calculate mentally. You must keep track of the place value of each digit in the partial products.

Note The symbols × and * are both used to indicate multiplication.

> **Example**
>
> $4 * 236 = ?$
>
> Estimate: You can round 236 to 200.
> Since 236 was rounded down, the product will be greater than $4 * 200 = 800$.
>
> Think of 236 as $200 + 30 + 6$.
> Multiply each part of 236 by 4.
>
> ```
>                    2  3  6
>                 *        4
>                 ---------
>     4 * 200 →   8  0  0
>     4 *  30 →   1  2  0
>     4 *   6 →      2  4
>                 ---------
> Add the partial products.
>                 9  4  4
> ```
>
> $4 * 236 = $ **944**
>
> The answer is reasonable because it is "in the hundreds" like the estimate of 800 and it is greater than 800.

Note If you can estimate and multiply to find the partial products using mental math, then you do not need to write the steps shown in green.

one hundred thirty-three

The Number System

Example

0.4 * 2.36 = ?

Step 1 Make an estimate.
  0.4 is close to 0.5, or $\frac{1}{2}$, and 2.36 rounds to 2.
  An estimated product is $\frac{1}{2}$ * 2 = 1.

Step 2 Multiply. Ignore the decimal points.
  Use partial-products multiplication as in the example on the previous page, 4 * 236 = 944.

Step 3 Use your estimate to place the decimal point in the product. Your estimate is 1. To have a product that is closest to 1, the decimal point should be placed before the 9 in 944.

So, 0.4 * 2.36 = **0.944.**

When solving multidigit multiplication problems, it can be helpful to make a diagram for identifying the partial products.

Example

43 * 26 = ?

Estimate: 43 rounds to 40, and 26 is close to 25.

An estimated product is 40 * 25 = 1,000.

Make a partial-products diagram. Rename 43 as 40 + 3. Rename 26 as 20 + 6. Draw a large rectangle. Divide the length of the rectangle into two sections. Label one section as 40 and the other as 3. Divide the width of the rectangle into two sections. Label one section as 20 and the other as 6.

	40	3
20	20 * 40	20 * 3
6	6 * 40	6 * 3

Find the product of the numbers in each section of the diagram. Add the partial products together to find the answer.

20 * 40 = 800   20 * 3 = 60
6 * 40 = 240   6 * 3 = 18

800 + 240 + 60 + 18 = 1,118

43 * 26 = **1,118**

This answer makes sense because it is close to the estimate of 1,000.

Check Your Understanding

Multiply using partial-products multiplication or a partial-products diagram. Estimate to check whether your answers are reasonable.

1. 179 * 4
2. 37 * 64
3. 2.8 * 4.6
4. 3.05 * 6.5
5. 0.52 * 3.03

Check your answers in the Answer Key.

The Number System

Multiplying Decimals Less Than 1

Sometimes when you multiply a decimal less than 1 by another decimal less than 1, making an estimate to place the decimal point can be difficult. You can use powers of 10 to place the decimal point.

Example

$0.2 * 0.041 = ?$

Multiply both factors by a power of 10 to make them whole numbers. Keep track of both powers of 10, or the total number of places the decimal point shifted in each factor.	$0.2 * 10^1 = 2$	Decimal point shifted 1 place to the right.
	$0.041 * 10^3 = 41$	Decimal point shifted 3 places to the right.
Multiply the whole numbers.	$2 * 41 = 82$	
Undo the multiplication by powers of 10 by dividing by powers of 10. Divide the product by the powers of 10 used to change the factors to whole numbers, or shift the decimal point to the left the total number of places it shifted when changing the factors to whole numbers.	$82 / (10^1 * 10^3) = 0.0082$	Decimal point shifted 4 places to the left in the product.

This is the same as $(10) * (10 * 10 * 10)$.

So, $0.2 * 0.041 = \mathbf{0.0082}$.

Here is another method for multiplying decimals when there are many decimal places in the factors and it is difficult to estimate the answer. This shortcut uses the information above.

Example

$0.2 * 0.041 = ?$

Step 1 Multiply as though both factors were whole numbers.

$2 * 41 = 82$

Step 2 Count the total number of places to the RIGHT of the decimal point for both factors.

0.2 has 1 decimal place.
0.041 has 3 decimal places.
There are 4 decimal places in all.

Step 3 Place the decimal point so that you have the same number of decimal places as the total in step 2. Insert 0s as needed.

0.0082.

So, $0.2 * 0.041 = \mathbf{0.0082}$.

Check Your Understanding

Multiply. Use an estimate to check whether the answer is reasonable.

1. $48 * 438 = ?$
2. $4.86 * 0.36 = ?$
3. $0.073 * 2.07 = ?$

Check your answers in the Answer Key.

The Number System

Lattice Multiplication

Lattice multiplication has been used for hundreds of years. The box with cells and diagonals is called a **lattice**. Each diagonal is the same as a place-value column.

Example

4 * 915 = ?

Estimate: 915 is close to 900. An estimated product is
4 * 900 = 3,600.

Step 1 Write 915 above the lattice.
Write 4 on the right side of the lattice.

Step 2 Find the products inside the lattice. Multiply 4 * 5. Then multiply 4 * 1. Then multiply 4 * 9. Write the answers as shown.

Step 3 Add the numbers along each diagonal, starting at the right.

Step 4 Compare the result with the estimate.
Read the answer: 4 * 915 = **3,660**.

So, 4 * 915 = **3,660**.

The answer is reasonable because it is close to 3,600.

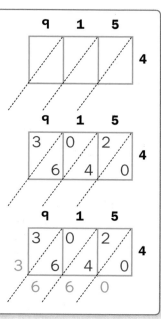

Example

45.5 * 3.06 = ?

Estimate: 45.5 is close to 45, and 3.06 is close to 3.
45 * 3 = 135

Step 1 Draw the lattice and write the factors, including the decimal points, at the top and right side. In the factor above the grid, the decimal point should be above a column line. In the factor on the right side of the grid, the decimal point should be to the right of a row line.

Step 2 Find the products inside the lattice.

Step 3 Add along the diagonals, moving from right to left.

Step 4 Locate the decimal point in the answer as follows. Slide the decimal point in the factor above the grid down along the column line. Slide the decimal point in the factor on the right side of the grid across the row line. When the decimal points meet, slide the decimal point down along the diagonal line. Write a decimal point at the end of the diagonal line.

Step 5 Compare the result with the estimate.
Read the answer: 45.5 * 3.06 = **139.230**.

So, 45.5 * 3.06 = **139.23**.

The answer is reasonable because it is close to the estimate of 135.

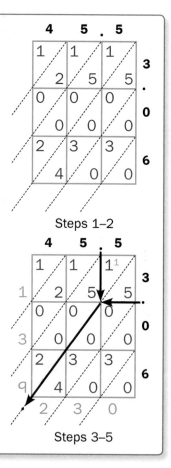

The Number System

U.S. Traditional Multiplication

You can use **U.S. traditional multiplication** to multiply. Use an estimate to check whether the answer is reasonable.

> **Example**
>
> 5 * 629 = ?
>
> **Estimate:** You can round 629 to 600.
> Since 629 was rounded down, the product will be greater than 5 * 600 = 3,000.
>
> **Step 1** Multiply the ones.
> 5 * 9 ones = 45 ones = 4 tens + 5 ones
> Write 5 in the 1s place below the line.
> Write 4 above the 2 in the 10s place.
>
> ```
>       4
>     6 2 9
>   *     5
>   -------
>         5
> ```
>
> **Step 2** Multiply the tens.
> 5 * 2 tens = 10 tens
> Remember the 4 tens from Step 1.
> 10 tens + 4 tens = 14 tens in all
> 14 tens = 1 hundred + 4 tens
> Write 4 in the 10s place below the line.
> Write 1 above the 6 in the 100s place.
>
> ```
>     1 4
>     6 2 9
>   *     5
>   -------
>       4 5
> ```
>
> **Step 3** Multiply the hundreds.
> 5 * 6 hundreds = 30 hundreds
> Remember the 1 hundred from Step 2.
> 30 hundreds + 1 hundred = 31 hundreds in all
> 31 hundreds = 3 thousands + 1 hundred
> Write 1 in the 100s place below the line.
> Write 3 in the 1,000s place below the line.
>
> ```
>     1 4
>     6 2 9
>   *     5
>   -------
>   3 1 4 5
> ```
>
> 5 * 629 = **3,145**
>
> The answer is reasonable because it is close to and greater than the estimate of 3,000.

The Number System

You can use U.S. traditional multiplication to multiply decimals.

Example

7.3 * 0.26 = **?**

Step 1 Make an estimate.

7.3 is close to 7, and 0.26 is close to 0.25, or $\frac{1}{4}$.
$7 * \frac{1}{4} = \frac{7}{4}$, or $1\frac{3}{4}$, so the product will be close to 1.75.

Step 2 Multiply as you would with whole numbers. One way to do this is to use U.S. traditional multiplication. Ignore the decimal points.

Multiply 73 by the 6 in 26 as if the problem were 73 * 6.

```
        1
        7  3
  *     2  6
  ─────────
     4  3  8
```

Think: What is the value of the 2 in 26? It's 20, so multiply 73 by the 2 in 26 as if the problem were 73 * 20.

```
           1
           7  3
  *        2  6
  ────────────
        4  3  8
     1  4  6  0
```

Add the two partial products to get the final answer.

```
           1
           7  3
  *        2  6
  ────────────
        4  3  8
  +  1  4  6  0
  ────────────
     1  8  9  8
```

73 * 26 = 1898

Step 3 Use the estimate to place the decimal point in the answer. The estimate was close to 1.75, so the decimal point should be placed between the 1 and the 8 in 1898.

7.3 * 0.26 = **1.898**

Check Your Understanding

Multiply. Use an estimate to check whether the answer is reasonable.

1. 77 * 86 **2.** 76 * 98 **3.** 7 * 648
4. 16.5 * 4.5 **5.** 4.03 * 17 **6.** 8.3 * 34.1

Check your answers in the Answer Key.

The Number System

Extended Division Facts

Numbers such as 10, 100, and 1,000 are called **powers of 10.**

The example box below shows how to use the following shortcut to divide a whole number that ends in zeros by a power of 10:

- Cross out zeros in the number, starting in the ones place.
- Cross out as many zeros as there are zeros in the power of 10.

Examples

90,000 / 1**0** = 9000~~0~~	63,000 / 1**0** = 6300~~0~~	860,000 / 1**0,000** = 86~~0000~~
90,000 / 1**00** = 900~~00~~	63,000 / 1**00** = 630~~00~~	7,000,000 / 1**00,000** = 70~~00000~~
90,000 / 1**,000** = 90~~000~~	63,000 / 1**,000** = 63~~000~~	

If you know basic division facts, you can use mental math to solve problems such as 540 / 9 and 18,000 / 3.

Examples

540 / 9 = **?**	18,000 / 3 = **?**
Think: 54 / 9 = 6	*Think:* 18 / 3 = 6
Then 540 / 9 is 10 times as much.	Then 18,000 / 3 is 1,000 times as much.
540 / 9 = 10 * 6 = **60**	18,000 / 3 = 1,000 * 6 = **6,000**

You can use a similar method to solve problems such as 18,000 / 30 and 32,000 / 400 mentally.

Examples

18,000 / 30 = **?**	32,000 / 400 = **?**
Think: 18,000 / 3 = 6,000	*Think:* 32,000 / 4 = 8,000
Then 18,000 / 30 is $\frac{1}{10}$ as much.	Then 32,000 / 400 is $\frac{1}{100}$ as much.
18,000 / 30 = $\frac{1}{10}$ * 6,000 = **600**	32,000 / 400 = $\frac{1}{100}$ * 8,000 = **80**

Check Your Understanding

Solve these problems using mental math.

1. 84,000 / 1,000
2. 56,000 / 8
3. 4,500 / 90
4. 45,000 / 900

Check your answers in the Answer Key.

The Number System

Dividing Decimals by Powers of 10

Here is one way to divide a decimal by a power of 10 greater than 1.

Example

45.6 / 1,000 = ?

Write the power of 10 in exponential notation.

$1,000 = 10 * 10 * 10 = 10^3$

Move the decimal point in the dividend. Use the exponent to find how many places to move the decimal point.

0.0 4 5.6

The decimal point moves 3 places to the left because 45.6 is divided by 10 three times. This makes each digit worth $\frac{1}{10^3}$ as much as it was worth before.

Sometimes you need to insert 0s to show the correct place value.

Dividing by 10 is the same as multiplying by $\frac{1}{10}$. Each time a number is multiplied by $\frac{1}{10}$, the original digits shift one place to the right, which moves the decimal point one place to the left.

So, 45.6 / 1,000 = **0.0456**.

Note Some powers of 10 greater than one:

$10^1 = 10$
$10^2 = 100$
$10^3 = 1,000$
$10^4 = 10,000$
$10^5 = 100,000$
$10^6 = 1,000,000$

Note When the dividend (the number you are dividing) does not have a decimal point, you must locate the decimal point before moving it. Note that in two of the examples below, 350 is written with a decimal point after the 0 (350 = 350.).

Examples

350 / 100 = ?
$100 = 10^2$
100 = 100.
1.0 0.
2 places
3.5 0.
350 / 100 = **3.50**

350 / 10,000 = ?
$10,000 = 10^4$
10,000 = 10000.
1.0 0 0 0.
4 places
0.0 3 5 0.
350 / 10,000 = **0.0350**

$290.50 / 1,000 = ?
$1,000 = 10^3$
1,000 = 1000.
1.0 0 0.
3 places
0.2 9 0.5 0
$290.50 / 1,000 = **$0.29**
(rounded to the nearest cent)

Check Your Understanding

Divide.

1. 79.8 / 10   **2.** 0.78 / 100   **3.** $360 / 1,000   **4.** 80 / 10,000

Check your answers in the Answer Key.

SRB
140   one hundred forty

The Number System

Interpreting Remainders in Division

Different symbols can be used to indicate division. For example, "94 divided by 6" can be written as $94 \div 6$, $6\overline{)94}$, $94 / 6$, and $\frac{94}{6}$.

- The number being divided is the **dividend.**
- The number that divides the dividend is the **divisor.**
- The answer to a division problem is the **quotient.**
- When numbers cannot be divided evenly, the answer includes a quotient and a **remainder.**

Four ways to show "123 divided by 4"
$123 \div 4 \rightarrow 30\ R3 \qquad 123 / 4 \rightarrow 30\ R3$ $4\overline{)123}\ \ 30\ R3 \qquad \frac{123}{4} \rightarrow 30\ R3$
123 is the dividend.   4 is the divisor.   30 is the quotient.   3 is the remainder.

The way you represent the quotient and remainder depends on the problem situation. There are several ways to think about remainders:

- Ignore the remainder. Use the quotient as the answer.
- Round the quotient up to the next whole number.
- Rewrite the remainder as a fraction or decimal. Use this fraction or decimal as part of the answer.

> **Example**
>
> Fran has 20 photos to place in a photo album. Three photos fit on a page. How many pages does she need for her photos?
>
> $20 / 3 \rightarrow 6\ R2$
>
> Round the quotient up to the next whole number. A seventh page is needed to include all the photos. The album will have 6 pages filled and another page only partially filled.
>
> Answer: 7 pages are needed

Note 6 R2 is not a number, so it is not entirely correct to write $20 / 3 = 6\ R2$. In *Everyday Mathematics,* an arrow is often used instead of an equal sign in number models for division with a remainder. The equal sign is correct when writing the remainder as a fraction or a decimal.

The Number System

To rewrite a remainder as a fraction, make the remainder the **numerator** of the fraction and the divisor the **denominator** of the fraction.

Problem	Answer	Remainder Written as a Fraction	Answer Written as a Mixed Number	Answer Written as a Decimal
375 / 4	93 R3	$\frac{3}{4}$	$93\frac{3}{4}$	93.75

Example

Suppose 3 friends share a 20-inch string of licorice. How long is each piece if the friends receive equal shares?

20 / 3 → 6 R2

If each friend receives 6 inches of licorice, 2 inches remain to be divided. Divide this 2-inch remainder into $\frac{1}{3}$-inch pieces. Each friend receives two $\frac{1}{3}$-inch pieces, or $\frac{2}{3}$ inch.

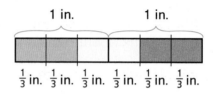

The remainder (2) has been rewritten as a fraction ($\frac{2}{3}$). Use this fraction as part of the answer.

Answer: Each friend will get a $6\frac{2}{3}$-inch piece of licorice.

Example

Lucy and her three brothers evenly split the cost of a present for their mother. The present cost $17. How much did each one pay?

$17 / 4 = ?$

$17 / 4 → 4 R1

You can use long division to rewrite the remainder as a decimal.

$17 / 4 = $4.25

Lucy and her three brothers each paid $4.25 for their mother's gift.

```
      4.25
   4)17.00
    -16
    ‾‾‾
      10
     - 8
    ‾‾‾
      20
     -20
    ‾‾‾
       0
```

The Number System

Division Methods

You can use the same methods for dividing whole numbers and decimals. The main difference is that with decimals you have to decide where to place the decimal point in the quotient. Here is one way to divide decimals.

- Estimate the quotient.
- Divide as if the divisor and dividend were whole numbers.
- Use your estimate to place the decimal point in the quotient.

Note You can estimate by using close-but-easier numbers or by rounding.

Partial-Quotients Division: Whole Number and Decimal Dividends

Partial-quotients division takes several steps to find the quotient. At each step, you find a partial answer (called a **partial quotient**). Then add the partial answers to find the quotient. Study the example below. To find the number of 6s in 1,010, first find partial quotients and then add them. Record the partial quotients in a column to the right of the original problem.

Example

1,010 / 6 = **?**

Estimate: 1,010 is close to 1,200.

The quotient will be less than 1,200 / 6 = 200.

```
        Write partial quotients in this column.
   6)1,010   ↓
    - 600  100   Think: How many 6s are in 1,010? At least 100.
    ─────         The first partial quotient is 100. 100 * 6 = 600
      410        Subtract 600 from 1,010. At least 50 [6s] are left in 410.
    - 300   50   The second partial quotient is 50. 50 * 6 = 300
    ─────        Subtract. At least 10 [6s] are left in 110.
      110
    -  60   10   The third partial quotient is 10. 10 * 6 = 60
    ─────        Subtract. At least 8 [6s] are left in 50.
       50
    -  48    8   The fourth partial quotient is 8. 8 * 6 = 48
    ─────
        2  168   Subtract. Add the partial quotients.
        ↑   ↑
   Remainder Quotient
```

The answer is **168 R2**. Record the answer as 6)1,010 with 168 R2 on top, or write 1,010 / 6 → 168 R2.

The answer is reasonable because it is close to and less than the estimate of 200.

one hundred forty-three SRB 143

The Number System

The example below shows different ways to find partial quotients for the same problem.

Example

381 / 4 = ?

Estimate: 381 is close to 400. An estimated quotient is 400 / 4 = 100.

One way:
```
4)381
-200 | 50
-----
 181
-120 | 30
-----
  61
- 40 | 10
-----
  21
- 20 |  5
-----
   1   95
```

A second way:
```
4)381
-200 | 50
-----
 181
-160 | 40
-----
  21
- 20 |  5
-----
   1   95
```

A third way:
```
4)381
-360 | 90
-----
  21
- 20 |  5
-----
   1   95
```

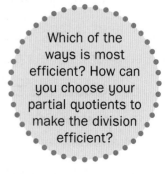

Which of the ways is most efficient? How can you choose your partial quotients to make the division efficient?

The answer, **95 R1,** is the same for all three ways. It is close to the estimate of 100.

You can use partial-quotients division to divide a decimal by a whole number.

Example

97.24 / 26 = ?

Step 1 Estimate: Round 97.24 to 100. 26 is close to 25. An estimated quotient is 100 / 25 = 4.

Step 2 Divide, ignoring the decimal point.
```
26)9,724
 -7800 | 300
 ------
   1924
  -1040 |  40
  ------
    884
   -780 |  30
   -----
    104
   -104 |   4
   -----
      0   374
```

Step 3 Decide where to place the decimal point in the answer. The estimate is 4. To have a quotient that is close to 4, the decimal point should be placed between the 3 and 7 in 374 to give 3.74.

9,724 / 26 = 374

97.24 / 26 = **3.74**

The Number System

Dividing Decimals by Decimals

When division problems have decimal divisors, sometimes estimating the quotient is difficult. For these types of problems, you can find an equivalent division problem that is easier to solve.

Just as an equivalent fraction expresses the same value with a different numerator and denominator, an **equivalent division problem** produces the same quotient using a different dividend and divisor. You can find an equivalent problem by multiplying both the dividend and divisor by the same amount.

Note You can use fractions to show division problems. The fraction $\frac{a}{b}$ is another way of saying a divided by b, $a \div b$, or a / b.

Example

2.84 / 0.004 = ?

Write the division problem as a fraction:

$$2.84 / 0.004 = \frac{2.84}{0.004}$$

Find an equivalent fraction with no decimals:

$$\frac{2.84 * 1{,}000}{0.004 * 1{,}000} = \frac{2{,}840}{4}$$

Write the equivalent fraction as a division problem:

$$\frac{2{,}840}{4} = 2{,}840 / 4 = 710$$

Since the two fractions $\frac{2.84}{0.004}$ and $\frac{2{,}840}{4}$ are equivalent, the quotient for 2.84 / 0.004 is the same as the quotient for 2,840 / 4.

2.84 / 0.004 = **710**

Note Multiplying the numerator and the denominator by the same number, 1,000, is like multiplying the fraction by $\frac{1{,}000}{1{,}000} = 1$.

To summarize, you can use these steps to divide a decimal by a decimal.

Step 1 Estimate.

You can estimate using the original problem or an equivalent problem.

Step 2 Write an equivalent problem with a whole-number divisor.

Use the multiplication rule to find an equivalent problem.

Step 3 Solve the equivalent problem using any division method.

The quotient for the equivalent problem is the same as the quotient for the original problem.

The Number System

Example

0.78 / 0.013 = **?**

Step 1 Estimate.

Step 2 Find an equivalent problem with a whole-number divisor.

One way to estimate is to use an equivalent problem.
0.78 * **1,000** = 780
0.013 * **1,000** = 13

Equivalent problem: 780 / 13

Estimate the quotient for the equivalent problem: 13 is close to 10.
An estimated quotient is 780 / 10 = 78.

Step 3 Solve the equivalent problem.

780 / 13 = 60
Since $\frac{0.78}{0.013} = \frac{780}{13}$, 0.78 / 0.013 = **60**.

0.78 / 0.013 = **60**

The answer is reasonable because it is close to the estimate of 78.

If the quotient does not come out evenly, remember that you can use the remainder to decide whether to round up or round down.

Example

0.73 / 0.03 = **?**

Step 1 Estimate.

69 / 3 is easy to do mentally and is close to 73 / 3.
An estimated quotient is $\frac{0.69 * 100}{0.03 * 100} = \frac{69}{3}$ or 23.

Step 2 Write an equivalent problem with a whole-number divisor.

0.73 * **100** = 73, and 0.03 * **100** = 3.
Equivalent problem: 73 / 3 = ?

Step 3 Solve the equivalent problem.

73 / 3 → 24 R1, or 73 / 3 = $24\frac{1}{3}$
Since $\frac{1}{3}$ is less than one-half, the quotient rounds down to 24.
0.73 / 0.03 is about 24.

0.73 / 0.03 ≈ 24 (The symbol ≈ means *is about equal to*.)

This quotient is reasonable because it is close to the original estimate of 23.

Check Your Understanding

Divide. Use an estimate to check whether your quotient is reasonable.

1. 4.02 / 0.5
2. 6.9 / 0.015
3. 210 / 0.25

Check your answers in the Answer Key.

The Number System

U.S. Traditional Long Division: Single-Digit Divisors

U.S. traditional long division is another method you can use to divide.

Example

Share $957 equally among 5 people.

Estimate: $957 is close to $1,000. An estimated quotient is $1,000 / 5 = $200.

Step 1 Share the [$100]s.

```
    1      ← Each person gets 1 [$100].
5)957
 −5        ← 1 [$100] each for 5 people
 ───
  4        ← 4 [$100]s are left.
```

Step 2 Trade 4 [$100]s for 40 [$10]s.
That makes 45 [$10]s in all.

```
    1
5)957
 −5
 ───
  45       ← 45 [$10]s are to be shared.
```

Step 3 Share the [$10]s.

```
   19      ← Each person gets 9 [$10]s.
5)957
 −5
 ───
  45
 −45       ← 9 [$10]s each for 5 people
 ───
   0       ← 0 [$10]s are left.
```

Step 4 Share the [$1]s.

```
  191      ← Each person gets 1 [$1].
5)957
 −5
 ───
  45
 −45
 ───
   07      ← 7 [$1]s are to be shared.
   −5      ← 1 [$1] each for 5 people
   ───
    2      ← 2 [$1]s are left.
```

$957 / 5 → $191 R$2

Each person gets $191; $2 are left over.

The answer is reasonable because it is close to the estimate of $200.

Check Your Understanding

Divide. Estimate to check whether your answers are reasonable.

1. 840 / 7   **2.** 6)984   **3.** 4)539   **4.** 5,280 / 6

Check your answers in the Answer Key.

one hundred forty-seven

The Number System

Short Division

Short division is a quick way to do division in which you multiply and subtract mentally. Short division works best with single-digit divisors. The following example compares long and short division.

Note The "leading" 0 in the quotient shown below can help you understand the division method. It should not be included in the answer.

Example

3,628 / 5 = **?**

Estimate: 3,628 is close to 3,500. An estimated quotient is 3,500 / 5 = 700.

Long Division	Short Division

Step 1 Start with the thousands.
 	
   0   ← There are not enough
5)3628   thousands to share 5 ways.
	Step 1 Start with the thousands.
	There are not enough thousands to share 5 ways.
	
	    0
	5 | 3 | 6 | 2 | 8

Step 2 So trade 3 thousands for 30 hundreds. Share the hundreds.
 	
  07   ← Each share gets 7 hundreds.
5)3628 ← 36 hundreds
−35   ← 7 hundreds * 5 shares
   1   ← 1 hundred is left.
	Step 2 So trade 3 thousands for 30 hundreds. Share the 36 hundreds.
	Each share gets 7 hundreds.
	Think: 5 * 7 = 35, and 36 − 35 = 1.
	1 hundred is left.
	
	   0 7
	5 | 3 | 6 | 2 | 8

Step 3 Trade 1 hundred for 10 tens. Share the tens.
 	
  072 ← Each share gets 2 tens.
5)3628
−35
  12 ← 10 tens + 2 tens
−10 ← 2 tens * 5 shares
   2 ← 2 tens are left.
	Step 3 Trade 1 hundred for 10 tens, and put the 1 before the 2 tens to make 12 tens. Share the 12 tens. Each share gets 2 tens.
	Think: 5 * 2 = 10, and 12 − 10 = 2.
	2 tens are left.
	
	   0 7 2
	5 | 3 | 6 | ¹2 | 8

Step 4 Trade 2 tens for 20 ones. Share the ones.
 	
  0725 ← Each share gets 5 ones.
5)3628
−35
  12
−10
  28 ← 20 ones + 8 ones
−25 ← 5 ones * 5 shares
   3 ← 3 ones are left.
	Step 4 Trade 2 tens for 20 ones, and put a 2 before the 8 ones to make 28 ones. Share the ones. Each share gets 5 ones.
	Think: 5 * 5 = 25, and 28 − 25 = 3.
	3 ones are left.
	
	   0 7 2 5 R3
	5 | 3 | 6 | ¹2 | ²8

3,628 / 5 → **725 R3** The answer is reasonable because it is close to the estimate of 700.

U.S. Traditional Long Division: Multidigit Divisors

You can use **U.S. traditional long division** to divide by larger numbers.

Example

Share $681 equally among 21 people.

Estimate: $681 is more than $600. Each person should get more than $600 / 20 = $30.

Make a table of easy multiples of the divisor.
This can help you decide how many to share at each step.

1 * 21	21
2 * 21	42
3 * 21	63
4 * 21	84
5 * 21	105
6 * 21	126
8 * 21	168
10 * 21	210

Double 21.
Add 2 * 21 and 1 * 21.
Double 2 * 21.
Halve 10 * 21.
Double 3 * 21.
Double 4 * 21.
Move the decimal point one place to the right.

Step 1 There are not enough [$100s] to share 21 ways, so trade 6 [$100s] for 60 [$10s]. Share the 68 [$10s].

```
        3      ← Each person gets 3 [$10s].
   21)681     ← There are 68 [$10s] to share.
    -63       ← 3 [$10s] * 21
      5       ← 5 [$10s] are left.
```

Step 2 Trade the 5 [$10s] for 50 [$1s]. Share the 51 [$1s].

```
       32     ← Each person gets 2 [$1s].
   21)681
    -63
      51     ← 50 [$1s] + 1 [$1]
     -42     ← 2 [$1s] * 21
       9     ← 9 [$1s] are left.
```

$681 / 21 → $32 R$9

The answer is reasonable because it is close to the estimate of $30.

The Number System

Example

7,720 / 25

Estimate: 7,720 is close to 7,500. An estimated quotient is 7,500 / 25 = 300.

Make a table of easy multiples of the divisor.

1 * 25	25
2 * 25	50
3 * 25	75
4 * 25	100
5 * 25	125
6 * 25	150
8 * 25	200
10 * 25	250

Double 25.
Add 2 * 25 and 1 * 25.
Double 2 * 25.
Halve 10 * 25.
Double 3 * 25.
Double 4 * 25.
Move the decimal point one place to the right.

Step 1 There are not enough thousands to share 25 ways, so trade the thousands for hundreds. Share the hundreds.

```
         3       ← Each share gets 3 hundreds.
   25)7720      ← 77 hundreds
      -75       ← 3 hundreds * 25 shares
      ───
        2       ← 2 hundreds are left.
```

Step 2 Trade the hundreds for tens. Share the tens.

```
        30       ← There are not enough
   25)7720          tens to share.
      -75
      ───
       22       ← 20 tens + 2 tens
```

Step 3 Trade the tens for ones. Share the ones.

```
       308      ← Each share gets 8 ones.
   25)7720
      -75
      ───
       220      ← 22 tens + 0 ones
      -200      ← 8 ones * 25 shares
      ────
        20      ← 20 ones are left.
```

7,720 / 25 → 308 R20

The answer is reasonable because it is close to the estimate of 300.

Check Your Understanding

Divide. Estimate to check whether your answers are reasonable.

1. 650 / 25   **2.** 3,495 / 18   **3.** 13)5,819   **4.** 48)5,286

Check your answers in the Answer Key.

The Number System

U.S. Traditional Long Division: Decimal Dividends

You can use **U.S. traditional long division** to divide money in dollars-and-cents notation.

Example

Share $5.29 equally among 3 people.

Estimate: $5.29 is less than $6. $6 / 3 = $2. Each person should get less than $2.00.

Step 1 Share the dollars.

```
      1        ← Each person gets 1 dollar.
   3)5.29
    -3         ← 1 dollar each for 3 people
    ---
     2         ← 2 dollars are left.
```

Step 2 Trade the dollars for dimes. Share the dimes.

```
      1.7      ← Each person gets 7 dimes. Write a decimal
   3)5.29        point to show amounts less than a dollar.
    -3
    ---
     2 2      ← 20 dimes + 2 dimes
    -2 1      ← 7 dimes each for 3 people
    ----
       1      ← 1 dime is left.
```

Step 3 Trade the dime for pennies. Share the pennies.

```
      1.76     ← Each person gets 6 pennies.
   3)5.29
    -3
    ---
     2 2
    -2 1
    ----
      19      ← 10 pennies + 9 pennies
     -18      ← 6 pennies each for 3 people
     ---
       1      ← 1 penny is left.
```

Each person gets $1.76. There is 1¢ left.

$5.29 / 3 → $1.76 R1¢

The answer is reasonable because it is a little less than the estimate of $2.00.

Note Another way to use U.S. traditional long division to divide with decimals:

Make an estimate. Ignore the decimal points and use U.S. traditional long division as if the divisor and dividend were whole numbers. Then use the estimate to place the decimal point.

Check Your Understanding

Divide. Estimate to check whether your answers are reasonable.

1. $7.26 / 6   **2.** 7)$8.61   **3.** 7)$5.62   **4.** $8.04 / 3

Check your answers in the Answer Key.

The Number System

You can use **U.S. traditional long division** to divide decimals that do not represent money.

Example

$3.97 / 5 = ?$

Estimate: 3.97 is close to 4, and $4/5 = \frac{4}{5}$. The answer will be a little less than 1.

Step 1 Trade the ones for tenths and share the tenths.

```
      .7       ← Each share gets 7 tenths. Write a decimal point in the quotient.
  5)3.97       ← 3 ones + 9 tenths = 39 tenths
   -3 5        ← 7 tenths * 5 = 35 tenths
   ----
      4        ← 4 tenths are left.
```

Step 2 Trade the remaining tenths for hundredths. Share the hundredths.

```
     .79       ← Each share gets 9 hundredths.
  5)3.97
   -35
   ----
     47        ← 4 tenths + 7 hundredths = 47 hundredths
    -45        ← 9 hundredths * 5 = 45 hundredths
    ----
      2        ← 2 hundredths are left.
```

At this point, you can either round 0.79 to 0.8 and write $3.97 / 5 \approx 0.8$, or you can continue dividing into the thousandths.

Step 3 Continue dividing into the thousandths. Attach a 0 to the end of 3.97. (Attaching 0s to the end of a decimal doesn't change its value.)

```
     .794      ← Each share gets 4 thousandths.
  5)3.970      ← 3.97 = 3.970
   -35
   ----
     47
    -45
    ----
     20        ← 2 hundredths + 0 thousandths = 20 thousandths
    -20        ← 4 thousandths * 5 = 20 thousandths
    ----
      0        ← No thousandths are left.
```

$3.97 / 5 = \mathbf{0.794}$

0.794 is a little less than 1, which matches the estimate, so the answer makes sense.

Check Your Understanding

Divide. Estimate to check whether your answers are reasonable.

1. $8.28 / 4$ **2.** $4)\overline{9.64}$ **3.** $6)\overline{8.67}$ **4.** $38.65 / 5$

Check your answers in the Answer Key.

The Number System

U.S. Traditional Long Division: Decimal Divisors

To use **U.S. traditional long division** to divide by a decimal number, such as 0.6 or 3.5, you can find an **equivalent problem** that has no decimal in the divisor. The answer to the equivalent problem is the same as the answer to the original problem.

Step 1 Think of the division problem as a fraction.

Step 2 Use the multiplication rule to find an equivalent fraction that has no decimal in the denominator.

Step 3 Think of the equivalent fraction as a division problem.

Step 4 Solve the division problem. The answer to the equivalent problem is the same as the answer to the original problem.

Example

$194 / 0.4 = ?$

Step 1 Think of the division problem as a fraction.

$$194 / 0.4 = \frac{194}{0.4}$$

Step 2 Find an equivalent fraction with no decimal in the denominator.

$$\frac{194 * 10}{0.4 * 10} = \frac{1{,}940}{4}$$

Step 3 Think of the equivalent fraction as a division problem.

$$\frac{1{,}940}{4} = 1{,}940 / 4$$

Estimate: 1,940 is close to 2,000. An estimated quotient is $2{,}000 / 4 = 500$.

Step 4 Solve the equivalent division problem.

```
    485
 4)1940
  -16
   ‾‾‾
    34
   -32
   ‾‾‾
    20
   -20
   ‾‾‾
     0
```

Because $\frac{1{,}940}{4}$ and $\frac{194}{0.4}$ are equivalent fractions, the division problems $1{,}940 / 4$ and $194 / 0.4$ are equivalent. So the answer to $1{,}940 / 4$ is the same as the answer to $194 / 0.4$.

$194 / 0.4 = \mathbf{485}$

The answer is reasonable because 485 is close to the estimate of 500.

The Number System

U.S. Traditional Long Division: Decimal Divisors and Dividends

Sometimes *both* the divisor (the number you are dividing by) and the dividend (the number being divided) are decimal numbers. To use **U.S. traditional long division** in these cases, you can first find an **equivalent problem** that has no decimal in the divisor. (Having a decimal part in the dividend is okay.) The answer to the equivalent problem is the same as the answer to your original problem.

Example

$3.78 / 0.7 = ?$

Step 1 Think of the division problem as a fraction.

$$3.78 / 0.7 = \frac{3.78}{0.7}$$

Step 2 Find an equivalent fraction with no decimal in the denominator.

$$\frac{3.78 * 10}{0.7 * 10} = \frac{37.8}{7}$$

Step 3 Think of the equivalent fraction as a division problem.

$$\frac{37.8}{7} = 37.8 / 7$$

Estimate: 37.8 is close to 35. An estimated quotient is $35 / 7 = 5$.

Step 4 Solve the division problem.

```
        5.4
    7)37.8
     -35
     ---
       2 8
      -2 8
      ----
         0
```

Because $\frac{37.8}{7}$ and $\frac{3.78}{0.7}$ are equivalent fractions, the division problems $37.8 / 7$ and $3.78 / 0.7$ are equivalent. So the answer to $37.8 / 7$ is the same as the answer to $3.78 / 0.7$.

$3.78 / 0.7 = \mathbf{5.4}$

The answer is reasonable because it is close to the estimate of 5.

Check Your Understanding

Divide. Estimate to check whether your answers are reasonable.

1. $784 / 0.7$ **2.** $36.9 / 1.5$ **3.** $4.68 / 0.03$ **4.** $3.05 / 0.005$

Check your answers in the Answer Key.

The Number System

Fractions

Fractions were invented thousands of years ago to name numbers that are between whole numbers. People likely first used these in-between numbers for making more precise measurements.

Today most rulers and other measuring tools have marks to name numbers that are between whole unit measures. Learning how to read these in-between marks is an important part of learning to use these tools. Here are some examples of measurements that use fractions: $1\frac{1}{4}$ inches, $\frac{2}{3}$ cup, $\frac{3}{4}$ hour, $\frac{9}{10}$ kilometer, and $13\frac{1}{2}$ pounds.

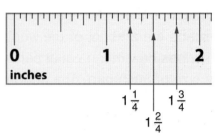

The $\frac{1}{4}$-inch marks between 1 and 2 are labeled.

You can use fractions to name parts of wholes. To understand the amount that a fraction represents, you need to know what the **whole** is, what the parts are, and how many of the parts are being considered.

The whole might be one single thing, like a stick of butter. The whole might be a distance, like an inch or a mile. The whole might also be a collection of things, like a box of crayons. The **whole** is sometimes called *one, one whole,* or *a unit.*

A fraction communicates an amount by describing how the parts are related to the whole. This means that a fraction like $\frac{1}{2}$ can communicate different amounts depending on how the whole is defined. For example, half an inch is much smaller than half a mile. Similarly, half a box of crayons could be many crayons or just a few crayons, depending on the number of crayons in the whole box.

There are many more crayons in half a box of 64 crayons than in half a box of 16.

Fractions can also be used to show division, ratios, rates, scales, and other relationships.

The Number System

Reading and Writing Fractions

Fractions can be written $\frac{a}{b}$, where *a* and *b* are two integers (*b* cannot be 0) separated by a fraction bar. For example, in the fraction $\frac{2}{3}$, 2 and 3 are both integers. Fractions can also be written in different ways. For example, two-thirds can be written as 2-thirds or 2 out of 3.

In fractions that name parts of wholes, the numerator and denominator together describe the amount of the whole that a fraction represents. The **denominator** describes how many equal parts it takes to make the whole, which determines the size of each part. The denominator cannot be 0. The **numerator** describes the number of equal-size parts that are being considered. When reading a fraction, say the numerator first. Then say the size of the equal parts represented by the denominator.

numerator ⟶ $\frac{2}{3}$ "two-thirds"
denominator ⟶

Fractions Equal To One

Some fractions represent numbers equal to one. If the numerator is equal to the denominator, the amount represented is equal to one whole.

> **Example**
>
> $\frac{4}{4}$ means 4 out of 4 equal parts. That's all of the parts of 1 whole.
>
> $\frac{4}{4} = 1$ whole
>
>
>
> $\frac{6}{6}$ means 6 out of 6 equal parts. That's all of the parts of 1 whole.
>
> $\frac{6}{6} = 1$ whole
>
>

The Number System

Fractions Greater Than One

If the numerator is larger than the denominator, the amount represented is greater than one whole. Fractions greater than one can also be written as **mixed numbers.** A mixed number has a whole-number part and a fraction part. In the mixed number $2\frac{3}{5}$, the whole-number part is 2 and the fraction part is $\frac{3}{5}$. A mixed number is equal to the sum of the whole-number part and the fraction part: $2\frac{3}{5} = 2 + \frac{3}{5}$. Mixed numbers are used in many of the same ways that fractions are used.

Note Fractions greater than one are sometimes called *improper fractions,* although there is nothing wrong with them. It is often easier to use fractions written with the numerator greater than the denominator than to use mixed numbers when writing number sentences to solve problems.

Example

Write $\frac{13}{5}$ as a mixed number.

You know $\frac{5}{5} = 1$ whole.

$\frac{5}{5} + \frac{5}{5} + \frac{3}{5} = \frac{13}{5}$

You can represent $\frac{13}{5}$ as two whole circles and three-fifths of another circle.

$2\frac{3}{5}$

$\frac{13}{5}$ can be written as $2\frac{3}{5}$.

one hundred fifty-seven **157**

The Number System

Meanings of Fractions

Fractions can have different meanings.

Parts of Regions
Fractions can be used to name part of a whole region:

$\frac{5}{8}$ of the circle is shaded.

Parts of Collections
Fractions can be used to mean part of a whole collection of objects:

$\frac{3}{10}$ of this collection of coins is quarters.

Points on a Number Line
Fractions can be used to mean distance from 0 on a number line. The whole is the distance from 0 to 1, and each whole can be divided into equal distances.

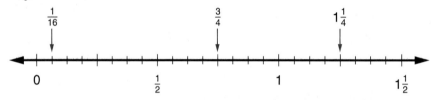

Division Notation
The fraction $\frac{a}{b}$ is another way of saying a divided by b. The division problem 24 divided by 3 can be written in any of these ways: $24 \div 3$, $3\overline{)24}$, $24/3$, or $\frac{24}{3}$. Similarly, 3 divided by 4 can be written in any of these ways: $3 \div 4$, $4\overline{)3}$, $3/4$, and $\frac{3}{4}$.

Ratios and Rates
Fractions can be used to compare quantities.

Ratios compare quantities. The quantities can have the same unit or different units. For example, DuSable School won 6 games and lost 14 games last year. The total number of games played was $6 + 14 = 20$ games. The fraction $\frac{6}{20}$ compares $\frac{\text{games won}}{\text{games played}}$. This fraction compares quantities with the same unit (games and games).

Fractions can be used to compare quantities with *different units*. This is a special kind of ratio called a **rate**. For example, Bill's car can travel about 35 miles on 1 gallon of gasoline. The fraction $\frac{35 \text{ miles}}{1 \text{ gallon}}$ compares quantities with different units (miles and gallons). At this rate, Bill's car can travel about 245 miles on 7 gallons of gasoline.

$\frac{35 \text{ miles}}{1 \text{ gallon}} = \frac{245 \text{ miles}}{7 \text{ gallons}}$

Uses of Fractions

Fractions have many uses in everyday life.

Scale Fractions can be used to compare the size of a drawing or model of an object to the size of the actual object.

The **map scale** shown here is given as 1 : 10,000. This means that every distance on the map is $\frac{1}{10,000}$ of the real-world distance. A 1-centimeter distance on the map stands for a real-world distance of 10,000 centimeters (100 meters).

Decimals Fractions can be used to rename decimals.

This coin is called a quarter because it is $\frac{1}{4}$ of a dollar. $\frac{1}{4}$ is equivalent to $\frac{25}{100}$. $\frac{25}{100}$ of a dollar is the same as 0.25 or $0.25.

1 quarter

1 dollar

Percents Fractions can be used to rename percents. Percent means *per hundred*, or *out of a hundred*. So 1% (one percent) has the same meaning as the fraction $\frac{1}{100}$ and the decimal 0.01.

Saving 50% means you will save half the cost. That's because 50% means $\frac{50}{100}$, or $\frac{1}{2}$.

"In-Between" Measures Fractions can name measures that are between whole-number measures.

Fractions can be used to name lengths measured with rulers, metersticks, yardsticks, and tape measures.

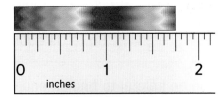

The ribbon measures $1\frac{3}{4}$ inches.

Fractions can be used to tell how far it is to a destination.

A scenic view is $\frac{3}{4}$ mile ahead.

Fractions can be used to name measures of volume.

The liquid in the measuring cup measures $\frac{1}{2}$ cup.

The Number System

Equivalent Fractions

Two or more fractions that name the same number are called **equivalent fractions.** Equivalent fractions are equal. One way to rename a fraction as an equivalent fraction is to *multiply* the numerator and denominator by the same number. Another way is to *divide* the numerator and the denominator by the same number.

> **Example**
>
> Rename $\frac{3}{4}$ as an equivalent fraction using multiplication.
>
> The rectangle is divided into 4 equal parts. 3 of the parts are blue. $\frac{3}{4}$ of the rectangle is blue.
>
> If each of the 4 parts is split into 2 equal parts, there are now 8 equal parts. 6 of them are blue. $\frac{6}{8}$ of the rectangle is blue.
>
> $\frac{3}{4}$ and $\frac{6}{8}$ both name the same amount of the rectangle that is blue. The number of parts in the rectangle was doubled. You can show this by multiplying both the numerator and the denominator by 2.
>
> $\frac{3}{4}$ is equivalent to $\frac{6}{8}$. $\frac{3}{4} = \frac{6}{8}$
>
> If each part in the rectangle is divided into 3 equal parts, the number of parts is tripled. You can show this by multiplying the numerator and the denominator 3.
>
> $\frac{3}{4}$ is equivalent to $\frac{9}{12}$. $\frac{3}{4} = \frac{9}{12}$
>
>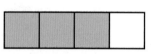
>
> $\frac{3}{4} = \frac{6}{8}$
>
>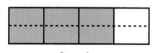
>
> $\frac{3 * 2}{4 * 2} = \frac{6}{8}$
>
>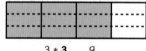
>
> $\frac{3 * 3}{4 * 3} = \frac{9}{12}$

> **Example**
>
> Rename $\frac{6}{12}$ as an equivalent fraction using division.
>
> $\frac{6}{12}$ of the region is green.   Divide the region into groups of 3. $\frac{2}{4}$ is green.
>
>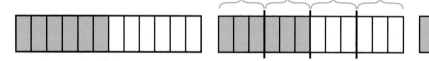
>
> $\frac{6}{12} = \frac{6 \div 3}{12 \div 3} = \frac{2}{4}$. So, $\frac{6}{12}$ is equivalent to $\frac{2}{4}$. $\frac{6}{12} = \frac{2}{4}$

It can be helpful to find an equivalent fraction with the smallest possible numerator and denominator. These fractions are in *simplest form* or in *lowest terms*. You can rename a fraction in simplest form by dividing its numerator and denominator by the **greatest common factor (GCF)** of the numerator and the denominator.

> **Example**
>
> Use the GCF to find an equivalent fraction for $\frac{18}{24}$.
>
> GCF (18, 24) = 6
>
> $\frac{18}{24} = \frac{(18 \div 6)}{(24 \div 6)} = \frac{3}{4}$ The fraction $\frac{3}{4}$ is in simplest form.

The Number System

Renaming Fractions Greater Than One

When the numerator of a fraction is greater than the denominator, the fraction is greater than one. You can rename fractions greater than one as mixed numbers or as whole numbers.

Different mixed-number names for a fraction can be useful for different reasons. For example, the name with the greatest whole-number part, such as $4\frac{2}{3}$, is useful for comparing the mixed number to whole numbers or for placing it on a number line. A name with a fraction part that is greater than one, such as $3\frac{5}{3}$, can be useful when you are subtracting mixed numbers.

One way to rename a fraction greater than one is to use **unit fractions** to make wholes.

Example

Rename $\frac{19}{4}$ as a mixed number.

Show 19 fourths of a circle.

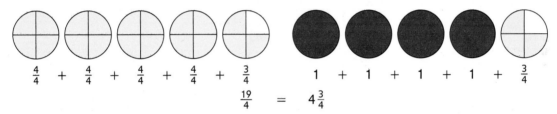

$\frac{4}{4} + \frac{4}{4} + \frac{4}{4} + \frac{4}{4} + \frac{3}{4}$ $\quad$ $1 + 1 + 1 + 1 + \frac{3}{4}$

$$\frac{19}{4} = 4\frac{3}{4}$$

- Trade a group of 4 fourths $\left(\frac{4}{4}\right)$ for 1 whole. Now you have 1 whole and 15 fourths. $1\frac{15}{4}$ is one mixed-number name for $\frac{19}{4}$.
- Trade another group of 4 fourths for 1 whole to get the name $2\frac{11}{4}$.
- Trade more groups to get $3\frac{7}{4}$ and $4\frac{3}{4}$.

$\frac{19}{4}$ can have different mixed-number names: $\frac{19}{4} = 1\frac{15}{4} = 2\frac{11}{4} = 3\frac{7}{4} = 4\frac{3}{4}$.

Another way to rename a fraction as a mixed number is to divide the numerator by the denominator. The remainder in the division problem can be rewritten as a fraction. The remainder is the numerator, and the divisor is the denominator. For example, to rename $\frac{37}{5}$ as a mixed number, divide 37 by 5: $37 \div 5 \to 7$ R2, or $7\frac{2}{5}$.

Example

Rename $\frac{19}{4}$ as a mixed number.

Think about $\frac{19}{4}$ as dividing 19 by 4. $19 \div 4 \to 4$ R3

- The quotient, 4, is the whole-number part of the mixed number. It tells the number of wholes in $\frac{19}{4}$.
- The remainder, 3, is the numerator of the fraction part of the mixed number. It tells how many fourths are left.

$\frac{19}{4} = 4\frac{3}{4}$

SRB
one hundred sixty-one 161

The Number System

Renaming Mixed Numbers

You can rename mixed numbers as fractions greater than one.

Example

Rename $3\frac{1}{2}$ as a fraction.

One way: Think about breaking apart the wholes.

If a circle is the whole, then $3\frac{1}{2}$ is three whole circles and $\frac{1}{2}$ of another circle.

$3\frac{1}{2}$

If you break apart each of the whole circles into halves, then you can see that $3\frac{1}{2} = \frac{7}{2}$.

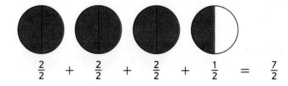

$\frac{2}{2} + \frac{2}{2} + \frac{2}{2} + \frac{1}{2} = \frac{7}{2}$

Another way: Think about the number line.

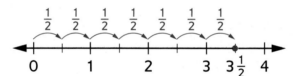

Count the halves. You count seven halves from 0 to $3\frac{1}{2}$, so $3\frac{1}{2} = \frac{7}{2}$.

Another way: Rename the whole number as a fraction with the same denominator as the fraction part, and add the fractions.

Rename 3 as $\frac{6}{2}$.

So, $3\frac{1}{2} = 3 + \frac{1}{2} = \frac{6}{2} + \frac{1}{2} = \frac{7}{2}$.

The Number System

You can rename many mixed numbers as other mixed numbers that have the same denominator.

Example

Write as many equivalent names as you can for $3\frac{2}{5}$ using the denominator 5.
Think about breaking apart each whole into fifths, one at a time.

$3\frac{2}{5} = 2\frac{7}{5} = 1\frac{12}{5} = \frac{17}{5}$

Note You can think of making wholes and breaking apart wholes as fair trades. Putting 4 fourths together to make a whole is like trading $\frac{4}{4}$ for one whole. This is a fair trade because $\frac{4}{4} = 1$. Breaking apart a whole into 5 fifths is like trading one whole for $\frac{5}{5}$. This is a fair trade because $1 = \frac{5}{5}$.

Check Your Understanding

Write each fraction as a mixed number.

1. $\frac{51}{4}$
2. $\frac{26}{3}$
3. $\frac{34}{5}$
4. $\frac{60}{16}$

Write each mixed number as a fraction.

5. $4\frac{3}{4}$
6. $3\frac{2}{3}$
7. $4\frac{5}{6}$
8. $1\frac{4}{3}$

9. Write as many equivalent names as you can for $6\frac{2}{3}$ using 3 as the denominator.

Check your answers in the Answer Key.

one hundred sixty-three

The Number System

Renaming Fractions as Decimals

Any fraction can be renamed as a decimal. Sometimes the decimal will end after a certain number of places. Decimals that end are called **terminating decimals**. Sometimes the decimal will have one or more digits that repeat in a pattern forever. Decimals that repeat in this way are called **repeating decimals**. The fraction $\frac{1}{2}$ is equal to the terminating decimal 0.5. The fraction $\frac{2}{3}$ is equal to the repeating decimal 0.6666….

One way to rename a fraction as a decimal is to remember the decimal name: $\frac{1}{2} = 0.5$, $\frac{3}{4} = 0.75$, $\frac{1}{8} = 0.125$, and so on. If you have memorized the decimal names for a few common fractions, then logical thinking can help you to rename many other common fractions as decimals. For example, if you know $\frac{1}{8} = 0.125$, then $\frac{3}{8} = 0.125 + 0.125 + 0.125 = 0.375$.

You can also find decimal names for fractions by using equivalent fractions, the Fraction-Stick Chart, division, or a calculator.

Using Equivalent Fractions

One way to rename a fraction as a decimal is to find an equivalent fraction with a denominator that is a power of 10, such as 10, 100, or 1,000. This method only works for some fractions.

> **Example**
>
> Rename $\frac{3}{5}$ as a decimal.
>
> The solid lines divide the square into 5 equal parts. Each of these parts is $\frac{1}{5}$ of the square. $\frac{3}{5}$ of the square is shaded.
>
> The dashed lines divide each fifth into 2 equal parts. Each of these parts is $\frac{1}{10}$, or 0.1, of the square. $\frac{6}{10}$, or 0.6, of the square is shaded.
>
> $\frac{3}{5} = \frac{6}{10} = 0.6$
>
>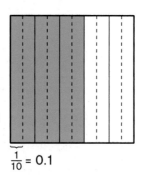
>
> $\frac{1}{10} = 0.1$

> **Check Your Understanding**
>
> Rename these fractions as decimals.
>
> 1. $\frac{3}{4}$ 2. $\frac{2}{5}$ 3. $\frac{3}{2}$ 4. $\frac{11}{20}$ 5. $\frac{3}{25}$
>
> Check your answers in the Answer Key.

The Number System

Using a Fraction-Stick Chart to Rename Fractions as Decimals

The Fraction-Stick Chart below, and also on page 376, can be used to rename fractions as decimals. Note that the result is usually only an approximation. You can use division or a calculator to obtain more precise approximations.

Example

Rename $\frac{2}{3}$ as a decimal.

Step 1 Locate $\frac{2}{3}$ on the "thirds" stick.

Step 2 Place one edge of a straightedge at $\frac{2}{3}$.

Step 3 Find where the straightedge crosses the numberline with decimals.

The straightedge crosses the numberline between 0.66 and 0.67.

So, $\frac{2}{3}$ is equal to about 0.66 or 0.67.

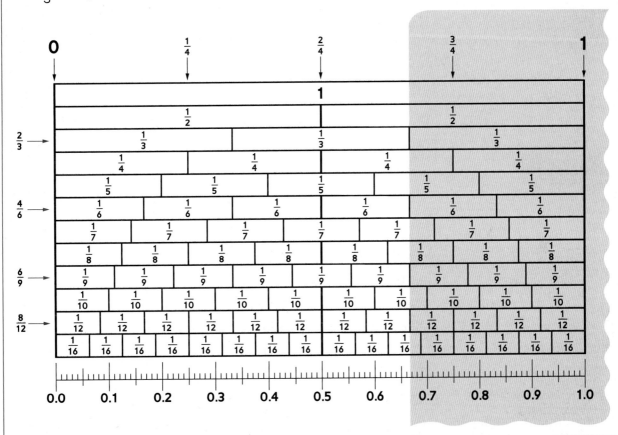

Check Your Understanding

Use the Fraction-Stick Chart to find an approximate decimal name for each fraction or mixed number.

1. $\frac{7}{10}$
2. $\frac{5}{8}$
3. $3\frac{1}{3}$
4. $\frac{9}{12}$
5. $1\frac{4}{9}$
6. $\frac{3}{7}$

Check your answers in the Answer Key.

one hundred sixty-five 165

The Number System

Using Division to Rename Fractions as Decimals

Fractions can be used to show division problems. For example, $\frac{7}{8}$ is another way to write $7 \div 8$. So one way to rename $\frac{7}{8}$ as a decimal is to divide 7 by 8. The following example illustrates how to rename a fraction as a decimal by dividing its numerator by its denominator.

Example

Use partial-quotients division to rename $\frac{7}{8}$ as a decimal.

Step 1 Estimate the quotient. It will be less than 1 but greater than $\frac{1}{2}$.

Step 2 Decide how many digits you want to the right of the decimal point. For measuring or solving everyday problems, two or three digits are usually enough. In this case rename $\frac{7}{8}$ as a decimal with three digits to the right of the decimal point.

Step 3 Rewrite the numerator with a 0 for each decimal place you want. Rewrite the numerator, 7, as 7.000.

Step 4 Use partial-quotients division to divide 7.000 by 8. Ignore the decimal point for now, and divide 7,000 by 8.

```
    8)7000
   - 6400 | 800
     ————
      600
    - 560 |  70
     ————
       40
     - 40 |   5
     ————
        0   875
```

Step 5 Use the estimate from Step 1 to place the decimal point in the quotient. Since $\frac{7}{8}$ is between $\frac{1}{2}$ and 1, the decimal point should be placed before the 8, to give 0.875.

So, $\frac{7}{8} = 0.875$.

This answer makes sense because it is less than 1 but greater than $\frac{1}{2}$, or 0.5.

Note This method will *always* work. Any fraction can be renamed as a decimal by dividing its numerator by its denominator. In this case, the fraction was renamed as a **terminating decimal**, and the answer worked out to exactly three decimal places. When the fraction is renamed as a **repeating decimal**, you will need to give a decimal name that is approximately equal to the fraction.

Renaming Decimals as Fractions

Any terminating decimal can be renamed as a fraction whose denominator is a power of 10. To change a terminating decimal to a fraction, use the place of the rightmost digit to help you write the denominator.

Examples

$0.4 = \frac{4}{10} = \frac{2}{5}$ $0.08 = \frac{8}{100} = \frac{4}{50} = \frac{2}{25}$ $0.25 = \frac{25}{100} = \frac{5}{20} = \frac{1}{4}$ $0.124 = \frac{124}{1,000} = \frac{62}{500} = \frac{31}{250}$

↓ tenths place   ↓ hundredths place   ↓ hundredths place   ↓ thousandths place

The Number System

Using a Calculator to Rename Fractions as Decimals

You can rename a fraction as a decimal by dividing the numerator by the denominator using a calculator.

Examples

Rename $\frac{3}{4}$ and $\frac{7}{8}$ as decimals.

Key in: 3 ÷ 4 =
Answer: 0.75
$\frac{3}{4} = 0.75$

Key in: 7 ÷ 8 =
Answer: 0.875
$\frac{7}{8} = 0.875$

In some cases, the decimal takes up the entire calculator display. If one or more digits repeat, you can write the decimal by writing the repeating digit or digits just once, and putting a bar above the digit or digits that repeat.

Examples

Fraction	Key in:	Calculator Display	Answer
$\frac{1}{3}$	1 ÷ 3 =	0.3333333	$0.\overline{3}$
$\frac{2}{3}$	2 ÷ 3 =	0.6666666 or 0.6666666667 (depending on the calculator)	$0.\overline{6}$
$\frac{1}{6}$	1 ÷ 6 =	0.1666666 or 0.1666666667 (depending on the calculator)	$0.1\overline{6}$
$\frac{4}{9}$	4 ÷ 9 =	0.4444444	$0.\overline{4}$
$\frac{6}{11}$	6 ÷ 11 =	0.5454545 or 0.5454545455 (depending on the calculator)	$0.\overline{54}$
$\frac{7}{12}$	7 ÷ 12 =	0.5833333	$0.58\overline{3}$

Note On some calculators, the final digit for a repeating decimal may not follow the pattern. For example, a calculator may show $\frac{2}{3} = 2 / 3 = 0.66666666667$. The digit 6 really does repeat forever, but this calculator has rounded the digit.

Check Your Understanding

Use a calculator to rename each fraction as a decimal.

1. $\frac{7}{8}$
2. $\frac{4}{12}$
3. $\frac{5}{12}$
4. $\frac{5}{16}$
5. $\frac{2}{9}$
6. $\frac{7}{6}$

Check your answers in the Answer Key.

The Number System

Comparing Fractions

You can use several strategies to compare fractions.

Use a Common Numerator	$\frac{5}{6} > \frac{5}{8}$ because sixths are larger than eighths of the same-size whole and there are 5 of each.	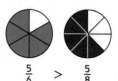 $\frac{5}{6} > \frac{5}{8}$
Compare to $\frac{1}{2}$ as a Benchmark	$\frac{3}{5} > \frac{4}{9}$ because $\frac{4}{9}$ is less than $\frac{1}{2}$ and $\frac{3}{5}$ is more than $\frac{1}{2}$.	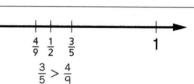 $\frac{3}{5} > \frac{4}{9}$
Compare to 1 as a Benchmark	Compare $\frac{3}{4}$ and $\frac{2}{3}$. They are both less than 1. But $\frac{3}{4}$ is $\frac{1}{4}$ away from 1, and $\frac{2}{3}$ is $\frac{1}{3}$ away from 1. So, $\frac{3}{4}$ is closer to 1 than $\frac{2}{3}$ is. That means $\frac{3}{4} > \frac{2}{3}$.	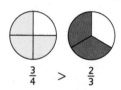 $\frac{3}{4} > \frac{2}{3}$
Use an Equivalent for One of the Fractions	To compare $\frac{2}{5}$ and $\frac{1}{4}$, change $\frac{1}{4}$ to $\frac{2}{8}$. Since $\frac{2}{5} > \frac{2}{8}$ (because fifths are larger than eighths), you know that $\frac{2}{5} > \frac{1}{4}$.	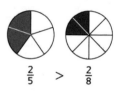 $\frac{2}{5} > \frac{2}{8}$
Use a Common Denominator	To compare $\frac{3}{5}$ and $\frac{5}{8}$, rename both fractions with a common denominator. $5*8 = 40$ is a common denominator. Since $\frac{24}{40} < \frac{25}{40}$, you know that $\frac{3}{5} < \frac{5}{8}$.	$\frac{3}{5} = \frac{3*8}{5*8} = \frac{24}{40}$ $\frac{5}{8} = \frac{5*5}{8*5} = \frac{25}{40}$
Convert to Decimals	To compare $\frac{13}{17}$ and $\frac{32}{41}$, use a calculator to convert both to decimals. Since $0.76... < 0.78...$, you know that $\frac{13}{17} < \frac{32}{41}$.	$\frac{13}{17} = 13 \boxed{\div} 17 \boxed{=} 0.7647059$ $\frac{32}{41} = 32 \boxed{\div} 41 \boxed{=} 0.7804878$

Check Your Understanding

Compare. Use > or <.

1. $\frac{5}{9} \square \frac{2}{5}$ 2. $\frac{14}{15} \square \frac{10}{11}$ 3. $\frac{12}{6} \square \frac{12}{5}$ 4. $\frac{4}{5} \square \frac{5}{7}$ 5. $\frac{17}{23} \square \frac{67}{93}$

Check your answers in the Answer Key.

The Number System

Finding Fractions Between Fractions

One way to find a fraction between fractions is to fold fraction strips into smaller pieces as you think about a number line.

Example

Find a fraction between $\frac{2}{5}$ and $\frac{3}{5}$.

Start with a fifths fraction strip. Fold the fraction strip to make smaller unit-fraction pieces. Think about the number line represented by the fraction strip.

This drawing shows what the fraction strip would look like if you fold each fifth in half. Dotted lines represent folds.

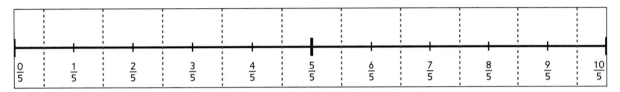

Make new tick marks on the number line at each fold. Each whole unit is now divided into 10 equal parts, so you can count by tenths to label the tick marks.

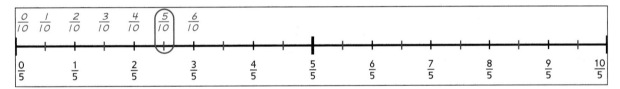

$\frac{5}{10}$ is a fraction between $\frac{2}{5}$ and $\frac{3}{5}$.

Check Your Understanding

1. Name a fraction between $\frac{3}{4}$ and $\frac{4}{4}$.
2. Name a fraction between $\frac{1}{3}$ and $\frac{2}{3}$.

Check your answers in the Answer Key.

The Number System

There are an infinite number of fractions between any two fractions. You cannot fold a fraction strip an infinite number of times. So another way to find fractions between fractions is to think about "zooming in" on a number line.

Example

Find a fraction between $\frac{23}{24}$ and $\frac{24}{24}$.

Start with part of a number line showing the fractions $\frac{23}{24}$ and $\frac{24}{24}$.

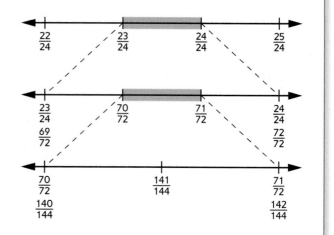

Zoom in between $\frac{23}{24}$ and $\frac{24}{24}$. Divide the interval between them into equal parts and rewrite each fraction as an equivalent fraction with a larger denominator. Here, the interval is divided into 3 equal parts, so $\frac{23}{24}$ and $\frac{24}{24}$ are each multiplied by $\frac{3}{3}$.

$$\frac{23}{24} * \frac{3}{3} = \frac{69}{72} \qquad \frac{24}{24} * \frac{3}{3} = \frac{72}{72}$$

Label the tick marks between $\frac{69}{72}$ and $\frac{72}{72}$.

$\frac{70}{72}$ and $\frac{71}{72}$ are two fractions that are between $\frac{23}{24}$ and $\frac{24}{24}$.

To find more fractions between $\frac{23}{24}$ and $\frac{24}{24}$, continue to zoom in on the number line between any two points in the interval.

If you zoom in between $\frac{70}{72}$ and $\frac{71}{72}$ and divide that interval into 2 equal parts, you can rename the fractions:

$$\frac{70}{72} * \frac{2}{2} = \frac{140}{144} \qquad \frac{71}{72} * \frac{2}{2} = \frac{142}{144}$$

Label the tick mark between them: $\frac{141}{144}$.

$\frac{141}{144}$ is another fraction between $\frac{23}{24}$ and $\frac{24}{24}$.

Remember that there are an infinite number of points between any two fractions. To find more points between $\frac{23}{24}$ and $\frac{24}{24}$ in the example above, you could continue zooming in on the number line. For example, you could find points between $\frac{141}{144}$ and $\frac{142}{144}$. Or, you could look for points in a different interval on the second number line. For example, you could look for points between $\frac{69}{72}$ and $\frac{70}{72}$. However, you will never be able to name them all.

The Number System

Fraction-Stick Chart

Each stick on the Fraction-Stick Chart represents 1 whole. Each stick (except the 1-stick) is divided into equal pieces. Each piece represents a fraction of 1 whole.

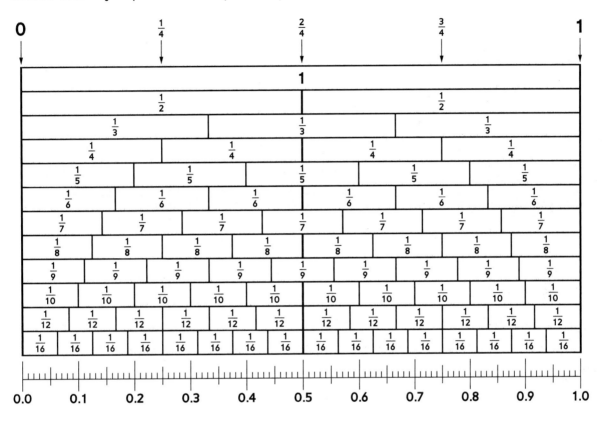

Locating a Fraction on the Fraction-Stick Chart

1. Select the stick shown by the denominator of the fraction.

2. Count the number of pieces shown by the numerator, starting at the left edge of the chart.

Example

Find $\frac{3}{4}$ on the Fraction-Stick Chart.

The "fourths" stick is equally divided into 4 pieces, each labeled $\frac{1}{4}$. This stick can be used to locate fractions whose denominators are 4. To locate the fraction $\frac{3}{4}$, count 3 pieces, starting at the left.

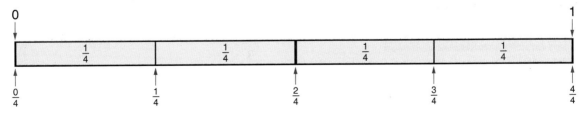

$\frac{3}{4}$ is located at the right edge of the third piece.

one hundred seventy-one **SRB 171**

The Number System

Finding Equivalent Fractions

Example

Find fractions that are equivalent to $\frac{2}{3}$.

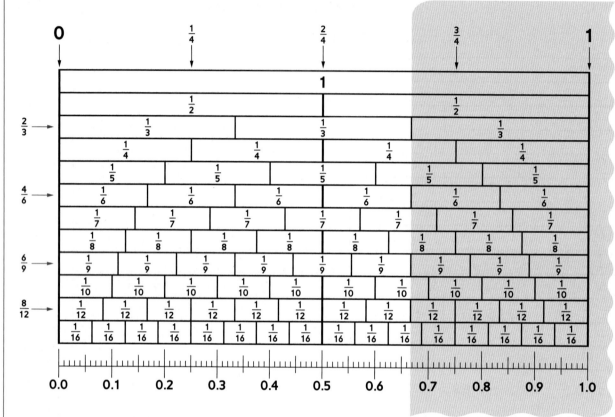

Place one edge of the straightedge at $\frac{2}{3}$. Find all pieces whose right edges touch the edge of the straightedge. The right edge of each piece shows the location of a fraction that is equivalent to $\frac{2}{3}$. $\frac{4}{6}$, $\frac{6}{9}$, and $\frac{8}{12}$ are equivalent to $\frac{2}{3}$.

Comparing Two Fractions

Example

Compare $\frac{4}{9}$ and $\frac{3}{8}$. Which is less?

Place one edge of a straightedge at $\frac{4}{9}$. Locate $\frac{3}{8}$ on the "eighths" stick. $\frac{3}{8}$ is to the left of $\frac{4}{9}$.

$\frac{3}{8}$ is less than $\frac{4}{9}$.

Check Your Understanding

Which fraction is less? Use the Fraction-Stick Chart to decide.

1. $\frac{1}{9}$ or $\frac{7}{8}$
2. $\frac{7}{9}$ or $\frac{8}{10}$
3. $\frac{3}{7}$ or $\frac{3}{6}$
4. $\frac{6}{7}$ or $\frac{13}{16}$
5. $\frac{11}{12}$ or $\frac{15}{16}$

Check your answers in the Answer Key.

The Number System

Common Denominators

To solve problems that involve fractions with different denominators, you may decide to rename the fractions so they have the same denominator. If two fractions have the same denominator, that denominator is called a **common denominator.**

There are several methods for renaming fractions so they have a common denominator.

Examples

Rename $\frac{3}{4}$ and $\frac{1}{6}$ with a common denominator.

Equivalent Fractions Method
List equivalent fractions for $\frac{3}{4}$ and $\frac{1}{6}$.

$\frac{3}{4} = \frac{6}{8} = \frac{9}{12} = \frac{12}{16} = \ldots$

$\frac{1}{6} = \frac{2}{12} = \frac{3}{18} = \frac{4}{24} = \ldots$

Both $\frac{3}{4}$ and $\frac{1}{6}$ can be renamed as fractions with the common denominator 12.

$\frac{3}{4} = \frac{9}{12}$ and $\frac{1}{6} = \frac{2}{12}$

The Multiplication Method
Multiply the numerator and the denominator of each fraction by the denominator of the other fraction.

$\frac{3}{4} = \frac{3 * 6}{4 * 6} = \frac{18}{24}$ $\frac{1}{6} = \frac{1 * 4}{6 * 4} = \frac{4}{24}$

Note The multiplication method gives what *Everyday Mathematics* calls the **quick common denominator.** The quick common denominator can be used with variables, so it is common in algebra.

Least Common Multiple Method
Find the least common multiple of the denominators.

Multiples of 4: 4, 8, **12**, 16, 20, …

Multiples of 6: 6, **12**, 18, 24, …

The least common multiple of 4 and 6 is 12.

Rename the fractions so that the denominator in each fraction is the least common multiple.

$\frac{3}{4} = \frac{3 * 3}{4 * 3} = \frac{9}{12}$ and $\frac{1}{6} = \frac{1 * 2}{6 * 2} = \frac{2}{12}$

This method gives what is known as the **least common denominator.**

Note The least common denominator is usually easier to use in complicated calculations, although finding it can often take more time.

Check Your Understanding

Rename each pair of fractions as fractions with a common denominator.

1. $\frac{1}{3}$ and $\frac{5}{6}$ 2. $\frac{3}{4}$ and $\frac{3}{5}$ 3. $\frac{7}{10}$ and $\frac{3}{2}$ 4. $\frac{1}{4}$ and $\frac{3}{10}$ 5. $\frac{4}{6}$ and $\frac{7}{8}$

Check your answers in the Answer Key.

The Number System

Using Unit Fractions to Solve Problems

A **unit fraction** is a fraction with 1 in its numerator, such as $\frac{1}{2}$, $\frac{1}{10}$, and $\frac{1}{25}$. If you know the part of a whole represented by a unit fraction, you can find the whole by multiplying the part of the whole by the denominator of the unit fraction.

Did You Know?

Levels of pollution are measured in parts per million (PPM). One PPM is one part out of 1,000,000, or $\frac{1}{1,000,000}$ of the whole.

Example

6 is $\frac{1}{4}$ of what number?

Since $\frac{1}{4}$ of the whole is 6,
$\frac{2}{4}$ must be 2 ∗ 6, or 12;
$\frac{3}{4}$ must be 3 ∗ 6, or 18;
and $\frac{4}{4}$ must be 4 ∗ 6, or 24.

So, 6 is $\frac{1}{4}$ of 24.

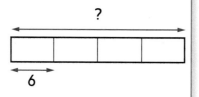

Example

Mark owns 2 white shirts. This is $\frac{1}{4}$ of the total number of shirts he owns. How many shirts does he own?

If 2 shirts are $\frac{1}{4}$ of the total number of shirts, then the total number of shirts is 4 times that number.

Since 4 ∗ 2 = 8, Mark owns a total of 8 shirts.

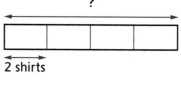

Unit fractions can be used even when a part is given as a fraction that is not a unit fraction.

Example

Sara lives 8 blocks from the library. This is $\frac{2}{3}$ of the distance from her home to school. How many blocks is it from Sara's home to school?

Step 1 Find $\frac{1}{3}$ of the distance to school. 8 blocks is $\frac{2}{3}$ of the distance to school. To find $\frac{1}{3}$ of the distance, divide 8 blocks by 2. 8 / 2 = 4 blocks

Step 2 Find the total distance to school. 4 blocks is $\frac{1}{3}$ of the distance to school. To find the total distance to school (or $\frac{3}{3}$ of the distance), multiply 4 blocks by 3. 3 ∗ 4 = 12 blocks

Sara lives 12 blocks from school.

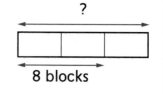

The Number System

Example

Ms. Partee spends $1,000 a month. This amount is $\frac{4}{5}$ of her monthly earnings. How much does she earn per month?

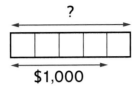

$1,000

Step 1 Find $\frac{1}{5}$ of her monthly earnings.
$1,000 is $\frac{4}{5}$ of her monthly earnings.

So to find $\frac{1}{5}$ of her earnings,
divide $1,000 by 4.

$1,000 / 4 = $250

Step 2 Find her total monthly earnings.
$250 is $\frac{1}{5}$ of her monthly earnings.
So to find her total earnings
(or $\frac{5}{5}$ of her earnings), multiply $250 by 5.
5 * $250 = $1,250

Ms. Partee earns $1,250 each month.

Check Your Understanding

Solve each problem.

1. Natalie collects movie posters. Her 4 posters from *Star Wars Episode III* are $\frac{1}{8}$ of her collection. How many posters does Natalie have in her whole collection?

2. If 12 counters are $\frac{3}{5}$ of a set, how many counters are in the whole set?

3. Mr. Hart spends $800 per month on rent. This is $\frac{3}{15}$ of his monthly earnings. How much does he earn per month?

Check your answers in the Answer Key.

The Number System

Estimating with Fractions

When working with fractions, you can use estimates to help you make sense of problem situations, approximate calculations, and check that answers are reasonable. One way to estimate with fractions is to think about visual representations, such as pictures, fraction strips, or number lines. Using a **benchmark,** a familiar reference point, can also help you estimate. Numbers such as 0, $\frac{1}{2}$, 1, $1\frac{1}{2}$, and 2 are often used as benchmarks because they are easy to picture in your head, or visualize, and compare with other fractions.

> **Example**
>
> Estimate: Will the sum of $\frac{7}{10} + \frac{3}{4}$ be greater than or less than 1?
>
> Visualize each fraction on a number line.
>
> $\frac{7}{10}$ is between $\frac{1}{2}$ and 1.
>
> $\frac{3}{4}$ is between $\frac{1}{2}$ and 1.
>
>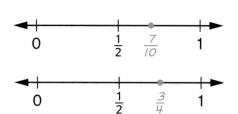
>
> Visualize finding the sum.
>
> Adding a number greater than $\frac{1}{2}$ to another number greater than $\frac{1}{2}$ will result in a sum greater than 1. So, $\frac{7}{10} + \frac{3}{4}$ is greater than 1.

> **Example**
>
> Estimate: Will the result of $2\frac{1}{4} - \frac{5}{6}$ be greater than or less than 1?
>
> You can visualize the starting amount, $2\frac{1}{4}$, as the blue area on the rectangles.
>
> You can think about the amount being subtracted as the area that is crossed out.
> $\frac{5}{6}$ is a little less than 1 whole, so about 1 whole will be subtracted. That will leave at least 1 other whole and the $\frac{1}{4}$ piece remaining.
>
> So, $2\frac{1}{4} - \frac{5}{6}$ will be greater than 1.
>
>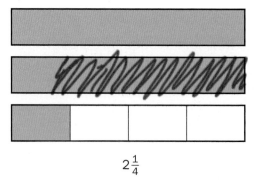

You can use benchmarks and mathematical reasoning to estimate the answers to fraction problems.

Example

Richard solved $1\frac{7}{8} - \frac{2}{5}$ and got $1\frac{5}{3}$. Is $1\frac{5}{3}$ a reasonable answer?

Make an estimate.

$1\frac{7}{8}$ is close to 2, and $\frac{2}{5}$ is close to $\frac{1}{2}$.

$2 - \frac{1}{2} = 1\frac{1}{2}$

So the difference of $1\frac{7}{8}$ and $\frac{2}{5}$ will be close to $1\frac{1}{2}$.

Compare your estimate to the answer you are checking.

The estimate is less than 2. Since $\frac{5}{3}$ is greater than 1, Richard's answer of $1\frac{5}{3}$ will be greater than 2. When compared to the estimate, the answer $1\frac{5}{3}$ does not make sense.

No, $1\frac{5}{3}$ is not a reasonable answer for $1\frac{7}{8} - \frac{2}{5}$. Richard should try the problem again.

When estimating sums and differences of mixed numbers, it is sometimes helpful to work with wholes and fractions separately.

Example

Estimate the sum of $2\frac{6}{10} + 3\frac{7}{8}$.

Add the wholes first.

$2 + 3 = 5$

Estimate the sum of the fractions using benchmarks.

$\frac{6}{10}$ is close to $\frac{1}{2}$. $\frac{7}{8}$ is close to 1.

$\frac{1}{2} + 1 = 1\frac{1}{2}$

Combine the sum of the wholes with your estimate for the sum of the fractions.

$5 + 1\frac{1}{2} = 6\frac{1}{2}$

The sum of $2\frac{6}{10} + 3\frac{7}{8}$ is about $6\frac{1}{2}$.

The Number System

Using Rounding to Estimate

Another way to estimate sums and differences of fractions is to round one or more of the numbers to whole numbers. This method can be especially helpful when you work with mixed numbers.

To round a mixed number to the nearest whole number, think: *What two whole numbers is this number between?* For example, $3\frac{1}{5}$ is between 3 and 4 because it is more than 3 wholes, but less than 4 wholes. Next look at the fraction part of the number you are rounding in order to determine which whole number is closer.

- If the fraction is greater than $\frac{1}{2}$, the number being rounded is closer to the higher whole number, so round up.

- If the fraction is less than $\frac{1}{2}$, the number being rounded is closer to the lower whole number, so round down.

- If the fraction is exactly $\frac{1}{2}$, use the problem situation to decide whether to round up or down. If there is nothing in the problem to help decide, many people use a rule to always round up.

Example

Estimate the sum of $14\frac{4}{5} + 15\frac{1}{3}$.

Round $14\frac{4}{5}$ to the nearest whole number.

$14\frac{4}{5}$ is between 14 and 15, and $\frac{4}{5}$ is greater than $\frac{1}{2}$.

So, $14\frac{4}{5}$ rounded to the nearest whole number is 15.

Next round $15\frac{1}{3}$ to the nearest whole number.

$15\frac{1}{3}$ is between 15 and 16, and $\frac{1}{3}$ is less than $\frac{1}{2}$.

So, $15\frac{1}{3}$ rounded to the nearest whole number is 15.

Add the rounded numbers: $15 + 15 = 30$.

The sum of $14\frac{4}{5} + 15\frac{1}{3}$ is about 30.

The Number System

Adding and Subtracting Fractions

To find the sum or difference of fractions that have the same denominator, add or subtract just the numerators. The denominator does not change. It often helps to represent the problem with a visual model such as a number line or a picture.

Examples

Find $\frac{1}{4} + \frac{2}{4}$.

$\frac{1}{4} + \frac{2}{4} = \frac{1+2}{4} = \frac{3}{4}$

Find $\frac{4}{5} - \frac{3}{5}$.

$\frac{4}{5} - \frac{3}{5} = \frac{4-3}{5} = \frac{1}{5}$

To add or subtract fractions that do not have the same denominator, you can rename the fractions as fractions with a common denominator.

Examples

Find $\frac{3}{4} + \frac{1}{8}$.

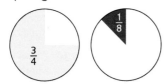

Estimate: $\frac{3}{4}$ is less than 1, and $\frac{1}{8}$ is less than $\frac{1}{4}$. $\frac{3}{4} + \frac{1}{8}$ will be a little less than 1.

Rename the fractions with a common denominator: $\frac{3}{4} = \frac{(3*2)}{(4*2)} = \frac{6}{8}$.

$\frac{3}{4} + \frac{1}{8} = \frac{6}{8} + \frac{1}{8} = \frac{7}{8}$

So, $\frac{3}{4} + \frac{1}{8} = \frac{7}{8}$.

$\frac{7}{8}$ is a little less than 1, so this answer makes sense.

Find $\frac{5}{6} - \frac{1}{4}$.

Estimate: $\frac{5}{6}$ is a little less than 1, and $\frac{1}{4}$ is less than $\frac{1}{2}$. $\frac{5}{6} - \frac{1}{4}$ will be close to $\frac{1}{2}$.

Rename the fractions with a common denominator: $\frac{5}{6} = \frac{(5*2)}{(6*2)} = \frac{10}{12}$ and $\frac{1}{4} = \frac{(1*3)}{(4*3)} = \frac{3}{12}$.

$\frac{5}{6} - \frac{1}{4} = \frac{10}{12} - \frac{3}{12} = \frac{7}{12}$

Subtract $\frac{3}{12}$.

So, $\frac{5}{6} - \frac{1}{4} = \frac{7}{12}$.

$\frac{7}{12}$ is close to the estimate of $\frac{1}{2}$, so this answer makes sense.

SRB
one hundred seventy-nine 179

The Number System

Adding Mixed Numbers

One way to add mixed numbers is to add the fractions and the whole numbers separately. You may need to find a common denominator of the fractions before you add them. You can rename the sum as an equivalent mixed number to make your answer easier to check.

Example

$1\frac{3}{8} + 2\frac{7}{8} = ?$

Estimate: $2\frac{7}{8}$ is close to 3. $1\frac{3}{8} + 3 = 4\frac{3}{8}$, so the sum should be close to $4\frac{3}{8}$.

Add the whole numbers.

$1\frac{3}{8}$
$+ 2\frac{7}{8}$
$\overline{3}$

Add the fractions.

$1\frac{3}{8}$
$+ 2\frac{7}{8}$
$\overline{3\frac{10}{8}}$

Rename the sum so that it is easier to check: $3\frac{10}{8} = 4\frac{2}{8}$.

$1\frac{3}{8} + 2\frac{7}{8} = 3\frac{10}{8}$, or $4\frac{2}{8}$ This makes sense because $4\frac{2}{8}$ is close to the estimate of $4\frac{3}{8}$.

When adding mixed numbers with unlike denominators, first consider whether you can solve the problem mentally. If you cannot, you can rename the fractions with a common denominator.

Note When adding mixed numbers, you will get the same answer whether you add the whole-number or fraction parts first.

Example

$4\frac{2}{5} + 1\frac{9}{10} = ?$

Estimate: $4 + 1$ is 5, and $\frac{2}{5} + \frac{9}{10}$ is more than one, so the sum is greater than 6.

Think: Can I solve this problem mentally? If not, rename the fractions using common denominators.
Notice that 10 is a multiple of 5, so use 10 as a common denominator.
Rewrite the problem with a common denominator. $4\frac{4}{10} + 1\frac{9}{10} = ?$

Add the fractions.

$4\frac{4}{10}$
$+ 1\frac{9}{10}$
$\overline{\frac{13}{10}}$

Add the whole numbers.

$4\frac{4}{10}$
$+ 1\frac{9}{10}$
$\overline{5\frac{13}{10} = 6\frac{3}{10}}$

Return to the original problem. $4\frac{2}{5} + 1\frac{9}{10} = 6\frac{3}{10}$

$6\frac{3}{10}$ is greater than 6, which matches the estimate.

The Number System

Subtracting Mixed Numbers

To subtract mixed numbers, subtract the fractions and whole numbers separately. You may need to rename the fractions with a common denominator before you subtract.

Example

$4\frac{3}{4} - 2\frac{1}{4} = ?$

Estimate: $4\frac{3}{4}$ is close to 5. $2\frac{1}{4}$ is close to 2. $5 - 2 = 3$, so the difference is about 3.

Subtract the fractions.

$$\begin{array}{r} 4\frac{3}{4} \\ -\ 2\frac{1}{4} \\ \hline \frac{2}{4} \end{array}$$

Subtract the whole numbers.

$$\begin{array}{r} 4\frac{3}{4} \\ -\ 2\frac{1}{4} \\ \hline \mathbf{2\frac{2}{4}} \end{array}$$

$4\frac{3}{4} - 2\frac{1}{4} = \mathbf{2\frac{2}{4}}$ The answer is reasonable because $2\frac{2}{4}$ is close to the estimate of 3.

When the fraction part of the mixed number you are taking away is greater than the fraction part of the starting number, you can rename the starting number as an equivalent mixed number with a larger fraction part before you subtract.

Example

$5\frac{1}{6} - 3\frac{5}{6} = ?$

Estimate: $5\frac{1}{6}$ is close to 5, and $3\frac{5}{6}$ is close to 4. $5 - 4 = 1$, so the difference is about 1.

$$\begin{array}{r} 5\frac{1}{6} \\ -\ 3\frac{5}{6} \\ \hline \end{array}$$

Show $5\frac{1}{6}$.

$\frac{1}{6}$ is less than $\frac{5}{6}$, so rename $5\frac{1}{6}$ as a mixed number with a larger fraction part.

Trade 1 whole for 6 sixths.

Break up 1 whole into 6 sixths. You now have 4 wholes and 7 sixths, or $4\frac{7}{6}$.

Rewrite the problem, then subtract $3\frac{5}{6}$.

$$\begin{array}{r} 5\frac{1}{6} \\ -\ 3\frac{5}{6} \\ \hline \end{array} \quad \begin{array}{r} 4\frac{7}{6} \\ -\ 3\frac{5}{6} \\ \hline 1\frac{2}{6} \end{array}$$

Take away $3\frac{5}{6}$. $1\frac{2}{6}$ is left.

$5\frac{1}{6} - 3\frac{5}{6} = \mathbf{1\frac{2}{6}}$ The answer makes sense because $1\frac{2}{6}$ is close to the estimate of 1.

The Number System

When you subtract mixed numbers with unlike denominators, first consider whether you can solve the problem mentally. If you can, solve it mentally. If you cannot, rename the mixed numbers with a common denominator.

Example

$5\frac{1}{3} - 2\frac{7}{8} = ?$

Estimate: $5\frac{1}{3}$ is close to 5, and $2\frac{7}{8}$ is close to 3. $5 - 3 = 2$, so the difference should be about 2.

Think: Can I solve this problem mentally? It may be difficult to subtract thirds and eighths mentally, so rename the fractions using a common denominator to solve the problem.

You can rename each fraction with a denominator of 24.

$\frac{1}{3} = \frac{(1 * 8)}{(3 * 8)} = \frac{8}{24}$ $\qquad$ $\frac{7}{8} = \frac{(7 * 3)}{(8 * 3)} = \frac{21}{24}$

Rewrite the problem with the renamed fractions. $\qquad 5\frac{8}{24} - 2\frac{21}{24} = ?$

Notice that the fraction part of the starting amount $\left(\frac{8}{24}\right)$ is less than the fraction part of the number being subtracted $\left(\frac{21}{24}\right)$. Rename the starting amount as an equivalent mixed number with a larger fraction part.

Trade 1 whole for $\frac{24}{24}$. $\qquad\qquad\qquad\qquad$ $5\frac{8}{24} = 4 + 1 + \frac{8}{24}$

$\qquad\qquad\qquad\qquad\qquad\qquad\qquad\qquad\qquad = 4 + \frac{24}{24} + \frac{8}{24}$

$\qquad\qquad\qquad\qquad\qquad\qquad\qquad\qquad\qquad = 4\frac{32}{24}$

Rewrite the problem. $\qquad\qquad\qquad\qquad\qquad 4\frac{32}{24} - 2\frac{21}{24} = ?$

Subtract the fractions. $\qquad\qquad\qquad\qquad$ Subtract the whole numbers.

$\quad 4\frac{32}{24}$ $\qquad\qquad\qquad\qquad\qquad\qquad\quad$ $4\frac{32}{24}$

$-\ 2\frac{21}{24}$ $\qquad\qquad\qquad\qquad\qquad\qquad -\ 2\frac{21}{24}$

$\quad\ \ \frac{11}{24}$ $\qquad\qquad\qquad\qquad\qquad\qquad\quad\ 2\frac{11}{24}$

Return to the original problem. $5\frac{1}{3} - 2\frac{7}{8} = \mathbf{2\frac{11}{24}}$

The answer is reasonable because it is close to the estimate of 2.

The Number System

You can subtract (or add) mixed numbers by renaming the mixed numbers as fractions greater than one.

Example

$4\frac{1}{6} - 2\frac{2}{3} = ?$

Estimate: $4\frac{1}{6}$ is close to 4, and $2\frac{2}{3}$ is close to 3. $4 - 3 = 1$, so the difference should be about 1.

Think: Can I solve this problem mentally? If not, rename the mixed numbers as fractions.

$4\frac{1}{6} = \frac{25}{6}$ $2\frac{2}{3} = \frac{8}{3}$

Rename the fractions with a common denominator. Use 6 as a common denominator, since 6 is a multiple of 3.

$\frac{8}{3} = \frac{(8 * 2)}{(3 * 2)} = \frac{16}{6}$

Rewrite the problem. $\frac{25}{6} - \frac{16}{6} = ?$

Subtract.   25 sixths − 16 sixths = 9 sixths

Rename the result as a mixed number. $\frac{9}{6} = \frac{6}{6} + \frac{3}{6}$
$= 1\frac{3}{6}$

Return to the original problem. $4\frac{1}{6} - 2\frac{2}{3} = \frac{9}{6}$, or $\mathbf{1\frac{3}{6}}$

This answer makes sense because it is close to the estimate of 1.

Check Your Understanding

Add and subtract. Estimate to check whether your answers are reasonable.

1. $2\frac{1}{8} + 7\frac{7}{8}$
2. $3\frac{4}{5} + 2\frac{1}{2}$
3. $6\frac{2}{3} + 3\frac{3}{4}$
4. $8\frac{9}{10} + 6\frac{4}{5}$
5. $5\frac{2}{5} - 1\frac{1}{3}$
6. $6 - 2\frac{7}{8}$
7. $4\frac{5}{9} - \frac{7}{9}$
8. $8\frac{1}{2} - 3\frac{2}{3}$

Check your answers in the Answer Key.

The Number System

Finding a Fraction of a Number

Many problems with fractions involve finding a fraction of a number.

Example

There are 24 students in Ms. Dunning's class. $\frac{1}{3}$ of the students are participating in a school performance. How many students in Ms. Dunning's class are participating in the performance?

Estimate: $\frac{1}{3}$ is less than $\frac{1}{2}$, so $\frac{1}{3}$ of 24 will be less than $\frac{1}{2}$ of 24. Half of 24 is 12, so the answer will be less than 12 students.

Note "$\frac{1}{3}$ of 24" has the same meaning as "$\frac{1}{3} * 24$." When you find a fraction of a number, you can replace the word *of* with a multiplication symbol. For example, $\frac{1}{6}$ of 18 means $\frac{1}{6} * 18$, just as 2 of 18 means $2 * 18$.

To find the number of students participating in the performance, find $\frac{1}{3}$ of 24.

Draw a picture to model the problem. Each X represents one student. Divide 24 Xs into 3 equal groups.

Each group has $\frac{1}{3}$ of the Xs. There are 8 Xs in each group. So, $\frac{1}{3}$ of 24 is 8.

Eight of Ms. Dunning's students are participating in the school performance. This answer is reasonable because it matches the estimate of less than 12 students.

Sometimes it can be helpful to find a unit fraction of a number in order to find another fraction of that number. Remember that a **unit fraction** is a fraction with a numerator of 1, such as $\frac{1}{2}$, $\frac{1}{4}$, or $\frac{1}{10}$.

Example

A jacket that sells for $45 is on sale for $\frac{2}{3}$ of the regular price. What is the sale price?
To find the sale price you have to find $\frac{2}{3}$ of $45.

Step 1 Find $\frac{1}{3}$ of 45. $45 \div 3 = 15$, so $\frac{1}{3}$ of 45 is 15.
Step 2 Use the answer from Step 1 to find $\frac{2}{3}$ of 45.
Since $\frac{1}{3}$ of 45 is 15, then $\frac{2}{3}$ of 45 is $2 * 15 = 30$.

The sale price is $30.
$\frac{2}{3}$ of $45 = $\frac{2}{3} * $45 = 30.

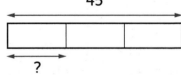

$45 \div 3 = 15$, so $\frac{1}{3}$ of 45 is 15.

The Number System

Finding a Fraction of a Fraction

Finding a fraction of a fraction is similar to finding a fraction of a whole number. You are still finding part of a given amount, except now the given amount is a fraction.

Using visual representations such as fraction strips can help you find a fraction of a fraction.

Example

Use fraction strips to find $\frac{1}{3}$ of $\frac{1}{2}$.

Estimate: $\frac{1}{3}$ of $\frac{1}{2}$ will be less than $\frac{1}{2}$.

Show $\frac{1}{2}$ using a halves fraction strip.

Fold each $\frac{1}{2}$ into 3 equal parts. (These folds are shown with red dashed lines.)

Each fold line represents $\frac{1}{3}$ of $\frac{1}{2}$.

Three equal parts in $\frac{1}{2}$ means each part is one-sixth of the whole.

$\frac{1}{3}$ of $\frac{1}{2}$ is $\frac{1}{6}$.

This solution makes sense because $\frac{1}{6}$ is less than $\frac{1}{2}$.

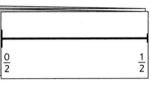
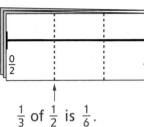

$\frac{1}{3}$ of $\frac{1}{2}$ is $\frac{1}{6}$.

Note The word *of* in a fraction of problem can be replaced with a multiplication symbol. For example, $\frac{1}{3}$ of $\frac{1}{2}$ means $\frac{1}{3} * \frac{1}{2}$, and $\frac{1}{3} * \frac{1}{2} = \frac{1}{6}$. Also, $\frac{3}{4}$ of $\frac{2}{3}$ means $\frac{3}{4} * \frac{2}{3}$, and $\frac{3}{4} * \frac{2}{3} = \frac{6}{12} = \frac{1}{2}$.

Using a number line is another way to find a fraction of a fraction.

Example

What is $\frac{3}{4}$ of $\frac{2}{3}$?

Estimate: Taking a fraction of $\frac{2}{3}$ will result in an amount less than $\frac{2}{3}$.

Start with a number line showing $\frac{2}{3}$.

Divide the distance between 0 and $\frac{2}{3}$ into 4 equal parts.

Name the tick mark $\frac{3}{4}$ of the distance to $\frac{2}{3}$:

Notice that each tick mark divides one third in half, so there are 6 equal parts in the whole. Rename each tick mark in sixths.

$\frac{3}{4}$ of $\frac{2}{3}$ is $\frac{3}{6}$, or $\frac{1}{2}$.

This answer makes sense because $\frac{1}{2}$ is less than $\frac{2}{3}$, which matches the estimate.

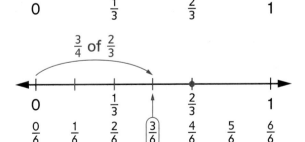

one hundred eighty-five SRB **185**

The Number System

Predicting the Size of Products

You can often predict the size of a product before you actually multiply. To predict the size of a product, think about the size of each factor.

When you multiply two numbers greater than one, the product is greater than both numbers.

Example

Predict the size of the product of $8 * 1\frac{1}{2}$.

Look at the size of each factor.	8 is a whole number greater than 1. $1\frac{1}{2}$ is a mixed number and is greater than 1.
Think about what multiplication means.	You can think of $8 * 1\frac{1}{2}$ as making 8 groups with $1\frac{1}{2}$ in each group. Just thinking of 2 groups of $1\frac{1}{2}$ gives 3, so 8 groups of $1\frac{1}{2}$ will certainly be more than $1\frac{1}{2}$.
	You can also think of $8 * 1\frac{1}{2}$ as making 1 full group of 8 and then adding on one-half of 8. Making more than 1 group of 8 means that the product will be greater than 8.

The product of $8 * 1\frac{1}{2}$ will be greater than each of the factors, 8 and $1\frac{1}{2}$.

When you multiply a given number by a number equal to one, the product is equal to the given number.

Example

Will the product of $\frac{1}{2} * \frac{3}{3}$ be greater than, equal to, or less than $\frac{1}{2}$?

Look at the size of each factor.	$\frac{1}{2}$ is a fraction less than 1. $\frac{3}{3}$ is a fraction equal to 1.
Think about what multiplication means.	Multiplying by a fraction is like finding a fraction of a number. Finding $\frac{1}{2}$ of $\frac{3}{3}$ is equivalent to finding $\frac{1}{2}$ of 1 whole, which results in $\frac{1}{2}$.

The product of $\frac{1}{2} * \frac{3}{3}$ will be equal to $\frac{1}{2}$.

The Number System

In general, when you multiply a given number (greater than zero) by a number less than one, the product is less than the given number.

Example

How will the product of $12 * \frac{2}{3}$ compare to 12?

Look at the size of each factor. 12 is a whole number greater than 1.
$\frac{2}{3}$ is a fraction less than 1.

Think about what multiplication means. $12 * \frac{2}{3}$ has the same product as $\frac{2}{3} * 12$. You can think of $\frac{2}{3} * 12$ as finding $\frac{2}{3}$ of 12. Taking a fractional part of 12 will result in a number less than 12.

The product of $12 * \frac{2}{3}$ will be less than 12.

Example

Complete the number sentences with >, <, or =. $\frac{1}{3} * \frac{3}{4} \square \frac{3}{4}$ $\frac{1}{3} * \frac{3}{4} \square \frac{1}{3}$

Look at the size of each factor. $\frac{1}{3}$ is a fraction less than 1.
$\frac{3}{4}$ is a fraction less than 1.

Think about what multiplication means. Think of $\frac{1}{3} * \frac{3}{4}$ as finding $\frac{1}{3}$ of $\frac{3}{4}$. Taking part of $\frac{3}{4}$ will result in an amount less than $\frac{3}{4}$.

$\frac{1}{3} * \frac{3}{4} = \frac{3}{4} * \frac{1}{3}$, so think of finding $\frac{3}{4}$ of $\frac{1}{3}$. Taking part of $\frac{1}{3}$ will result in an amount less than $\frac{1}{3}$.

So, $\frac{1}{3} * \frac{3}{4} < \frac{3}{4}$, and $\frac{1}{3} * \frac{3}{4} < \frac{1}{3}$.

Check Your Understanding

Complete each number sentence with >, <, or =.

1. $\frac{9}{8} * 16 \square 16$
2. $\frac{6}{8} * \frac{3}{4} \square \frac{3}{4}$
3. $1\frac{1}{2} * 2\frac{1}{3} \square 1\frac{1}{2}$
4. $\frac{9}{12} * \frac{2}{2} \square \frac{9}{12}$

Check your answers in the Answer Key.

one hundred eighty-seven   SRB 187

The Number System

Multiplying Fractions and Whole Numbers

People sometimes think that "multiplication makes things bigger." But multiplication involving fractions can result in products that are smaller than at least one of the factors. For example, $10 * \frac{1}{2} = 5$.

Number-Line Model

One way to multiply a fraction and a whole number is to think about hops on a number line. The whole number tells how many hops to make, and the fraction tells how long each hop should be. For example, to find $5 * \frac{2}{3}$, imagine taking 5 hops on a number line, each $\frac{2}{3}$ of a unit long.

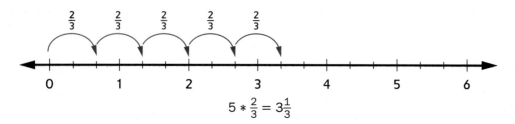

$5 * \frac{2}{3} = 3\frac{1}{3}$

Addition Model

You can use addition to multiply a fraction and a whole number. For example, to find $4 * \frac{2}{3}$, draw 4 representations of $\frac{2}{3}$. Then add all of the fractions.

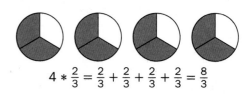

$4 * \frac{2}{3} = \frac{2}{3} + \frac{2}{3} + \frac{2}{3} + \frac{2}{3} = \frac{8}{3}$

Area Model

Think of the problem $\frac{2}{3} * 4$ as "What is $\frac{2}{3}$ of an area that has 4 square units?"

You can draw a rectangle made up of 4 squares, each with an area of 1 square unit. The rectangle has an area of 4 square units.

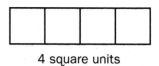

4 square units

Divide the rectangle into 3 equal strips and shade 2 strips ($\frac{2}{3}$ of the area). The shaded area equals $\frac{8}{3}$ (8 small rectangles, each with an area of $\frac{1}{3}$ square unit).

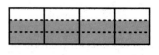

$\frac{2}{3} * 4 = \frac{8}{3}$

So, $\frac{2}{3}$ of 4 square units equals $\frac{8}{3}$ square units. $\frac{2}{3}$ of $4 = \frac{2}{3} * 4 = \frac{8}{3}$.

Check Your Understanding

Predict whether the product will be greater than or less than the whole number factor. Then multiply.

1. $5 * \frac{3}{4}$
2. $\frac{2}{3} * 6$
3. $4 * \frac{4}{5}$
4. $6 * \frac{2}{5}$
5. $\frac{3}{4} * 6$

Check your answers in the Answer Key.

The Number System

Multiplying Fractions

You can use an area model to multiply two fractions.

Example

$\frac{3}{4} * \frac{2}{3} = ?$

Estimate: Taking a fraction of $\frac{2}{3}$ will result in an amount less than $\frac{2}{3}$.

Shade $\frac{2}{3}$ of a rectangle blue.

Shade $\frac{3}{4}$ of the $\frac{2}{3}$ a darker blue.

Name the area that is shaded in the darker color. There are 6 parts shaded out of 12 equal parts in the whole rectangle, so the shaded area is $\frac{6}{12}$ of the whole.

$\frac{3}{4} * \frac{2}{3} = \frac{6}{12}$ This answer makes sense because $\frac{6}{12} = \frac{1}{2}$, which is less than $\frac{2}{3}$.

Multiplication of Fractions Property

The problem above is an example of this general pattern: To multiply fractions, multiply the numerators and multiply the denominators. The pattern can be expressed as follows:

$\frac{a}{b} * \frac{c}{d} = \frac{a * c}{b * d}$ (b and d may not be 0.)

Examples

Multiply $\frac{3}{4} * \frac{2}{3}$.

$\frac{3}{4} * \frac{2}{3} = \frac{3 * 2}{4 * 3} = \frac{6}{12} = \frac{1}{2}$

Multiply $\frac{5}{4} * \frac{6}{7}$.

$\frac{5}{4} * \frac{6}{7} = \frac{5 * 6}{4 * 7} = \frac{30}{28} = \frac{15}{14} = 1\frac{1}{14}$

You can use the Multiplication of Fractions Property to multiply a whole number and a fraction. First, rename the whole number as a fraction with 1 in the denominator.

Examples

Multiply $5 * \frac{2}{3}$.

$5 * \frac{2}{3} = \frac{5}{1} * \frac{2}{3} = \frac{5 * 2}{1 * 3} = \frac{10}{3} = 3\frac{1}{3}$

Multiply $\frac{2}{11} * 4$.

$\frac{2}{11} * 4 = \frac{2}{11} * \frac{4}{1} = \frac{2 * 4}{11 * 1} = \frac{8}{11}$

one hundred eighty-nine

The Number System

Multiplying Mixed Numbers

Multiplying Mixed Numbers Using Partial Products

You can use **partial products** to multiply mixed numbers by whole numbers, fractions, or mixed numbers. Each part of the mixed number must be multiplied by each part of the other factor. Partial-products diagrams can help you find the partial products. To make a partial-products diagram:

- First decide how to decompose the mixed numbers as whole numbers and fractions.
- Draw a rectangular grid and label the outside edges with the decomposed mixed numbers.
- Find the outside numbers that intersect in each section of the grid and use them as factors to record a number model in that section.
- Add the partial products to find the total.

The example below shows how to use a partial-products diagram to multiply a mixed number by a fraction.

> **Example**
>
> $4 * 2\frac{3}{4} = ?$
>
> Estimate: Since $2\frac{3}{4}$ is greater than 2, the product of $4 * 2\frac{3}{4}$ will be greater than 8.
>
> Use a partial-products diagram.
>
> - 4 is a whole number. $2\frac{3}{4}$ can be decomposed into $2 + \frac{3}{4}$.
> - Draw a rectangular grid and label the edges with the decomposed factors.
> - Record number models in each section of the grid: $2 * 4 = \mathbf{8}$ and $\frac{3}{4} * 4 = \mathbf{3}$.
> - Add the partial products: $\mathbf{8 + 3 = 11}$.
>
> $4 * 2\frac{3}{4} = \mathbf{11}$
>
> This answer of 11 makes sense because it is more than 8, which matches the estimate.

The Number System

The example below shows how to use a partial-products diagram to multiply a mixed number and a fraction.

Example

$5\frac{1}{2} * \frac{2}{3} = ?$

Estimate: $\frac{2}{3}$ is a fraction less than 1. So, $\frac{2}{3}$ of $5\frac{1}{2}$ should be less than $5\frac{1}{2}$.

Use a partial-products diagram.

- $5\frac{1}{2}$ can be decomposed into $5 + \frac{1}{2}$.
- Draw a grid and partition it into 2 sections. Label the edges with the decomposed factors.
- Record number models in each section of the grid: $5 * \frac{2}{3} = \frac{10}{3}$ and $\frac{1}{2} * \frac{2}{3} = \frac{2}{6}$.
- Rename the partial products with a common denominator, then add.

$\frac{2}{6} = \frac{1}{3}$

$\frac{10}{3} + \frac{2}{6} = \frac{10}{3} + \frac{1}{3} = \frac{11}{3} = 3\frac{2}{3}$

$5\frac{1}{2} * \frac{2}{3} = \mathbf{3\frac{2}{3}}$

This answer makes sense because it is less than $5\frac{1}{2}$, which matches the estimate.

This example shows how to use a partial-products diagram to multiply two mixed numbers.

Example

$7\frac{1}{2} * 2\frac{3}{5} = ?$

Estimate: $7\frac{1}{2}$ is close to 7, and $2\frac{3}{5}$ is close to 3. $7 * 3 = 21$, so the product should be close to 21.

Use a partial-products diagram.

- $7\frac{1}{2} = 7 + \frac{1}{2}$, and $2\frac{3}{5} = 2 + \frac{3}{5}$.
- Draw a grid and partition it into four sections. Label the edges with the decomposed factors.
- Record number models in each section of the grid:

  $7 * 2 = \mathbf{14}$

  $\frac{1}{2} * 2 = \mathbf{1}$

  $7 * \frac{3}{5} = \frac{21}{5}$

  $\frac{1}{2} * \frac{3}{5} = \frac{3}{10}$

- Rename the partial products with a common denominator, then add.

  $\frac{21}{5} = \frac{42}{10}$, or $\mathbf{4\frac{2}{10}}$

  $14 + 1 + 4\frac{2}{10} + \frac{3}{10} = 19\frac{5}{10}$

$7\frac{1}{2} * 2\frac{3}{5} = \mathbf{19\frac{5}{10}}$

This makes sense because $19\frac{5}{10}$ is reasonably close to the estimate of 21.

SRB 191

The Number System

Multiplying Mixed Numbers by Renaming Them as Fractions

Another way to multiply with mixed numbers is to rename any whole or mixed numbers as fractions, and then multiply the fractions.

Example

Find $5 * 2\frac{1}{4}$.

Estimate: $2\frac{1}{4}$ is a little more than 2. $5 * 2 = 10$, so the answer will be a little more than 10.

Rename any whole or mixed numbers as fractions. $5 = \frac{5}{1}$ $2\frac{1}{4} = \frac{9}{4}$

Rewrite the problem, then multiply. $\frac{5}{1} * \frac{9}{4} = \frac{(5 * 9)}{(1 * 4)} = \frac{45}{4}$, or $11\frac{1}{4}$

$5 * 2\frac{1}{4} = 11\frac{1}{4}$

The answer is reasonable because it is a little more than 10, which matches the estimate.

Example

Solve $3\frac{1}{4} * 1\frac{5}{6}$.

Estimate: $3\frac{1}{4}$ is close to 3, and $1\frac{5}{6}$ is close to 2. $3 * 2 = 6$, so the product will be close to 6.

Rename the mixed numbers as fractions. $3\frac{1}{4} = \frac{13}{4}$ $1\frac{5}{6} = \frac{11}{6}$

Rewrite the problem, then multiply. $\frac{13}{4} * \frac{11}{6} = \frac{(13 * 11)}{(4 * 6)} = \frac{143}{24}$

Rename the product as a mixed or whole number. $143 \div 24 \rightarrow 5\ R23$, or $5\frac{23}{24}$

$3\frac{1}{4} * 1\frac{5}{6} = 5\frac{23}{24}$

The answer is reasonable because it is close to 6, which matches the estimate.

Check Your Understanding

Estimate and multiply using the method of your choice.

1. $\frac{1}{4} * 1\frac{1}{2}$
2. $2\frac{2}{3} * 5$
3. $3\frac{2}{5} * 2\frac{1}{2}$

Check your answers in the Answer Key.

The Number System

Division of Fractions

Dividing a number by a fraction often gives a quotient that is larger than the dividend. For example, $4 \div \frac{1}{2} = 8$. To understand why, it's helpful to think about what division means.

Equal Groups

A division problem like $a \div b = ?$ asks "How many bs are there in a?" For example, the problem $6 \div 3 = ?$ asks, "How many 3s are in 6?" The figure at the right shows that there are two groups of 3 in 6, so $6 \div 3 = 2$.

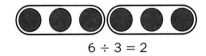

$6 \div \frac{1}{3} = ?$ asks, "How many $\frac{1}{3}$s are in 6?" The figure at the right shows that there are 18 thirds in 6, so $6 \div \frac{1}{3} = 18$.

> **Example**
>
> A cup of rice is about $\frac{1}{2}$ pound. About how many cups of rice are in 5 pounds?
>
>
>
> $\frac{1}{2}$ lb + $\frac{1}{2}$ lb + $\frac{1}{2}$ lb + $\frac{1}{2}$ lb + $\frac{1}{2}$ lb + $\frac{1}{2}$ lb + $\frac{1}{2}$ lb + $\frac{1}{2}$ lb + $\frac{1}{2}$ lb + $\frac{1}{2}$ lb = 5 lb
>
> Find how many $\frac{1}{2}$s are in 5, which is the same as $5 \div \frac{1}{2}$.
> There are about 10 cups of rice in 5 pounds.

Missing Factors

A division problem is equivalent to a multiplication problem with a missing factor.
A problem like $6 \div \frac{1}{2} = \square$ is equivalent to $\frac{1}{2} * \square = 6$.
$\frac{1}{2} * \square = 6$ is the same as asking "$\frac{1}{2}$ of what number equals 6?"
Since $\frac{1}{2} * 12 = 6$, you know that $6 \div \frac{1}{2} = 12$.

> **Example**
>
> Solve $10 \div \frac{2}{3} = \square$.
>
> This problem is equivalent to $\frac{2}{3} * \square = 10$, which means "$\frac{2}{3}$ of what number equals 10?"
>
> The diagram shows that $\frac{2}{3}$ of the missing number is 10.
> Since $\frac{2}{3}$ of the missing number is 10, $\frac{1}{3}$ must be 5.
> Since $\frac{1}{3}$ of the missing number is 5, the missing number must be $3 * 5 = 15$.
> So, $\frac{2}{3}$ of 15 = 10, which means that $\frac{2}{3} * 15 = 10$.
> $10 \div \frac{2}{3} = 15$
>
>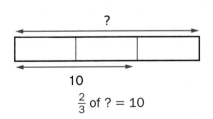

The Number System

Equal-Sharing Division

In equal-sharing division problems, you can think of the dividend as an amount that is split into a number of equal shares. For example, splitting $\frac{1}{3}$ of a watermelon among four people is an equal-sharing division problem. You can model this situation with the expression $\frac{1}{3} \div 4$.

Problems in which a fraction is divided by a whole number can usually be interpreted as equal-sharing division. One way to solve equal-sharing division problems is to use visual models, such as pictures or number lines.

Example

Share $\frac{1}{2}$ loaf of bread equally among 3 friends. What part of a whole loaf is each share?

Estimate: $\frac{1}{2}$ loaf is being split 3 ways, so the amount that each person gets should be small compared to $\frac{1}{2}$.

Solve this equal-sharing division problem by finding $\frac{1}{2} \div 3$.
Draw a picture to model the problem.
Use a rectangle to represent a loaf of bread.
Split one-half of the loaf of bread into 3 equal parts.
Each person's share is one of the equal parts.
One share fits into the whole 6 times, so one share is $\frac{1}{6}$ of a whole loaf.
Each person gets $\frac{1}{6}$ loaf of bread.
This makes sense because $\frac{1}{6}$ loaf is small compared to $\frac{1}{2}$ loaf.

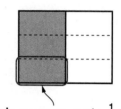

Each person gets $\frac{1}{6}$ loaf.

You can divide a fraction by a whole number by thinking of finding a fraction of a number. You can use what you know about fraction multiplication to solve fraction division problems.

Examples

$\frac{1}{2} \div 3 = ?$

$\frac{1}{2} \div 3$ is like finding $\frac{1}{3}$ of $\frac{1}{2}$. Remember that $\frac{1}{3}$ of $\frac{1}{2}$ is the same as $\frac{1}{3} * \frac{1}{2}$.

$\frac{1}{2} \div 3 = \frac{1}{3} * \frac{1}{2} = \frac{1 * 1}{3 * 2} = \frac{1}{6}$ So, $\frac{1}{2} \div 3 = \boldsymbol{\frac{1}{6}}$.

You can check division problems using multiplication. For example, you can check that $6 \div 2 = 3$ by multiplying $3 * 2 = 6$. You can check division problems with fractions in a similar way.

Example

Use multiplication to check that $\frac{1}{2} \div 3 = \frac{1}{6}$.

Think: Does $\frac{1}{6} * 3 = \frac{1}{2}$? Do 3 [$\frac{1}{6}$s] make $\frac{1}{2}$?

Yes, $\frac{1}{6} * 3 = \frac{3}{6}$, which is the same as $\frac{1}{2}$.

The Number System

Common Denominators

One way to solve a fraction division problem is to rename both the dividend and the divisor as fractions with a common denominator. Then divide the numerators and the denominators.

Example

Find $6 \div \frac{2}{3}$.

Rename 6 as $\frac{18}{3}$.

Divide the numerators and the denominators.

$$6 \div \frac{2}{3} = \frac{18}{3} \div \frac{2}{3}$$
$$= \frac{18 \div 2}{3 \div 3}$$
$$= \frac{9}{1}, \text{ or } 9$$

So, $6 \div \frac{2}{3} = 9$.

To see why this method works, imagine putting the 18 thirds into equal groups that have $\frac{2}{3}$ in each group. There would be 9 groups. There are 9 two-thirds in $\frac{18}{3}$, or 6.

Example

Julia has 6 pounds of modeling clay. She wants to put it in packages that hold $\frac{3}{4}$ of a pound each. How many packages can she make?

To solve $6 \div \frac{3}{4}$, rename 6 as $\frac{24}{4}$.

Then divide.

$$6 \div \frac{3}{4} = \frac{24}{4} \div \frac{3}{4}$$
$$= \frac{24 \div 3}{4 \div 4}$$
$$= \frac{8}{1}, \text{ or } 8$$

The 24 fourths can be put into 8 groups of $\frac{3}{4}$ each.

So Julia can make 8 packages of modeling clay.

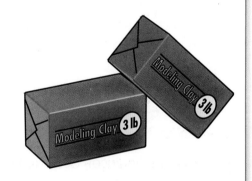

Check Your Understanding

Solve. Then write a division number model for each problem.

1. Regina has 9 pizzas. If each person can eat $\frac{1}{2}$ of a pizza, how many people can Regina serve?

2. Selena has 10 yards of plastic strips for making bracelets. She needs $\frac{1}{3}$ yard for each bracelet. How many bracelets can she make?

3. 7 is $\frac{1}{4}$ of a number. What is the number?

Check your answers in the Answer Key.

The Number System

Division of Fractions and Mixed Numbers

The **reciprocal** of a fraction $\frac{a}{b}$ is the fraction $\frac{b}{a}$. For example, the reciprocal of $\frac{3}{4}$ is $\frac{4}{3}$.

If n is a whole number, $n = \frac{n}{1}$. So the reciprocal of n is $\frac{1}{n}$. For example, the reciprocal of 9 is $\frac{1}{9}$. 0 can be written as $\frac{0}{1}$, but 0 does not have a reciprocal, because $\frac{1}{0}$ is not defined. (Division by 0 is never allowed.)

The product of a number and its reciprocal is always 1. $\frac{a}{b} * \frac{b}{a} = 1$

Examples

$\frac{4}{7} * \frac{7}{4} = \frac{4*7}{7*4} = \frac{28}{28} = 1$ $\frac{120}{3} * \frac{3}{120} = \frac{360}{360} = 1$ $29 * \frac{1}{29} = \frac{29}{1} * \frac{1}{29} = \frac{29}{29} = 1$

Reciprocals are useful when dividing fractions.

Division of Fractions Property

To find the quotient of two fractions, multiply the first fraction by the reciprocal of the second fraction. $\frac{a}{b} \div \frac{c}{d} = \frac{a}{b} * \frac{d}{c}$

Examples

$\frac{4}{5} \div \frac{2}{3} = \frac{4}{5} * \frac{3}{2}$
$= \frac{12}{10} = 1\frac{2}{10}$, or $1\frac{1}{5}$

$2\frac{3}{4} \div 1\frac{1}{3} = \frac{11}{4} \div \frac{4}{3}$
$= \frac{11}{4} * \frac{3}{4} = \frac{33}{16}$, or $2\frac{1}{16}$

Examples

Roger has $6\frac{1}{4}$ pounds of ground hamburger. He wants to make 5-ounce hamburger patties. How many can he make?

1 pound = 16 ounces So, 1 ounce = $\frac{1}{16}$ pound, and
5 ounces = $\frac{5}{16}$ pound.
Think: How many $\frac{5}{16}$s are in $6\frac{1}{4}$? This is a division problem: $6\frac{1}{4} \div \frac{5}{16}$.
Rename the mixed number $6\frac{1}{4}$ as $\frac{25}{4}$ and use the Division of Fractions Property:
$\frac{25}{4} \div \frac{5}{16} = \frac{25}{4} * \frac{16}{5} = \frac{400}{20} = 20$
Roger can make 20 hamburger patties.

Check Your Understanding

Divide.

1. $\frac{3}{5} \div \frac{1}{4}$ 2. $5 \div \frac{5}{3}$ 3. $\frac{1}{7} \div \frac{3}{7}$ 4. $2\frac{2}{3} \div 4$ 5. $3\frac{1}{2} \div 1\frac{1}{4}$

Check your answers in the Answer Key.

Expressions and Equations

Expressions and Equations

Algebra

Algebra uses mathematical statements to describe patterns, express numerical relationships, and model real-world situations. Algebra is like arithmetic, but algebra uses letters, blanks, question marks, and other symbols to stand for quantities that vary or are unknown. By learning how to write and use algebraic statements, you will expand the number and kinds of tools you can use to solve problems.

> **Did You Know?**
>
> Our word *algebra* comes from the Arabic word *al-jabru*, which means "restoration." Algebra was known as the "science of restoration and balancing."

The origins of algebra can be traced back thousands of years to ancient Egypt and Babylon. In early times, algebra involved solving number problems for which one or more of the numbers was not known. The objective was to find these missing numbers, called the "unknowns." Words were used for the unknowns, as in "Five plus some number equals eight."

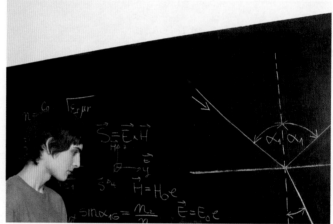

Algebra helps in solving many types of complex problems.

Then, in the late 1500s, François Viète began using letters to stand for unknown quantities, as in $5 + x = 8$. Viète's invention made solving number problems much easier and led to an explosion of discoveries in mathematics and science that has continued into modern times.

Letters, blanks, or other symbols that are used to stand for unknown numbers are called variables. **Variables** are used in several other ways.

Variables Can Be Used to State the Properties of the Number System

Properties of a number system are rules that are true for all numbers. For example, the Commutative Property of Addition states that the order of the addends does not change the sum in an addition problem. You can use variables to state this property more efficiently: For all numbers a and b, $a + b = b + a$. You were introduced to this property as the "turn-around rule" to help you learn addition facts.

Many additional properties are listed on pages 231–232. All of these properties can be stated using variables.

Expressions and Equations

Variables Can Be Used to Express Rules or Functions

Function machines and "What's My Rule?" tables have rules that tell you how to get the *out* numbers from the *in* numbers. These rules can be written using variables. For example, a function machine might have the rule "double and add 1." This rule can be written using variables, as $y = 2 * x + 1$ or as $y = 2x + 1$. The rule can also be graphed on a coordinate grid, as shown at the right.

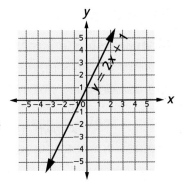

Variables Can Be Used in Formulas

Formulas are used in everyday life, in science, in business, and in many other situations as a compact way of expressing relationships. For example, the formula $d = r * t$ expresses the relationship between a distance (*d*), the rate at which a distance is covered (*r*), and the time (*t*) it takes to cover the distance.

miles (*d*)	50	100	150	?
hours (*t*)	1	2	3	4

$d = r * t$ $r = 50$ miles per hour

Variables Can Be Used in Computers and Calculators

Variables are used in computer spreadsheets, making it possible to evaluate formulas quickly and efficiently. For example, in the spreadsheet at the right, cell C2 has been selected, and the display bar shows "=A2*B2." Changing the values of the variables, or the values in cells A2 and B2, will change the value shown in cell C2.

Variables are also used in writing computer programs. Computer programs are made up of a series of "commands" similar to number sentences that contain variables. Some calculators, especially graphing calculators, use variables to name calculator key functions and to specify procedures.

C2   fx =A2*B2

	A	B	C
	Length (cm)	Width (cm)	Area (sq cm)
2	2.0	3.0	6.0
3	2.7	4.0	10.8
4	3.0	5.1	15.3
5	4.6	7.0	32.2
6	6.5	12.0	78
7	8.0	5.1	40.8

This spreadsheet uses a formula to find the area of rectangles.

Check Your Understanding

1. How far will a car travel in $2\frac{1}{2}$ hours if the average speed is 50 miles per hour?
2. Use the rule $y = 2 * x + 1$ and the graph above to find the value of *y* when $x = 2$.

Check your answers in the Answer Key.

Expressions and Equations

Expressions

An **expression** is a mathematical phrase that may include numbers, variables, operation symbols, and grouping symbols. Unlike a number sentence, an expression does not include a relational symbol such as <, >, or =.

Numerical Expressions

Expressions that do not contain variables are called *numerical expressions.*

Examples

Numerical Expressions:

3 + 4 $1.45 * 10^4$

(2 * 8) / 5   246

$(3^2 - 4) * 6$ $(24 * 75) + 4 * (8^3 - 27)$

Algebraic Expressions

Expressions that contain variables are called **algebraic expressions.**

Consider the following statement: "Marcy read four more books last year than Gina." You can't tell how many books Marcy read unless you know the number of books Gina read. There are many possibilities:

- If Gina read 8 books, then Marcy read 12 books.
- If Gina read 27 books, then Marcy read 31 books, and so on.

One way to represent the number of books Marcy read is to write an algebraic expression with a variable that represents the number of books Gina read. For example, if G represents the number of books Gina read, then G + 4 represents the number of books Marcy read.

Examples

Write an algebraic expression to answer each question.

Statement	Algebraic Expression
Mary is 5 years older than her sister Carla. How old is Mary?	If Carla is C years old, then Mary is C + 5 years old.
Mrs. Roth weighs 30 pounds less than Mr. Roth. How much does Mrs. Roth weigh?	If Mr. Roth weighs R pounds, then Mrs. Roth weighs R − 30 pounds.
Anna bought a box of crayons with 8 crayons per box for each of her nieces. How many crayons did she buy all together?	If Anna has N nieces, then she bought a total of 8 * N crayons.
Claude earned $6 an hour and was paid an additional $3.50 for his lunch. How much did Claude get paid?	If Claude worked H hours, he got paid (6 * H) + 3.50 dollars.

Expressions and Equations

Algebraic expressions can be used to represent relationships among numbers.

Example
Write an algebraic expression for each phrase.

17 less than y	$y - 17$
4 more than $\frac{1}{3}$ of h	$4 + (\frac{1}{3})h$
The product of 6 and the quantity of $5 + p$	$6 * (5 + p)$
19 divided by the sum of 9 and w	$19 / (9 + w)$

Note When you read expressions with parentheses, you can use "the quantity of" to tell what is inside the parentheses. For example, you can read the expression $7 * (x - 9)$ as "seven times the quantity of x minus 9."

Mathematicians follow several conventions when writing algebraic expressions.

- When two or more variables or a variable and a number are multiplied, you do not need to write the multiplication symbol between them. For example, $a * b$ can be written as ab and $4 * x$ can be written as $4x$.

- The product of a number and a variable is written with the number, called a **coefficient,** first. For example, an expression to represent the cost of some number of pencils (n) if each pencil costs 15 cents is $15n$. The coefficient in this expression is 15, and it is written before the variable n.

- The multiplication symbol can be dropped before or between parentheses. For example, $5 * (7 - h)$ can be written as $5(7 - h)$ and $(4 + y) * (y - 6)$ can be written as $(4 + y)(y - 6)$. The Commutative Property of Multiplication is used when renaming $(n + 1) * 6$ as $6(n + 1)$.

- Division problems can be written using fraction notation. For example, $(d + 2) \div 7$ can be written as $\frac{d + 2}{7}$.

Expressions and Equations

Evaluating Expressions

To evaluate an expression is to find its numerical value. To **evaluate a numerical expression,** simply carry out the operations in the correct order. To evaluate an algebraic expression, first replace each variable with its value and then carry out the operations in the correct order. Page 203 gives more information about the correct order of operations for expressions and number sentences.

Examples

Evaluate $(24 - 6) * 2 = ?$

The parentheses tell you to subtract $24 - 6$ first.	$(24 - 6) * 2$
	$18 * 2$
Then multiply by 2.	36

$(24 - 6) * 2 = 36$

Evaluate $24 - (6 * 2) = ?$

The parentheses tell you to multiply $6 * 2$ first.	$24 - (6 * 2)$
	$24 - 12$
Then subtract.	12

$24 - (6 * 2) = 12$

Example

Evaluate $(3 * C) + H$ if $C = 4$ and $H = 5$.

Replace C with 4 and replace H with 5.
The result is $(3 * 4) + 5$. $(3 * 4) + 5$
The parentheses tell you to multiply $3 * 4$ first. $12 + 5$
Then add. 17

If $C = 4$ and $H = 5$, then $(3 * C) + H$ is 17.

Check Your Understanding

Write an algebraic expression for each situation using the suggested variable.

1. Mark is m inches tall. If Audrey is 3 inches shorter than Mark, what is Audrey's height?
2. It takes Herman H minutes to do his homework. If it takes Sue twice as long, how long does it take her?
3. Dawn went on R rides at the amusement park. If the park charges a $3 admission fee and $0.50 per ride, how much did Dawn spend?

Write an algebraic expression for each phrase.

4. twice the quantity of $4 - m$
5. 46 more than the product of 7 and q
6. one-fourth of the sum of 15 and x
7. the quantity of $8 + n$ divided by 12

Check your answers in the Answer Key.

Expressions and Equations

Order of Operations

In many everyday situations, the order in which you do things is important. When you bake a cake, for example, you crack the eggs before adding them to the batter. In mathematics, you need to do many operations in a certain order.

Rules for the Order of Operations
1. Do the operations inside **parentheses** or other **grouping symbols** following Rules 2–4. Work from the innermost set of grouping symbols outward.
2. Calculate all expressions with **exponents**.
3. **Multiply** and **divide** in order from left to right.
4. **Add** and **subtract** in order from left to right.

Example

Evaluate. $5 * 4 - 6 * 3 + 2 = ?$

Multiply first.	$5 * 4 - 6 * 3 + 2$
Subtract next.	$20 - 18 + 2$
Then add.	$2 + 2$
	4

$5 * 4 - 6 * 3 + 2 = \mathbf{4}$

Example

Evaluate. $5^2 + (3 * 4 - 2) / 5$

Do the operations inside parentheses first.	$5^2 + (3 * 4 - 2) / 5$
Calculate exponents next.	$5^2 + 10 / 5$
Divide.	$25 + 10 / 5$
Then add.	$25 + 2$
	27

$5^2 + (3 * 4 - 2) / 5 = 27$

Check Your Understanding

Evaluate each expression.

1. $28 - 15 / 3 + 8$
2. $1 + (5 * 10) / 4$
3. $10 * 6 / 2 - 30$
4. $10 * (12 / 6 + 4) / 12 + 1$
5. $3[4^2 * (9 - 4)]$
6. $8 + 2^3 * (24 / 6)$

Check your answers in the Answer Key.

Expressions and Equations

The Distributive Property

You have been using the **Distributive Property** to help you calculate and solve problems for a long time. For example, when you solve 40 * 57 with partial products, you think of 57 as 50 + 7 and multiply each part by 40. The Distributive Property says: 40 * (50 + 7) = (40 * 50) + (40 * 7).

The Distributive Property can be illustrated by finding the area of a rectangle.

40 * 50 → 2 0 0 0
40 * 7 → 2 8 0
40 * 57 → 2 2 8 0

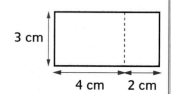

```
      5   7
*     4   0
  2 0 0   0
      2 8 0
  2 2 8   0
```

> **Example**
>
> Show how the Distributive Property works by finding the area of the rectangle in two different ways.
>
> **Method 1** Find the total width of the rectangle and multiply that by the height.
>
> A = 3 cm * (4 cm + 2 cm)
>
> = 3 cm * 6 cm
>
> = 18 cm²
>
> **Method 2** Find the area of each smaller rectangle, and then add these areas.
>
> A = (3 cm * 4 cm) + (3 cm * 2 cm)
>
> = 12 cm² + 6 cm²
>
> = 18 cm²
>
> Both methods show that the area of the rectangle is 18 cm².
>
> 3 * (4 + 2) = (3 * 4) + (3 * 2)
>
> This is an example of the Distributive Property of Multiplication over Addition.

The Distributive Property of Multiplication over Addition can be stated in two ways:

a * (x + y) = (a * x) + (a * y)

(x + y) * a = (x * a) + (y * a)

Expressions and Equations

The Distributive Property of Multiplication over Subtraction can also be stated in two ways:

$a * (x - y) = (a * x) - (a * y)$

$(x - y) * a = (x * a) - (y * a)$

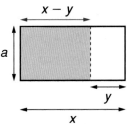

Example

Show how the Distributive Property of Multiplication over Subtraction works by finding the area of the shaded part of the rectangle in two different ways.

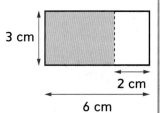

Method 1 Multiply the width of the shaded rectangle by its height.

$A = 3 \text{ cm} * (6 \text{ cm} - 2 \text{ cm})$
$= 3 \text{ cm} * 4 \text{ cm}$
$= 12 \text{ cm}^2$

Method 2 Subtract the area of the unshaded rectangle from the entire area of the whole rectangle.

$A = (3 \text{ cm} * 6 \text{ cm}) - (3 \text{ cm} * 2 \text{ cm})$
$= 18 \text{ cm}^2 - 6 \text{ cm}^2$
$= 12 \text{ cm}^2$

Both methods show that the area of the shaded part of the rectangle is 12 cm².
$3 * (6 - 2) = (3 * 6) - (3 * 2)$
This is an example of the Distributive Property of Multiplication over Subtraction.

Check Your Understanding

Use the Distributive Property to solve the problems.

1. $6 * (100 + 40)$
2. $(35 - 15) * 6$
3. $4 * (80 - 7)$
4. Use a calculator to verify that $1.23 * (456 + 789) = (1.23 * 456) + (1.23 * 789)$.

Check your answers in the Answer Key.

SRB
two hundred five 205

Expressions and Equations

Equivalent Expressions

Equivalent expressions are expressions that represent the same number, or when simplified, have the same **terms.** To find an equivalent expression for a given expression, you can use properties of numbers and operations. More information about these properties can be found on pages 231–232.

When you evaluate a numerical expression, you are **simplifying** it, or writing it as an equivalent expression in simpler form.

Note In an expression or equation, a **term** is a number, a variable, or a product of a number and one or more variables. For example, in the expression $7x + y + 4$, the terms are $7x$, y, and 4. $7x$ and y are **variable terms** and 4 is a constant term, or **constant,** because it has no variable part and will not change.

Examples

Simplify each numerical expression to find equivalent expressions.

Expression	$4 * (3^2 + 1)$	$25 - 10 * 2 + 2$
Equivalent Expressions	$4 * (9 + 1)$ $4 * 10$ 40	$25 - 20 + 2$ $5 + 2$ 7

Note The name-collection boxes you used in earlier grades are collections of equivalent numerical expressions. The expressions in this name-collection box are all equivalent expressions representing the number 15.

15
$5 * 3$
$20 - 5$
$\frac{3}{4} * 20$
$11 + 4$

Algebraic expressions are equivalent when they consist of exactly the same terms and operations in simplified form. To simplify an algebraic expression, combine **like terms,** or terms that have exactly the same unknown or unknowns. For example, $2k$ and $3k$ are like terms because they have the same unknown, k. They can be combined to equal $5k$ ($2k + 3k = 5k$). The Distributive Property, along with the Commutative and Associative Properties, can be especially useful for simplifying algebraic expressions.

Examples

Find equivalent expressions by simplifying these algebraic expressions.

$12x + 5 + 3x$
$= 12x + 3x + 5$
$= 15x + 5$

$25 + 63g + 134 + 18g$
$= 25 + 134 + 63g + 18g$
$= (25 + 134) + (63 + 18) * g$
$= 159 + 81g$

$7(n + 11) - 4n$
$= 7n + 77 - 4n$
$= 7n - 4n + 77$
$= (7 - 4) * n + 77$
$= 3n + 77$

Check Your Understanding

Which of the following expressions are equivalent to $4y + 24 + y - 10$?

a. $5y + 34$ **b.** $5y + 14$ **c.** $4y + 14$ **d.** $5y + 24 - 10$

Check your answers in the Answer Key.

Expressions and Equations

Number Sentences

Number sentences are made up of mathematical symbols.

Mathematical Symbols							
Digits	Variables	Operation Symbols		Relation Symbols		Grouping Symbols	
0, 1, 2, 3, 4, 5, 6, 7, 8, 9	n x y z a b c d C M P ? ☐	+ − × or * / or ÷	plus minus times divided by	= ≠ < > ≤ ≥	is equal to is not equal to is less than is greater than is less than or equal to is greater than or equal to	( ) [ ]	parentheses brackets

A number sentence must contain numbers (or variables) and a relation symbol. It may or may not contain operation symbols and grouping symbols.

Number sentences that contain the = symbol are called **equations.** Number sentences that contain any one of the symbols ≠, <, >, ≤, or ≥ are called **inequalities.**

If a number sentence does not contain variables, then it is always possible to tell whether it is true or false.

> **Example**
>
> **Equations**
> $3 + 3 = 8$ False $3 + 3 = 6$, not 8
> $(24 + 3) / 9 = 3$ True $27 / 9 = 3$
> $100 = 9^2 + 9$ False $9^2 + 9 = 90$, and 90 is not equal to 100.
>
> **Inequalities**
> $\frac{4}{5} - \frac{2}{3} < \frac{1}{2}$ True $\frac{4}{5}$ is close to 1, and $\frac{2}{3}$ is close to 1, so $\frac{4}{5} - \frac{2}{3}$ is less than $\frac{1}{2}$.
>
> $27 \neq 72$ True   27 is not equal to 72.
> $19 < 19$ False   19 is not less than itself.
> $16 * 4 \geq 80 \div 3$ True   64 is greater than or equal to $26\frac{2}{3}$.

> **Check Your Understanding**
>
> True or false?
>
> **1.** $32 - 14 = 18$ **2.** $4 * 7 < 30$ **3.** $0 = \frac{5}{5}$
> **4.** $25 + 5 \leq 5 * 6$ **5.** $50 - 12 = 7 * 2^2$ **6.** $84 \neq 84$
>
> Check your answers in the Answer Key.

Expressions and Equations

Open Sentences

In some number sentences, one or more numbers may be missing. In place of each missing number is a letter, a question mark, or some other symbol. These number sentences are called **open sentences**. A symbol, usually a letter, used in place of a missing number is called a **variable**.

Note If the same letter appears more than once in a single equation, it must represent the same number each time.

Some open sentences are always true. For example, $9 + y = y + 9$ is true if you replace y with any number. Some open sentences are always false. For example, $C - 1 > C$ is false if you replace C with any number.

For many open sentences, you cannot tell whether the sentence is true or false until you know which number replaces the variable. For example, $5 + x = 12$ is an open sentence in which x stands for some number.

- If you replace x with 3 in $5 + x = 12$, you get the number sentence $5 + 3 = 12$, which is false.
- If you replace x with 7 in $5 + x = 12$, you get the number sentence $5 + 7 = 12$, which is true.

If a number used in place of a variable makes the number sentence true, this number is called a **solution of the open sentence**. For example, the number 7 is a solution of the open sentence $5 + x = 12$ because the number sentence $5 + 7 = 12$ is true. Finding the solution(s) of an open number sentence is called solving the **equation** or solving the **inequality**.

Many simple equations, such as $5 + x = 12$, have just one solution, but inequalities may have many solutions. For example, 9, 3.5, $2\frac{1}{2}$, and -8 are all solutions of the inequality $x < 10$. In fact, no matter how many solutions you name, there are always more numbers that are less than 10. This means that $x < 10$ has an *infinite*, or unlimited, number of solutions.

Mathematicians often use *set notation* to record all possible solutions for an equation or inequality. Set notation lists all possible solutions inside brackets { }. You can use set notation for the solutions to the equation and inequality above.

- Since 7 is the only solution to the equation $5 + x = 12$, you can record the solution set as {7}.
- Since any number less than 10 is a solution of the inequality $x < 10$, the solution set can be recorded as {all numbers <10}.

Expressions and Equations

Examples

Use set notation to record the solution set.

$9 + y = y + 9$ — Since this open sentence is always true, the solution set includes all numbers. It can be recorded as {all numbers}.

$c - 1 > c$ — Since this open sentence is always false, the solution set is the **empty set**, or null set. It is recorded as { }, or ø.

$|b| = 4$ — Since $|-4|$ and $|4|$ both equal 4, the solution set can be recorded as $\{-4, 4\}$.

Number Models

In *Everyday Mathematics*, a number sentence or an expression that describes a situation is called a **number model.** Often, two or more number models can fit the same situation. For example, suppose that you had $20, spent $8.50, and had $11.50 left. The number model $20 − $8.50 = $11.50 fits this situation. The number model $20 = $8.50 + $11.50 also fits. Number models can be useful in solving problems.

Note Number models are not always number sentences. The expression $20 − $8.50 is not a number sentence, but it is another model that fits the situation.

Example

Write two number models that fit the following problem:

Juan is saving money for a bicycle that costs $119. He has $55. How much more does he need?

Two possible number models:
$119 = $55 + x
$119 − $55 = x

The first number model suggests counting up to find how much more Juan needs. The second number model suggests subtracting to find the answer.

Check Your Understanding

Find the solution of each equation.

1. $6 + c = 20$
2. $42 = 6 * z$
3. $(2 * f) + 5 = 26$

Write a number model that fits each problem.

4. Hunter used a $20 bill to pay for a video game that cost $12.49. How much change did he get?

5. Eve earns $10 a week babysitting. How many weeks will it take her to earn $90?

Check your answers in the Answer Key.

Expressions and Equations

Inequalities

An **inequality** is a number sentence that contains one of these symbols: ≠, <, >, ≥, or ≤. An inequality that contains a variable is an open sentence. Any number substituted for the variable that makes the inequality true is called a **solution** of the inequality.

Many inequalities have an infinite number of solutions, so it is often impossible to list all the solutions. Instead, the set of solutions, called the **solution set,** is either described or shown on a number-line graph.

Note Mathematicians often use *set notation* to record all possible solutions for an inequality. Set notation lists all possible solutions inside brackets. See pages 208–209 for examples.

Example

Describe and graph the solution set of $x + 3 > 10$.

The inequality $x + 3 > 10$ is an open sentence.
100 is a solution of $x + 3 > 10$ because $100 + 3 > 10$ is true.
2 is not a solution of $x + 3 > 10$ because $2 + 3 > 10$ is not true.

Any number that is less than or equal to 7 is not a solution of $x + 3 > 10$.
For example, 6 is not a solution of $x + 3 > 10$ because $6 + 3$ is not greater than 10.

Any number greater than 7 is a solution of $x + 3 > 10$.
Using set notation, you can write the solution set as {all numbers > 7}.
The graph of the solution set of $x + 3 > 10$ looks like this:

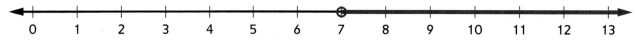

The shaded part of the number line tells you that any number greater than 7 is a solution (for example, 7.1, 8, 10.25). Notice the open circle at 7. This tells you that 7 is not part of the solution set.

Example

Graph the solution set of $y - 3 \leq 1$.

Any number less than or equal to 4 is a solution. For example, 4, 2.1, and −6 are all solutions.
Using set notation, you can write the solution set as {all numbers ≤ 4}.
Notice the shaded circle at 4. This tells you that 4 is one of the solutions.

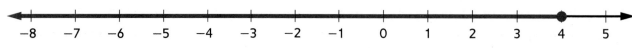

Check Your Understanding

Describe the solution set in Problems 1 and 2. Graph the solution set in Problem 3.

1. $n - 6 < 2$
2. $6 + c > 6$
3. $y \leq 8$

Check your answers in the Answer Key.

Expressions and Equations

When graphing a solution set for an inequality that represents a real-world situation, you often need to consider whether there are any constraints. *Constraints* are limitations to the solution set based on the context. For example, there are some situations where only whole numbers or counting numbers make sense as possible solutions. In other situations, there might be a reasonable minimum or maximum number that is not included in the problem description.

Example

Write a number sentence and graph the solution set to represent the number of people allowed on the elevator at a time.

If n represents the number of people allowed on the elevator at a time, $n \leq 8$.

Without the real world situation, you might graph $n \leq 8$ like this:

However, there are not an infinite number of solutions to this problem. For example, you can't have $2\frac{1}{2}$ people or -4 people on the elevator. There can only be whole numbers of people, and the minimum possible number of people on an elevator is 0. The graph for the solution set to this problem should look like this:

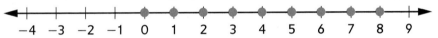

Example

In Michigan, bass must be at least 14 inches long in order to be kept by the people who catch them. This rule allows bass to reach adulthood and lay new eggs before being caught. It helps maintain a healthy population of fish living in Michigan's rivers and lakes. The longest bass ever caught was 32 inches long.

Jason is going bass fishing in Michigan. Write a number sentence and graph the solution set to represent the length of a bass that Jason might bring home.

Think: What length of bass can Jason keep? Are there any real-world constraints?

- If b represents the length of a bass that Jason can keep, then $b \geq 14$ inches because of Michigan's fishing rules.
- Bass will not grow infinitely long. Assuming that Jason doesn't catch a bass as long as or longer than the record of 32 inches, you can write another inequality to represent this situation more precisely: $b < 32$ inches.
- The solution set should be continuous (include all the numbers in the interval on the number line) because there are infinitely many possible lengths of fish.

By combining these two inequalities, you can graph the solution set:

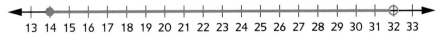

two hundred eleven 211

Expressions and Equations

Formulas

Formulas express relationships between quantities. (A **quantity** is a number with a unit, often a measurement or count.) Quantities in a formula are represented by variables.

Example

What is the formula for the area of a parallelogram?

The variable A stands for the area, b for the length of the base, and h for the height of the parallelogram.

The formula for the area of a parallelogram is $A = b * h$.

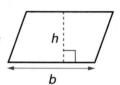

Variables in Formulas

Often the symbol for a variable is the first letter or other important letter of the quantity it represents.

Examples

Write each formula using variables.

Area of a parallelogram = base length * height $A = b * h$

Volume of a prism = area of the base * height $V = B * h$

A letter variable can have different meanings depending on whether it is capital or lowercase.

Example

The area of the shaded region in the figure at the right can be found by using the formula $A = S^2 - s^2$.

Note that S stands for the length of the side of the larger square and s stands for the length of the side of the smaller square.

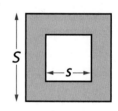

Evaluating a Formula

To *evaluate* a formula means to find the value of one variable in the formula when the values of the other variables are given.

Example

Evaluate the formula for the area of the shaded region in the figure above when $S = 4$ cm and $s = 2$ cm.

$A = S^2 - s^2$
$A = (4 \text{ cm})^2 - (2 \text{ cm})^2$
$A = 16 \text{ cm}^2 - 4 \text{ cm}^2 = 12 \text{ cm}^2$

The area of the shaded region is 12 cm^2.

Expressions and Equations

Units in Formulas

It is important that the **units** (hours, inches, meters, and so on) in a formula are consistent. For example, an area formula will not give a correct result if one measurement is in millimeters and another is in centimeters.

Example

Find the area of the rectangle.

First change millimeters to centimeters or centimeters to millimeters. Then use the formula $A = b * h$.

Change centimeters to millimeters: 5 cm = 50 mm $A = 7 \text{ mm} * 50 \text{ mm} = 350 \text{ mm}^2$

Or change millimeters to centimeters: 7 mm = 0.7 cm $A = 0.7 \text{ cm} * 5 \text{ cm} = 3.5 \text{ cm}^2$

The area of the rectangle is 350 mm², or 3.5 cm².

If something travels at a constant speed, the distance formula is $d = r * t$. d is the distance traveled; t is the travel time; and r is the rate of travel (speed). If the value of r is in miles per hour, the value of t should be in hours. If the value of r is in meters per second, t should be in seconds.

Example

Find the distance if $r = 50$ miles per hour and $t = 2$ hours.
Use the formula $d = r * t$.

If $r = 50$ miles per hour and $t = 2$ hours,
then $d = 50$ miles per hour $* 2$ hours $= 100$ miles.

Did You Know?

If an object is dropped over an open space, then $d = 16 * t^2$ and $s = 32 * t$. t is the time in seconds since the object started falling; d is the distance traveled (in feet); and s is the speed of the object (in feet per second).

Check Your Understanding

1. Find the area of rectangle ABCD.
2. If a car is traveling 20 feet per second, how far will it travel in one minute?
3. If $S = 9$ meters and $s = 3$ meters, what is the area of the shaded region in the square at the right?

Express the following relationship with a formula.

4. The interest (i) earned on $1,000 deposited in a savings account is equal to $1,000 times the rate of interest (r) times the length of time the money is left in the account (t).

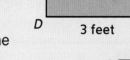

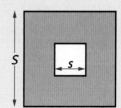

Check your answers in the Answer Key.

two hundred thirteen

Expressions and Equations

Using Trial and Error to Solve Equations

When you substitute a number for a variable in an equation, it can make the number sentence either true or false. If the number you substitute makes the number sentence true, then that number is a **solution** of the equation. One way to solve an equation is to try several test numbers until you find a solution. This method is called **trial and error.**

Note To *substitute* means to replace one thing with another. In an equation, it often means to replace variables with numbers.

Each test number can help you get closer to an exact solution. Comparing the results of each trial can help you decide how to adjust your test number for the next trial.

- A number that makes the values of the expressions in the equation nearly correct is close to a solution, but is not a solution.
- If one number makes a value too large and another number makes it too small, then the solution will likely be between those two numbers.

Example

Solve for y.

$9 + y = 16 - y$

Start with the test number 4. Substitute 4 for y in the equation.

Think: Does $9 + 4$ equal $16 - 4$? No. $13 > 12$, so 4 is not a solution, but 13 is close to 12.

Try a number smaller than 4 so that the sum of $9 + y$ will be smaller.

Substitute 3 for y in the equation.

Think: Does $9 + 3$ equal $16 - 3$? No. $12 < 13$, so 3 is not a solution.

Try a number greater than 3 so that the difference of $16 - y$ will be smaller.

Try $y = 3.5$ because it is greater than 3 and less than 4.

$9 + 3.5 = 16 - 3.5$

$12.5 = 12.5$

$y = 3.5$ is a solution of this equation.

Expressions and Equations

Sometimes it is difficult to find an exact solution with trial and error, but you can approximate a solution. A table can help you keep track of numbers you have tested and the results.

Example

Find an approximate solution for this equation: $4k + k^2 = 30$.

	k	$4k$	k^2	$4k + k^2$	Compare $(4k + k^2)$ to 30
You can start by substituting 3 for k.	3	12	9	21	21 < 30
The sum is less than 30, so try $k = 4$.	4	16	16	32	32 > 30
The sum is greater than 30, so you know $k > 3$ and $k < 4$. Try $k = 3.5$.	3.5	14	12.25	26.25	26.25 < 30
26.25 < 30, so try a larger number.	3.8	15.2	14.44	29.64	29.64 < 30
Keep adjusting your guess to get closer and closer to the solution.	3.9	15.6	15.21	30.81	30.81 > 30
	3.85	15.4	14.8225	30.2225	30.2225 > 30

The solution will be very close to 3.85. $k \approx 3.85$.

(The symbol $\approx$ means *is about equal to*.)

Expressions and Equations

Using Bar Models to Solve Equations

Bar models are diagrams that can help you solve equations. Bar models are useful when solving equations with variables on one or both sides of the equation. To make a bar model:

- Draw a rectangle with 2 layers. Write one side of the equation in the top layer and the other side in the bottom layer.
- Line up any equivalent quantities.
- Create new sections to match the remaining amount in one layer with the remaining amount in the other layer.
- Continue aligning equivalent expressions until a number in one layer lines up with a single copy of the variable in the other.

Examples

Solve the equations using bar models.

$3a = 16 + a$

3a	
16	a

a	a	a
16		a

a	a	a
8	8	a

Write one side of the equation in the top layer of the rectangle, and the other side in the bottom layer.

$3a$ can be written as $a + a + a$. One a in the top layer is equal to one a in the bottom layer. So, 16 is equivalent to $a + a$.

Divide 16 into 2 equal parts. $8 + 8 = 16$, so each a must equal 8.

Check: Substitute 8 for a in the original problem. $\quad 3 * 8 = 16 + 8$

This number sentence is true, so $a = 8$ is a solution. $\quad\quad\quad 24 = 24$

$2q + 16 = 3q + 5$

2q	16	
3q		5

2q	16	
2q	q	5

2q	11	5
2q	q	5

Write each side of the equation in one layer of the rectangle.

Rename $3q + 5$ as $2q + q + 5$ to line up $2q$ in the top with $2q$ in the bottom. This leaves $16 = q + 5$.

Break apart 16 into $11 + 5$. 5 in the top row aligns with 5 in the bottom row, so q must equal 11.

Check: Substitute 11 for q in the original problem. $\quad 2 * 11 + 16 = 3 * 11 + 5$

This number sentence is true, so $q = 11$ is a solution. $\quad\quad 22 + 16 = 33 + 5$

$\quad\quad\quad\quad\quad\quad\quad\quad\quad\quad\quad\quad\quad\quad\quad\quad\quad\quad\quad\quad\quad 38 = 38$

Expressions and Equations

Pan-Balance Problems and Solving Equations

When objects are placed in a pan balance, one of three things happens: the left side is heavier than the right, the right side is heavier than the left, or the sides balance. If two different kinds of objects are placed in the pan balance and the sides balance, then you know the two sides have equal weights. You can find the weight of one kind of object in terms of the other.

> **Example**
>
> The pan balance at the right has 2 balls and 6 marbles in one pan, and 1 ball and 8 marbles in the other pan. How many marbles weigh as much as 1 ball?
>
> **Step 1** If 1 ball is removed from each pan, the pan balance will remain balanced. One ball and 6 marbles will be left in the pan on the left and 8 marbles will be left in the pan on the right.
>
> **Step 2** If 6 marbles are removed from each pan, the pan balance will remain balanced. One ball will be left in the pan on the left and 2 marbles will be left in the pan on the right.
>
> 1 ball weighs as much as 2 marbles.

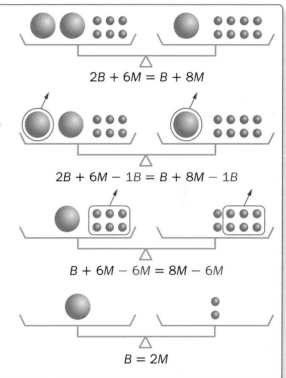

$2B + 6M = B + 8M$

$2B + 6M - 1B = B + 8M - 1B$

$B + 6M - 6M = 8M - 6M$

$B = 2M$

When solving a pan-balance problem, the pans must balance after each step. If you *always do the same thing to the objects in both pans*, then the pans will remain balanced. For example, you might remove the same number of the same kind of object from each pan as in Step 1 and Step 2 in the example above.

You can think of pan-balance problems as models of equations. Suppose that B stands for the weight of 1 ball and M stands for the weight of 1 marble. The pan-balance problem in the example above can then be expressed by the equation $2B + 6M = B + 8M$.

Note Here are some other ways to do the same thing to the objects in both pans:
- Double the number of each kind of object in each pan.
- Remove half of each kind of object from each pan.

two hundred seventeen

Expressions and Equations

You can use pan balances to solve equations. Use a sequence of pan balances to show changes to an equation. As long as you perform the same operations on both sides of the equal sign, you will have equivalent equations.

Example

Solve using a pan-balance model.

$4p = 2p + 12$

Subtract $2p$.

Divide by 2.

$p = 6$

Note Equivalent equations have the same solution set. For example, $2 + x = 7$ and $5 + x = 10$ are equivalent equations because the solution set for each is $x = 5$.

You can replace an expression on one side of the pan balance with an equivalent expression and not affect the balance of the equation. This creates an equivalent equation with the same solution.

Example

Solve using a pan-balance model.

$6(x + 2) = (5 - x) * 3$

Use the Distributive Property to write equivalent expressions.

Subtract 12.

Add $3x$.

Divide by 9.

$x = \frac{3}{9}$

Check Your Understanding

Write an equation for each pan-balance problem.

1.

2.

Solve.

3. $6n + 10 = 16$

4. $30 = 9w - 15$

5. $4(y + 3) = 28$

Check your answers in the Answer Key.

Expressions and Equations

Using Inverse Operations to Solve Equations

Many equations with just one unknown can be solved using the **inverse-operations strategy.** This strategy involves reversing operations in an equation by using their inverses. For example, subtraction reverses addition and division reverses multiplication.

If the unknown appears on both sides of the equal sign, use inverse operations to change the equation to an equation with the unknown appearing on only one side. You may also have to use inverse operations to move all the constants to the other side of the equal sign.

This method works best if you reverse addition and subtraction before reversing multiplication and division. Remember, just like with the pan-balance method, any operation performed on one side of the equation must be done on the other side.

Note A **constant** is just a number, such as 3 or 7.5 or $\frac{1}{2}$. Constants don't change, or vary, the way variables do.

Example

Solve $3y + 10 = 7y - 6$.

Step	Operation	Equation
1. Remove the unknown term (the variable term) from the left side of the equation.	Subtract $3y$ from each side.	$3y + 10 = 7y - 6$ $3y - 3y + 10 = 7y - 3y - 6$ $10 = 4y - 6$
2. Remove the constant term from the right side of the equation.	Add 6 to both sides.	$10 = 4y - 6$ $10 + 6 = 4y - 6 + 6$ $16 = 4y$
3. Change the $4y$ term to a $1y$ term. (Remember: $1y$, $1 * y$, and y all mean the same thing.)	Divide both sides by 4.	$16 = 4y$ $16 / 4 = 4y / 4$ $4 = y$

Check: Substitute the solution, 4, for y in the original equation.

$$3y + 10 = 7y - 6$$
$$3 * 4 + 10 = 7 * 4 - 6$$
$$12 + 10 = 28 - 6$$
$$22 = 22$$

Since $22 = 22$ is true, the solution of 4 is correct.

So, $y = 4$.

Each step in the above example produced a new equation that looks different from the original equation. Even though these equations look different, they all have the same solution set (which is {4}). Equations that have the same solution set are called **equivalent equations.**

Expressions and Equations

Like terms are terms that have exactly the same unknown or unknowns. The terms 4x and 2x are like terms because they both contain x. The terms 6 and 15 are like terms because they both contain no variables; 6 and 15 are both constants.

If an equation has parentheses, or if the unknown or constants appear on both sides of the equal sign, you can simplify the equation by **simplifying the expressions** on both sides of the equation.

- If an equation has parentheses, use the Distributive Property or other properties to write an equation without parentheses.

- If an equation has two or more like terms on one side of the equal sign, combine the like terms. To *combine like terms* means to rewrite the sum or difference of like terms as a single term. For example, $4y + 7y = 11y$, and $7y - 4y = 3y$.

Example

Solve $5(b + 3) - 3b + 5 = 4(b - 1)$.

Operation	Equation
1. Use the Distributive Property to remove the parentheses.	$5(b + 3) - 3b + 5 = 4(b - 1)$ $5b + 15 - 3b + 5 = 4b - 4$
2. Combine like terms on each side of the equation.	$(5b - 3b) + (15 + 5) = 4b - 4$ $2b + 20 = 4b - 4$
3. Subtract $2b$ from both sides.	$2b - 2b + 20 = 4b - 2b - 4$ $20 = 2b - 4$
4. Add 4 to both sides.	$20 + 4 = 2b - 4 + 4$ $24 = 2b$
5. Divide both sides by 2.	$24 / 2 = 2b / 2$ $12 = b$

Note Reminder: $5(b + 3)$ means the same as $5 * (b + 3)$.

Check Your Understanding

1. Check that 12 is the solution of the equation in the example above.

Solve.

2. $5x - 7 = 1 + 3x$ **3.** $5 * (s + 12) = 10 * (3 - s)$ **4.** $3(9 + b) = 6(b + 3)$

Check your answers in the Answer Key.

Expressions and Equations

"What's My Rule?" Problems

Imagine a machine that works like this: When a number (the *input*, or "in" number) is dropped into the machine, the machine changes the number according to a rule. A new number (the *output*, or "out" number) comes out the other end.

This machine adds 5 to any "in" number. Its rule is "+ 5."

- If 4 is dropped in, 9 comes out.
- If 7 is dropped in, 12 comes out.
- If 53 is dropped in, 58 comes out.
- If 1.6 is dropped in, 6.6 comes out.

"In" and "out" numbers can be displayed in table form as shown at the right.

To solve a "What's My Rule?" problem, you need to find the missing information. In the following examples, the solutions (the missing information) appear in color.

in	out
x	$x + 5$
4	9
7	12
53	58
1.6	6.6

Example

Find the "out" numbers.
Rule: subtract 7 from "in"

in	out	
z	$z - 7$	
9	2	$9 - 7 = 2$
27	20	$27 - 7 = 20$

Find the "in" numbers.
Rule: multiply "in" by 2

in	out	
w	$w * 2$	
4	8	$4 * 2 = 8$
24	48	$24 * 2 = 48$

Find the rule.
Rule: raise "in" to the second power

in	out	
r	r^2	
2	4	$2^2 = 4$
5	25	$5^2 = 25$

Check Your Understanding

Solve these "What's My Rule?" problems.

1. *Rule:* divide "in" by 3

in	out
n	$n / 3$
9	
36	

2. *Rule:* subtract 4 from "in"

in	out
k	$k - 4$
	7
	24

3. *Rule:* ?

in	out
x	
4	120
10	300

Check your answers in the Answer Key.

Expressions and Equations

Rules, Tables, and Graphs

Many problems can be solved by showing relationships between variables with rules, tables, or graphs.

Example

Sara earns $6 per hour. Use a rule, a table, and a graph to find how much Sara earns in $2\frac{1}{2}$ hours.

Rule: If h stands for the number of hours worked, and e stands for her earnings, then $e = \$6 * h$.

$$e = \$6 * h = \$6 * 2\frac{1}{2} = \$15$$

Table:

Time (hours)	Earnings ($)
h	$6 * h$
0	0
1	6
2	12
3	18

Think of $2\frac{1}{2}$ hours as 2 hours + $\frac{1}{2}$ hour. For 2 hours, Sara earns $12. For $\frac{1}{2}$ hour, Sara earns half of $6, or $3. In all, Sara earns $12 + $3 = $15. Note that $2\frac{1}{2}$ hours is halfway between 2 hours and 3 hours, so her earnings are halfway between $12 and $18, which is $15.

Graph:

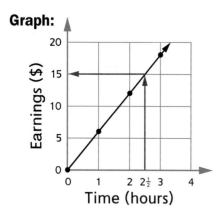

To draw the graph, plot several of the number pairs from the table. For example, plot (0, 0), (1, 6), (2, 12), and (3, 18). Plot each number pair as a point on the coordinate grid. The straight line connecting these points is the graph.

To use the graph, first find $2\frac{1}{2}$ hours on the horizontal axis. Then go straight up to the line for Sara's earnings. Turn left and go across to the vertical axis. You will end up at the same answer as you did when you used the table, $15.

Sara earns $15 in $2\frac{1}{2}$ hours.

Check Your Understanding

1. Christie types about 40 words per minute. Use the graph to find how many words she can type in 12.5 minutes.

2. Daniel earns $4.50 an hour. He worked 7 hours. Use the rule to find how much he earned. (e stands for earnings; h stands for number of hours worked.) Rule: $e = \$4.50 * h$

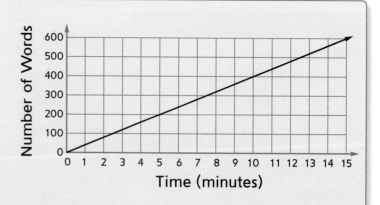

Check your answers in the Answer Key.

Expressions and Equations

Independent and Dependent Variables

Formulas and the rules in "What's My Rule?" problems often use equations to relate two or more variables. For example:

- The formula for the area A of a square with a given side length s is $A = s^2$.
- An equation for the number of miles m a car travels in a given number of hours h at a constant speed of 45 miles per hour is $45 * h = m$.
- An equation for the "What's My Rule?" function machine at the right is $y = x + 5$.

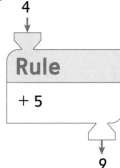

Each of these equations has an independent variable and a dependent variable. An **independent variable** is one whose value does not rely on any other variable. A **dependent variable** is one whose value depends on the value of another variable.

Examples

Identify the independent and dependent variables in the equations above.

| In $A = s^2$, the area depends on the side length. So s is independent and A is dependent. | In $45 * h = m$, the miles traveled depends on the number of hours. So h is independent and m is dependent. | In $y = x + 5$, the "out" variable depends on the value of the "in" variable. So x is independent and y is dependent. |

The situation for which an equation is written determines the independent variable or variables. The independent variables in the area and distance equations are the given values: side length s for area and hours h for distance. In "What's My Rule?" problems, the independent variable is the "in" number.

The next example shows that sometimes it is not clear which variable or variables are independent.

Example

Judd wants to build an 84 square foot rectangular deck w feet wide and l feet long. So, $84 = l * w$. Which variable in this equation is the independent variable?

Because Judd can choose a value for either l or w, both are independent.

Once a value for either l or w is chosen, the other variable is dependent on that value to make the equation true.

Expressions and Equations

By convention, values of independent variables in equations are listed in the left column of a table and graphed on the horizontal axis of a graph.

$A = s^2$

s	A
1	1
2	4
3	9

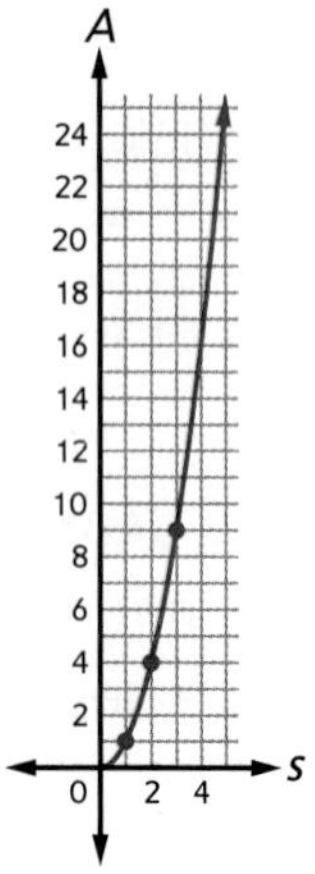

$45 * h = m$

h	m
1	45
2	90
3	135

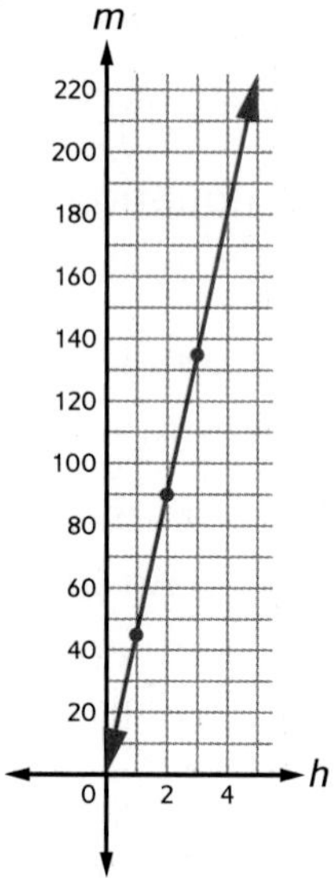

$y = x + 5$

x	y
1	6
2	7
3	8

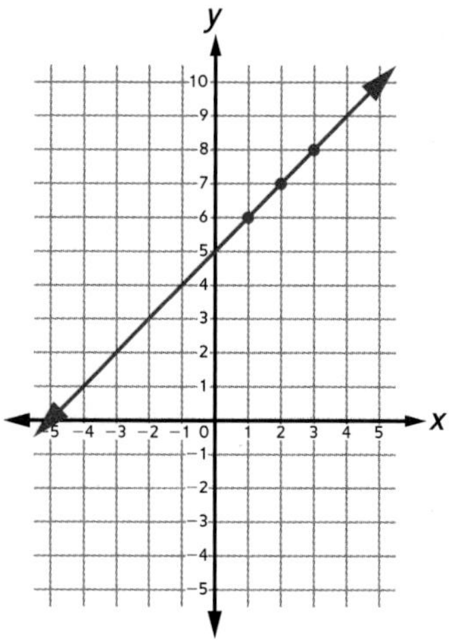

It is possible to have more than one independent variable in an equation. The formula for the volume V of a prism with base area B and height h is $V = B * h$. Both B and h are independent variables and V is dependent on both.

Check Your Understanding

Identify the independent and dependent variables in each situation. Then write an equation relating them.

1. The rule in a "What's My Rule?" problem is "multiply by $\frac{1}{2}$."
2. The volume V of a cube with edge length s

Check your answers in the Answer Key.

Expressions and Equations

Representing Patterns with Algebra

Many rules involving numbers can be described with the help of variables, such as n or $\square$. For example, the rule "The square of a number is the number multiplied by itself" can be expressed as $n^2 = n * n$, where n stands for any number. This is sometimes called a general pattern. A special case of this general pattern can be given by replacing the variable n with any number. For example, $3^2 = 3 * 3$ and $\left(\frac{2}{3}\right)^2 = \frac{2}{3} * \frac{2}{3}$ are special cases of the general pattern $n^2 = n * n$.

Example

Use variables to describe the general pattern for the following special cases.

Special cases: $8 / 8 = 1$ $0.5 / 0.5 = 1$ $2\frac{3}{4} / 2\frac{3}{4} = 1$

Step 1 Write everything that is the same for all the special cases.
Use blanks for the parts that change.   ___ / ___ = 1

Step 2 Fill in the blanks. Use a variable for the number that varies.
(Use 2 different variables if there are 2 different numbers that vary.) $b / b = 1$

The general pattern is $b / b = 1$. (Division by 0 is not allowed, so b may not equal 0.)

You can generalize a numerical pattern by writing an expression with variables.

Example

Look at the pattern below. Write an expression to represent the number of tiles in each figure.

      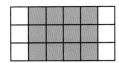

Figure 1   Figure 2   Figure 3   Figure 4   ...   Figure f ...

Special cases: How many tiles are in Figure 1? In Figure 2? In Figure 3?
Figure 1 has 1 column of 3 blue tiles and 2 columns of 3 white tiles: $(1 * 3) + (2 * 3) = 3 + 6 = 9$
Figure 2 has 2 columns of 3 blue tiles and 2 columns of 3 white tiles: $(2 * 3) + (2 * 3) = 6 + 6 = 12$
Figure 3 has 3 columns of 3 blue tiles and 2 columns of 3 white tiles: $(3 * 3) + (2 * 3) = 9 + 6 = 15$
General pattern: Write an algebraic expression for the number of tiles that will be in Figure f.
Think: What is the same for all of the special cases?
Each figure has the same number of blue columns as its figure number, f, and each column has 3 tiles. The number of blue tiles in each figure is $(f * 3)$.
Each figure also has 2 columns of 3 white tiles. There are $(2 * 3)$ white tiles in each figure.
Add the blue and white tiles together to find the total number of tiles in Figure f.

The number of tiles in Figure f can be generalized by the expression
$(f * 3) + (2 * 3) = 3f + 6$.

Expressions and Equations

How to Balance a Mobile

A *mobile* is a sculpture constructed of rods and other objects that are suspended in midair by wire, twine, or thread. The rods and objects are connected in such a way that the sculpture is balanced when it is suspended. The rods and objects move independently when they are stirred by air currents.

The simplest mobile consists of a single rod with two objects hanging from it. The point at which the rod is suspended is called the *fulcrum*. The fulcrum may be at any point on the rod. Objects may be hung at the ends of the rod or at points between the ends of the rod and the fulcrum.

Suppose the fulcrum is the center point of the rod and you hang one object on each side of the fulcrum.

Let W = the weight of one object

D = the distance of this object from the fulcrum

w = the weight of the second object

d = the distance of this object from the fulcrum

The mobile will balance if $W * D = w * d$.

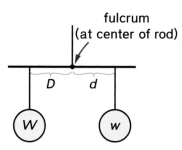

Example

The mobile below is balanced. What is the missing distance, x?

Replace the variables in the formula $W * D = w * d$ with the values shown in the diagram. Then solve the equation.

$W = 6 \qquad D = 7 \qquad w = 10 \qquad d = x$

Solution:
$W * D = w * d$
$6 * 7 = 10 * x$
$42 = 10 * x$
$4.2 = x$

So the missing distance to the fulcrum is 4.2 units. $x = 4.2$ units

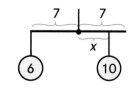

Check Your Understanding

Decide whether the mobiles are in balance.

1.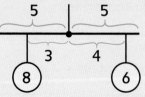

2. (mobile with distances 8 and 8 from fulcrum; 4 and 7 shown; weights 11 and 6)

Check your answers in the Answer Key.

Expressions and Equations

Suppose the fulcrum is *not* the center point of the rod and you hang one object on each side of the fulcrum.

Let R = the weight of the rod

L = the distance from the center of the rod to the fulcrum

W = the weight of the object that is on the same side of the fulcrum as the center of the rod

D = the distance of this object from the fulcrum

w = the weight of the other object

d = the distance of this object from the fulcrum

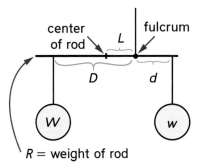

The mobile will balance if $(W * D) + (R * L) = w * d$.

Example

The mobile below is balanced. What is the missing weight, 5x?

Replace the variables in the formula $(W * D) + (R * L) = w * d$ with the values shown in the diagram. Then solve the equation.

R = 25   L = 2   W = 7   D = 10   w = 5x   d = 6

Solution : $(W * D) + (R * L) = w * d$

$(7 * 10) + (25 * 2) = 5x * 6$

$70 + 50 = 30x$

$120 = 30x$

$4 = x$

Since x = 4, 5x = 5 * 4 = 20.

So the weight of the object suspended to the right of the fulcrum is 20 units.

Check Your Understanding

In Problems 1 and 2, decide whether the mobiles are balanced. In Problem 3, find the weight of the object on the left of the fulcrum if the mobile is balanced.

1.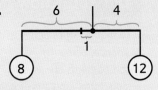
weight of rod = 5

2.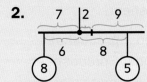
weight of rod = 4

3.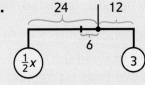
weight of rod = 2

Check your answers in the Answer Key.

two hundred twenty-seven

Expressions and Equations

Solving Problems with Computer Spreadsheets and Formulas

Spreadsheets: A History

Everyday Rentals—Debit Statement for May 1964

Company	Type	Invoice #	Invoice Amount	Amount Paid	Balance Due
Electric	Utility	2704-3364	342.12	100.00	242.12
Gas	Utility	44506-309	129.43	50.00	79.43
Phone	Utility	989-2209	78.56	78.56	0.00
Water	Utility	554-2-1018	13.12		13.12
NW Bank	Mortgage	May 1964	1,264.00	1,264.00	0.00
Waste Removal	Garbage	387-219	23.00		23.00
NW Lumber	Supplies	e-318	239.47	50.00	189.47
Total			2,089.70	1,542.56	547.14

Above is a copy of a financial record for Everyday Rentals Corporation for May 1964. A financial record often had more columns of figures than would fit on one sheet of paper, so accountants taped several sheets together. They folded the sheets for storage and spread them out to read or make entries. Such sheets came to be called *spreadsheets*.

Note that the "Balance Due" column and the "Total" row are calculated from other numbers in the spreadsheet. Before they had computers, accountants wrote spreadsheets by hand. If an accountant changed a number in one row or column, several other numbers would have to be erased, recalculated, and reentered.

For example, when Everyday Rentals Corporation pays the $23 owed to Waste Removal, the accountant must enter that amount in the "Amount Paid" column. That means the total of the "Amount Paid" column must be changed as well. That's not all—making a payment changes the amount in the "Balance Due" column and the total of the "Balance Due" column. One entry requires three other changes to update, or revise, the spreadsheet.

When personal computers were developed, **spreadsheet programs** were among the first applications. Spreadsheet programs save time by making changes automatically. Suppose the record at the top of this page is on a computer spreadsheet. If a spreadsheet is programmed correctly, when the accountant enters the payment of $23, the computer recalculates all of the numbers that are affected by that payment.

Did You Know?

In mathematics and science, computer spreadsheets are used to store large amounts of data and to perform complicated calculations. People use spreadsheets at home to keep track of budgets, payments, and taxes.

Expressions and Equations

Spreadsheets and Computers

A **spreadsheet program** enables you to use a computer to evaluate formulas quickly and efficiently. On a computer screen, a spreadsheet looks like a table. Each **cell** in the table has an **address** made up of a letter and a number. The letter identifies the column, and the number identifies the row in which the cell is found. For example, cell B3 is in column B, row 3.

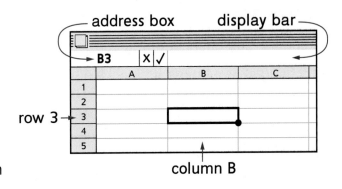

To enter information in a cell, you can use a computer mouse to click on the cell and the address of the cell will appear in the **address box.** Then type the information you want to enter in the cell and the information will appear in the **display bar.**

There are three kinds of information that may be entered in a spreadsheet.

- **Labels** (many consist of words, numbers, or both): These are used to display information about the spreadsheet, such as headings for columns and rows. Numbers in labels are never used in calculations. When a label is entered from the keyboard, it is stored in its address and shown in its cell on the screen.

- **Numbers** (those not included in labels): These are used in calculations. When a number is entered from the keyboard, it is stored in its address and appears in its cell on the screen.

- **Formulas:** These tell the computer what calculations to make on numbers in other cells. When a formula is entered from the keyboard, it is stored in its address but is *not* shown in its cell on the screen. Instead, a number is shown in the cell. This number is the result of applying the formula to the numbers in other cells.

Example

Study the spreadsheet at the right.

All the entries in row 1 and column A are *labels*.

The entries in cells B3 though B6 and cells C3 through C6 are *numbers* that are not labels. They were entered from the keyboard and are used in calculations.

Cells D3 through D10 also display numbers, but they were not entered from the keyboard. Instead, a *formula* was entered in each of these cells. The computer program used these formulas to calculate the numbers that appear in column D.

The numbers in column D are the results of calculations.

	A	B	C	D
D3	x ✓ =B3*C3			
	A	B	C	D
1	item name	unit price	quantity	totals
2				
3	pencils	0.29	6	1.74
4	graph paper	1.19	2	2.38
5	ruler	0.50	1	0.50
6	book	5.95	1	5.95
7				
8	Subtotal			10.57
9	tax 7%			0.74
10	Total			11.31

two hundred twenty-nine **229**

Expressions and Equations

Example

Study the spreadsheet at the right.

The *address box* shows that cell D3 has been selected; the *display bar* shows "= B3∗C3." This stands for the formula D3 = B3 ∗ C3. It is not necessary to enter D3, since D3 is already identified as the address of the cell. This formula is stored in the computer; it is not shown in cell D3.

D3	x ✓ =B3∗C3		
A	B	C	D
item name	unit price	quantity	totals
pencils	0.29	6	1.74
graph paper	1.19	2	2.38
ruler	0.50	1	0.50
book	5.95	1	5.95
Subtotal			10.57
tax 7%			0.74
Total			11.31

When the formula is entered, the program multiplies the number in cell B3 (0.29) by the number in cell C3 (6) and displays the product (1.74) in cell D3.

Suppose that you clicked on cell C3 and changed the 6 to an 8. The entry in cell D3 would change automatically to 2.32 (= 0.29 ∗ 8). At the same time, the entries in cells D8, D9, and D10 would also change automatically. The entry in cell D8 is the result of a calculation involving the entry in cell D3. The entry in cell D8 is used to calculate the entry in cell D9. And the entries in cells D8 and D9 are used to calculate the entry in cell D10.

Check Your Understanding

The following spreadsheet gives budget information for a class picnic.

Class Picnic ($$)

	A	B	C	D
1		budget for class picnic		
2				
3	quantity	food items	unit price	cost
4	6	packages of hamburgers	2.79	16.74
5	5	packages of hamburger buns	1.29	6.45
6	3	bags of baby carrots	3.12	9.36
7	3	quarts of macaroni salad	4.50	13.50
8	4	gallons of lemonade	1.69	6.76
9			subtotal	52.81
10			8% tax	4.23
11			total	57.04

Use the spreadsheet to answer the following questions.

1. What kind of information is shown in column B?
2. What information is shown in cell C7?
3. Which cell shows the title of the spreadsheet?
4. What information is shown in cell A5?
5. Which occupied cells do not hold labels or formulas?
6. Which column holds formulas?

Check your answers in the Answer Key.

Expressions and Equations

Properties of Numbers and Operations

The following properties are true for all numbers. The variables a, b, c, and d stand for any numbers (except 0 if the variable stands for a divisor).

Properties	Examples
Commutative Property The sum or product of two numbers is the same, regardless of the order of the numbers. $a + b = b + a$ $a * b = b * a$	$7 + 8 = 8 + 7 = 15$ $\frac{3}{4} * \frac{4}{5} = \frac{4}{5} * \frac{3}{4} = \frac{3}{5}$
Associative Property The sum or product of three or more numbers is the same, regardless of how the numbers are grouped. $a + (b + c) = (a + b) + c$ $a * (b * c) = (a * b) * c$	$(7 + 5) + 8 = 7 + (5 + 8)$ $\quad 12 \; + 8 = 7 + \quad 13$ $\quad\quad 20 = 20$ $2\frac{1}{2} * (2 * 3) = (2\frac{1}{2} * 2) * 3$ $2\frac{1}{2} * \quad 6 \; = \quad\quad 5 \; * 3$ $\quad\quad 15 = 15$
Distributive Property When a number a is multiplied by the sum or difference of two other numbers, the number a is "distributed" to each of these numbers. $a * (b + c) = (a * b) + (a * c)$ $a * (b - c) = (a * b) - (a * c)$	$5 * (8 + 2) = (5 * 8) + (5 * 2)$ $5 * \quad 10 \; = \quad 40 \; + \quad 10$ $\quad\quad 50 = 50$ $2 * (8 - 3) = (2 * 8) - (2 * 3)$ $2 * \quad 5 \; = \quad 16 \; - \quad 6$ $\quad\quad 10 = 10$
Addition Property of Zero (Identity Property of Addition) The sum of any number and 0 is equal to the original number. $a + 0 = 0 + a = a$	$5.37 + 0 = 5.37$ $0 + 6 = 6$
Multiplication Property of One (Identity Property of Multiplication) The product of any number and 1 is equal to the original number. $a * 1 = 1 * a = a$	$\frac{2}{3} * 1 = \frac{2}{3}$ $1 * 19 = 19$
Opposites Property The opposite of a number, a, is usually written $-a$. In *Everyday Mathematics*, this is sometimes written as OPP(a) for the opposite of a. If a is a positive number, then OPP(a) is a negative number. If a is a negative number, then OPP(a) is a positive number. If $a = 0$, then OPP(a) = 0. Zero is the only number that is its own opposite.	OPP(8) = -8 OPP($-\frac{3}{4}$) = $\frac{3}{4}$ OPP(-7) = 7 OPP(0) = 0

Expressions and Equations

Properties	Examples
Opposite of Opposites Property The opposite of the opposite of a number is equal to the original number. $OPP(OPP(a)) = OPP(-a) = a$	$OPP(OPP(\frac{2}{3})) = OPP(-\frac{2}{3}) = \frac{2}{3}$ $OPP(OPP(-9)) = OPP(9) = -9$
Equivalent Fractions Property If the numerator and denominator of a fraction are multiplied or divided by the same number, the resulting fraction is equal to the original fraction. $\frac{a}{b} = \frac{a * c}{b * c}$ $\frac{a}{b} = \frac{a \div c}{b \div c}$	$\frac{2}{3} = \frac{2 * 5}{3 * 5} = \frac{10}{15}$ $\frac{6}{8} = \frac{6 \div 2}{8 \div 2} = \frac{3}{4}$
Addition and Subtraction of Fractions Properties The sum or difference of fractions with like denominators is the sum or difference of the numerators over the denominator. $\frac{a}{c} + \frac{b}{c} = \frac{a+b}{c}$ $\frac{a}{c} - \frac{b}{c} = \frac{a-b}{c}$ To add or subtract fractions with unlike denominators, you can rename the fractions so that they have a common denominator. $\frac{a}{b} + \frac{c}{d} = \frac{ad + bc}{bd}$ $\frac{a}{b} - \frac{c}{d} = \frac{ad - bc}{bd}$	$\frac{3}{5} + \frac{1}{5} = \frac{3+1}{5} = \frac{4}{5}$ $\frac{5}{6} - \frac{1}{6} = \frac{5-1}{6} = \frac{4}{6} = \frac{2}{3}$ $\frac{2}{3} + \frac{1}{5} = \frac{10}{15} + \frac{3}{15} = \frac{10+3}{15} = \frac{13}{15}$ $\frac{2}{3} - \frac{1}{4} = \frac{8}{12} - \frac{3}{12} = \frac{8-3}{12} = \frac{5}{12}$
Multiplication of Fractions Property The product of two fractions is the product of the numerators over the product of the denominators. $\frac{a}{b} * \frac{c}{d} = \frac{a * c}{b * d}$	$\frac{5}{8} * \frac{3}{4} = \frac{5 * 3}{8 * 4} = \frac{15}{32}$
Division of Fractions Property The quotient of two fractions is the product of the dividend and the reciprocal of the divisor. $\frac{a}{b} \div \frac{c}{d} = \frac{a}{b} * \frac{d}{c} = \frac{a * d}{b * c}$	$9 \div \frac{2}{3} = 9 * \frac{3}{2} = \frac{27}{2}$, or $13\frac{1}{2}$ $\frac{5}{6} \div \frac{1}{4} = \frac{5}{6} * \frac{4}{1} = \frac{20}{6}$, or $3\frac{1}{3}$
Powers of a Number Property If a is any number and b is a positive whole number, then a^b is the product of a used as a factor b times. $a^b = \underbrace{a * a * a * \ldots * a}_{b \text{ factors}}$ a^0 is equal to 1.	$5^2 = 5 * 5 = 25$ $(\frac{2}{3})^4 = \frac{2}{3} * \frac{2}{3} * \frac{2}{3} * \frac{2}{3} = \frac{16}{81}$ $4^0 = 1$

Geometry

Geometry

What is Geometry?

Geometry is the mathematical study of space and objects in space. It deals with the shape and size of objects and the position of objects in space.

Plane geometry concerns 2-dimensional objects (figures) on a flat surface (a plane). Figures studied in plane geometry include points, lines, line segments, angles, polygons, and circles.

Solid geometry is the study of 3-dimensional objects. These objects are often simplified versions of familiar everyday things. For example, **rectangular prisms** are suggested by boxes, cylinders by food cans, and spheres by balls.

Graphs, coordinates, and coordinate grids are also a part of geometry. The branch of geometry dealing with figures on a coordinate grid is called *coordinate geometry,* or *analytic geometry.* Analytic geometry combines algebra with geometry.

A *transformation* is an operation that changes an object. Reflections (flips), translations (slides), and rotations (turns) are familiar operations in transformation geometry.

A world globe is 3-dimensional and shows Earth correctly. Flat maps are 2-dimensional; all flat views of earth have some distortions.

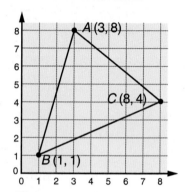

Triangles are 2-dimensional objects with three vertices (points), three sides (line segments), and three angles. The location of this triangle is given by the coordinates of its points on the coordinate grid.

The image of the butterfly's left wing is a reflection of the image of the butterfly's right wing.

Geometry

Geometry originated in ancient Egypt and Mesopotamia as a practical tool for surveying and constructing buildings. (The word *geometry* comes from the Greek words *ge*, meaning *Earth*, and *metron*, meaning *to measure*.) Around 300 BCE, the Greek mathematician Euclid gathered the geometric knowledge of his time into a book known as the *Elements*. Euclid's *Elements* is one of the great achievements of human thought. It begins with ten unproven statements, called postulates and common notions. In modern wording, one of the postulates reads, "Through a given point not on a given line, there is exactly one line parallel to the given line."

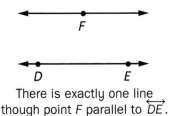
There is exactly one line though point F parallel to $\overleftrightarrow{DE}$.

Euclid used logic to deduce several hundred propositions (theorems) from the postulates and common notions—for example, "The sum of any two sides of a triangle is greater than the remaining side."

Mathematicians began to develop other forms of geometry in the seventeenth century, beginning with Rene Descartes' analytic geometry (1637). Descartes invented the **coordinate** system, in which points on a plane are identified by numbers. The numbers indicate how far the point is from each of two axes. This system gave mathematicians a way to connect algebra and geometry.

The problem of perspective in drawings and paintings led to *projective geometry.*

In the nineteenth century, mathematicians explored the results of changing Euclid's postulate about parallel lines quoted above. In non-Euclidean geometry, there are either no lines or at least two lines parallel to a given line through a given point.

Topology, a modern branch of geometry, deals with properties of geometric objects that do not change when their shapes are changed.

Even though the rails in this track are parallel and always the same distance apart, they appear to meet at a common point on the horizon. This is one of the laws of perspective.

Geometry

Angles

An **angle** is formed by two rays, two line segments, or a ray and line segment that share the same endpoint. The symbol for an angle is ∠.

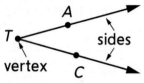
angle formed by 2 rays
names: ∠T, or ∠ATC, or ∠CTA

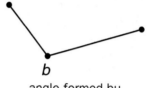
angle formed by 2 segments
name: ∠b

angle formed by 1 ray and 1 segment

The endpoint where the rays or segments meet is called the **vertex** of the angle. The rays or segments are called the sides of the angle.

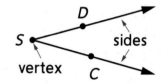

Angle Measures

The size of an angle is the amount of turning about the vertex from one side of the angle to the other. Angles are measured in degrees. A **degree** is a unit of measure for the size of an angle. A full turn about a vertex makes an angle that measures 360 degrees. You can use a **protractor** to measure angles. There are two types: full-circle protractors and half-circle protractors.

The **degree symbol** ° is often used in place of the word *degrees*. The measure of ∠S above is 30 degrees, or 30°.

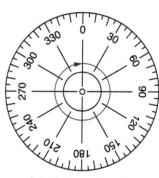
full-circle protractor

Classifying Angles

Angles may be classified according to size.

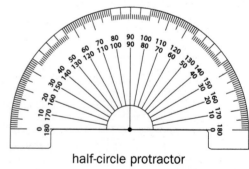
half-circle protractor

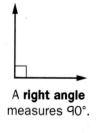

A **right angle** measures 90°.

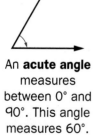

An **acute angle** measures between 0° and 90°. This angle measures 60°.

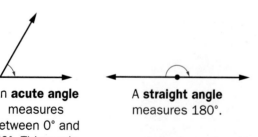

A **straight angle** measures 180°.

An **obtuse angle** measures between 90° and 180°. This angle measures 120°.

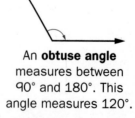

A **reflex angle** measures between 180° and 360°. This angle measures 300°.

SRB

Geometry

Line Segments, Rays, Lines, and Angles

Figure	Symbol	Name and Description
• A	A	**point:** a location in space
B, C (endpoints)	$\overline{BC}$ or $\overline{CB}$	**line segment:** a straight path between two points called its endpoints
N, M (endpoint)	$\overrightarrow{MN}$	**ray:** a straight path that goes on forever in one direction from an endpoint
S, T	$\overleftrightarrow{ST}$ or $\overleftrightarrow{TS}$	**line:** a straight path that goes on forever in both directions
vertex T, S, P	∠T or ∠STP or ∠PTS	**angle:** two rays or line segments with a common endpoint, called the vertex
A, B, C, D	$\overleftrightarrow{AB} \parallel \overleftrightarrow{CD}$	**parallel lines:** lines in the same plane that never cross or meet and are everywhere the same distance apart
	$\overline{AB} \parallel \overline{CD}$	**parallel line segments:** segments that are parts of lines that are parallel
R, E, D, S	none	**intersecting lines:** lines that cross or meet
	none	**intersecting line segments:** segments that cross or meet
B, F, E, C	$\overleftrightarrow{BC} \perp \overleftrightarrow{EF}$	**perpendicular lines:** lines that intersect at right angles
	$\overline{BC} \perp \overline{EF}$	**perpendicular line segments:** segments that intersect at right angles

The painted dividing lines at this intersection illustrate parallel and perpendicular line segments. Parallel lines are often used to separate a street into traffic lanes.

Did You Know?

Lane markings for roads were invented by Dr. June McCarroll in 1924. By 1939, lane markings were officially standardized throughout the United States.

two hundred thirty-seven

Geometry

Polygons

A **polygon** is a flat, **2-dimensional** figure made up of line segments called sides. A polygon can have any number of sides, as long as it has at least three. The **interior** (inside) of a polygon is not a part of the polygon.

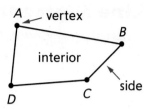

- The sides of a polygon are connected end to end and make one closed path.
- The sides of a polygon do not cross.

Each endpoint where two sides meet is called a **vertex.** The plural of vertex is *vertices*.

Figures That Are Polygons

4 sides, 4 vertices   3 sides, 3 vertices   7 sides, 7 vertices

Figures That Are NOT Polygons

All sides of a polygon must be line segments. Curved lines are not line segments.   The sides of a polygon must form a closed path.   A polygon must have at least 3 sides.   The sides of a polygon must not cross.

The prefix in a polygon's name tells the number of sides it has.

Prefixes

tri- 3   quad- 4   penta- 5   hexa- 6   hepta- 7

triangle   quadrilateral   pentagon   hexagon   heptagon
3 sides   4 sides   5 sides   6 sides   7 sides

octa- 8   nona- 9   deca- 10   dodeca- 12

octagon   nonagon   decagon   dodecagon
8 sides   9 sides   10 sides   12 sides

Geometry

Convex Polygons

A **convex** polygon is a polygon in which all the sides are pushed outward. The polygons below are all convex.

triangle

quadrilateral

pentagon

hexagon

octagon

Concave (Nonconvex) Polygons

A **concave,** or nonconvex, polygon is a polygon in which at least two sides are pushed in. The polygons below are all concave.

quadrilateral

pentagon

hexagon

octagon

Side or Angle Markings

When the markings on the sides of a polygon are the same, it means the sides are the same length. When the markings on angles of a polygon are the same, it means that the angles have the same measure.

In quadrilateral EFGH, side $\overline{EF}$ has the same length as side $\overline{GH}$, and side $\overline{HE}$ has the same length as side $\overline{FG}$. ∠E has the same angle measure as ∠G, and ∠F has the same angle measure as ∠H.

Regular Polygons

A polygon is a **regular polygon** if the sides all have the same length and the angles inside the figure are all the same measure. A regular polygon is always convex. The polygons below are all regular.

equilateral triangle

square

regular pentagon

regular hexagon

regular octagon

regular nonagon

Check Your Understanding

1. What is the name of a polygon that has
   a. 6 sides?   b. 4 sides?   c. 8 sides?
2. a. Draw a convex heptagon.   b. Draw a concave decagon.

Check your answers in the Answer Key.

Geometry

Triangles

Triangles have fewer sides and angles than any other polygon. The prefix *tri-* means *three*. All triangles have three vertices, three sides, and three angles.

For the triangle shown here:

- The vertices are the points B, C, and A.
- The sides are $\overline{BC}$, $\overline{BA}$, and $\overline{CA}$.
- The angles are ∠B, ∠C, and ∠A.

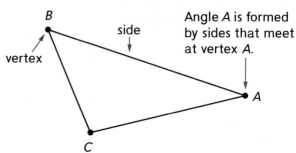

The symbol for triangle is △. Triangles have three-letter names. You name a triangle by listing each letter name for the vertices. The triangle above has six possible names: △BCA, △BAC, △CAB, △CBA, △ABC, and △ACB.

Triangles may be classified according to the length of their sides.

A **scalene triangle** is a triangle with sides that all have different lengths.

An **isosceles triangle** is a triangle with at least two sides that have the same length.

An **equilateral triangle** is a triangle with sides that all have the same length.

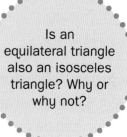
Is an equilateral triangle also an isosceles triangle? Why or why not?

Triangles may also be classified according to the size of their angles.

An **acute triangle** is a triangle whose angles are all acute.

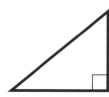
A **right triangle** is a triangle with one right angle.

An **obtuse triangle** is a triangle with one obtuse angle.

Check Your Understanding

1. Draw and label an equilateral triangle named △JKL. Write the five other possible names for this triangle.
2. Draw an isosceles triangle.
3. Draw a right scalene triangle.

Check your answers in the Answer Key.

Triangle Hierarchy

In mathematics, a **hierarchy** can be used to classify shapes.

This triangle hierarchy shows how triangles can be classified into **categories** and **subcategories**.

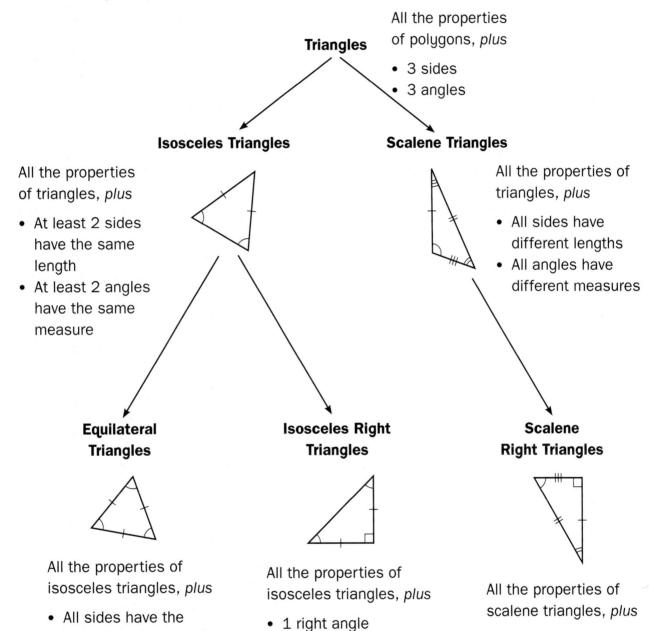

Geometry

Quadrilaterals

A **quadrilateral** is a polygon with four sides. The prefix *quad-* means *four*. All quadrilaterals have four vertices and four sides. Another name for quadrilateral is **quadrangle**.

For the quadrilateral shown here:

- The vertices are *R*, *S*, *T*, and *U*.
- The sides are $\overline{RS}$, $\overline{ST}$, $\overline{TU}$, and $\overline{UR}$.
- The angles are ∠*R*, ∠*S*, ∠*T*, and ∠*U*.

A quadrilateral is named by listing the letter names for the vertices in order. The quadrilateral above has eight possible names:

RSTU, RUTS, STUR, SRUT, TURS, TSRU, URST, UTSR

Some quadrilaterals have two pairs of parallel sides. These quadrilaterals are called **parallelograms**.

Reminder: Two sides are parallel if they never meet, no matter how far they are extended in either direction.

Figures That Are Parallelograms

Opposite sides are parallel in each figure.

Figures That Are NOT Parallelograms

No parallel sides   Only 1 pair of parallel sides   3 pairs of parallel sides not a quadrilateral

> **Did You Know?**
>
> In mathematics, there are times when concepts or ideas are defined differently. In *Everyday Mathematics*, a trapezoid is defined as a quadrilateral that has *at least* one pair of parallel sides. Using this definition, parallelograms are also trapezoids. Other sources define a trapezoid as a quadrilateral with *exactly* one pair of parallel sides. Using this definition, parallelograms are not trapezoids since parallelograms have more than one pair of parallel sides.

Check Your Understanding

1. Draw and label a quadrilateral named *EFGH* that has exactly one pair of parallel sides.
2. Is *EFGH* a parallelogram?

Check your answers in the Answer Key.

Geometry

Quadrilateral Hierarchy

This quadrilateral hierarchy shows one way to classify quadrilaterals:

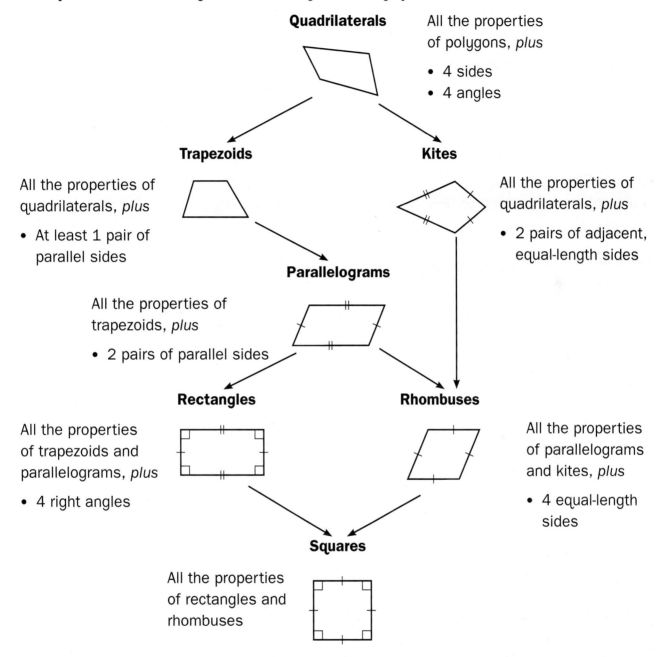

This quadrilateral hierarchy gives a lot of information about the properties of quadrilaterals and the relationships among quadrilaterals. For example:

- Kites are a subcategory of quadrilaterals, so kites have all of the attributes of quadrilaterals.

- Rhombuses are a subcategory of parallelograms, kites, trapezoids, and quadrilaterals, so rhombuses have all of the attributes of parallelograms, kites, trapezoids, and quadrilaterals. Therefore, a rhombus is also a parallelogram, kite, trapezoid, and quadrilateral.

two hundred forty-three 243

Geometry

Geometric Solids

Polygons and circles are flat, **2-dimensional** figures. The surfaces they enclose have area, but they do not have any thickness and do not take up any volume.

Three-dimensional shapes have length, width, *and* thickness. They take up volume. Cans, boxes, balls, books, and toys are all examples of 3-dimensional shapes.

A **geometric solid** is the surface or surfaces that surround a 3-dimensional shape. The surfaces of a geometric solid may be flat, curved, or both. Despite its name, a geometric solid is hollow; it does not include the points within its interior.

- A flat surface of a solid is called a **face**.
- A curved surface of a solid does not have a special name.

Examples

Describe the surfaces of each geometric solid.

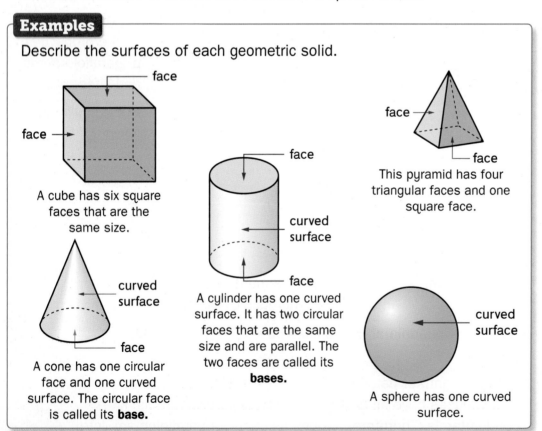

A cube has six square faces that are the same size.

A cone has one circular face and one curved surface. The circular face is called its **base**.

A cylinder has one curved surface. It has two circular faces that are the same size and are parallel. The two faces are called its **bases**.

This pyramid has four triangular faces and one square face.

A sphere has one curved surface.

SRB
244   two hundred forty-four

Geometry

The **edges** of a geometric solid are the line segments or curves where surfaces meet.

A corner of a geometric solid is called a **vertex** (plural *vertices*).

A vertex is usually a point at which edges meet. The vertex of a cone is an isolated corner completely separated from the edge of the cone.

A cone has one edge and one vertex. The vertex opposite the circular base is called the **apex.**

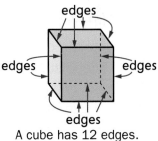
A cube has 12 edges.

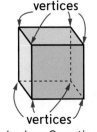
A cube has 8 vertices.

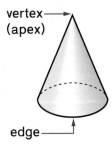

The pyramid shown below has eight edges and five vertices. The vertex opposite the rectangular base is called the **apex.**

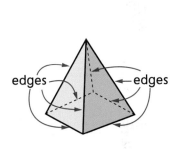
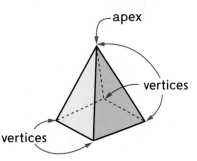

Did You Know?

Euler's Theorem is a formula that tells how the number of faces, edges, and vertices in a polyhedron are related. Let F, E, and V denote the number of faces, edges, and vertices of a polyhedron. Then $F + V - E = 2$. (Polyhedrons are defined on page 246.)

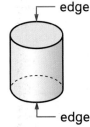

A cylinder has two edges. It has no vertices.

A sphere has no edges and no vertices.

Check Your Understanding

1. a. How are cylinders and cones alike?   **b.** How do they differ?

2. a. How are pyramids and cones alike?   **b.** How do they differ?

Check your answers in the Answer Key.

two hundred forty-five

Geometry

Polyhedrons

A **polyhedron** is a geometric solid whose surfaces are all formed by polygons. These surfaces are the faces of the polyhedron. A polyhedron does not have any curved surfaces.

Two important groups of polyhedrons are shown below. These are **pyramids** and **prisms.**

Pyramids

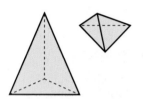
triangular pyramids

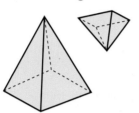
rectangular pyramids

pentagonal pyramid

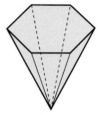
hexagonal pyramid

Prisms

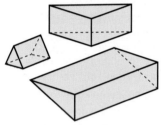
triangular prisms

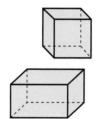
rectangular prisms

hexagonal prism

Many polyhedrons are not pyramids or prisms. Some examples are shown below.

Polyhedrons That Are NOT Pyramids or Prisms

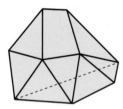
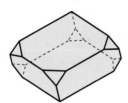
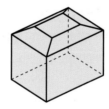

> **Check Your Understanding**
>
> **1. a.** How many faces does a rectangular pyramid have?
> **b.** How many faces have a rectangular shape?
>
> **2. a.** How many faces does a triangular prism have?
> **b.** How many faces have a triangular shape?
>
> Check your answers in the Answer Key.

Geometry

Pyramids

All of the geometric solids below are **pyramids.**

   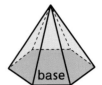

triangular pyramid   square pyramid   pentagonal pyramid   hexagonal pyramid

The shaded face of each pyramid is called the **base** of the pyramid.

- The polygon that forms the base can have any number of sides.
- The faces that are not the base are all triangles.
- The faces that are not the base all meet at a common vertex, called the **apex.**

The shape of the base is used to name the pyramid. If the base is a triangle, the pyramid is called a *triangular pyramid*. If the base is a square, the pyramid is called a *square pyramid*.

The great pyramids of Giza were built near Cairo, Egypt, around 2600 BCE. They have square bases, so they are square pyramids.

The largest of the Giza pyramids covers about 55,000 square meters at its base and is 137 meters high.

Example

The hexagonal pyramid shown here has seven faces—six triangular faces and one hexagonal face.

It has 12 edges. Six edges surround the hexagonal base. The other six edges meet at the apex (tip) of the pyramid.

It has seven vertices. Six of the vertices are on the hexagonal base. The remaining vertex is the apex of the pyramid.

The apex is the vertex opposite the base of the pyramid.

Check Your Understanding

1. a. How many faces does a triangular pyramid have?
   b. How many edges?
   c. How many vertices?
2. What is the name of a pyramid that has 10 edges?
3. a. How are pyramids and prisms alike?
   b. How are they different?

Check your answers in the Answer Key.

Geometry

Prisms

All of the geometric solids below are **prisms.**

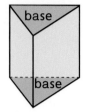 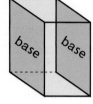 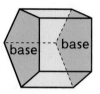

triangular prism   rectangular prism   pentagonal prism   hexagonal prism

The two shaded faces of each prism are called **bases.**

- The bases have the same size and shape.
- The bases are parallel. This means that the bases will never meet, no matter how far they are extended.
- The other faces connect the bases and are all shaped like parallelograms.

Note Notice that the edges connecting the bases of a prism are parallel to each other.

The shape of its bases is used to name a prism. If the bases are shaped like triangles, it is called a **triangular prism.** If the bases are shaped like rectangles, it is called a **rectangular prism.** Rectangular prisms have three possible pairs of bases.

Did You Know?

A triangular prism made of glass may be used to separate light into colors. The band of separated colors is called the *spectrum*.

The number of faces, edges, and vertices that a prism has depends on the shape of the base.

Example

The triangular prism shown here has five faces—three rectangular faces and two triangular bases. It has nine edges and six vertices.

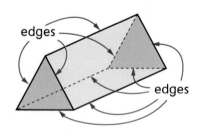 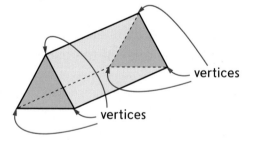

Check Your Understanding

1. a. How many faces does a pentagonal prism have?
   b. How many edges?
   c. How many vertices?

Check your answers in the Answer Key.

SRB
248   two hundred forty-eight

Geometry

Regular Polyhedrons

A polyhedron is **regular** if:

- It is convex.
- Each face is a regular polygon.
- The faces all have the same size and shape.

There are only five kinds of regular polyhedrons.

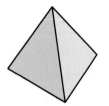 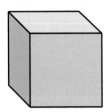   

| regular tetrahedron (pyramid) (4 faces) | cube (prism) (6 faces) | regular octahedron (8 faces) | regular dodecahedron (12 faces) | regular icosahedron (20 faces) |

The pictures below show a **net** of each regular polyhedron with its faces unfolded. There is more than one way to unfold each polyhedron.

regular tetrahedron
(4 equilateral triangles)

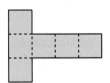

cube
(6 squares)

regular octahedron
(8 equilateral triangles)

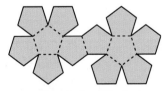

regular dodecahedron
(12 regular pentagons)

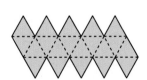

regular icosahedron
(20 equilateral triangles)

Check Your Understanding

1. **a.** How many edges does a regular rectangular polyhedron have?
   b. How many vertices?
2. **a.** How are regular tetrahedrons and regular octahedrons alike?
   b. How are they different?

Check your answers in the Answer Key.

Geometry

Length

Length is the measure of the distance between two points. Length is a 1-dimensional measurement.

Metric Units

In the metric system, units of length include meters (m), decimeters (dm), centimeters (cm), millimeters (mm), and kilometers (km). Length in the metric system is usually measured with a meterstick, the centimeter side of a ruler, or a metric tape measure.

Metric Units of Length
1 kilometer (km) = 1,000 meters (m)
1 meter = 10 decimeters (dm)
1 meter = 100 centimeters (cm)
1 meter = 1,000 millimeters (mm)

You can estimate lengths by using the lengths of common objects and distances that you know. These are called personal references. Some examples of personal references for metric units of length are given below.

The point of a thumbtack is about 1 millimeter thick.

Personal References for Metric Units of Length	
About 1 millimeter	About 1 centimeter
• Thickness of a thumbtack point • Thickness of a dime	• Width of a fingertip • Thickness of a crayon
About 1 meter	About 1 kilometer
• One big step (for an adult) • Width of a front door	• Length of 10 football fields (including the end zones) • 1,000 big steps (for an adult)

Note The personal references for a meter can also be used for a yard. One yard equals 36 inches, while one meter is about 39.37 inches. A meter is often called a "fat yard," or one yard plus one hand width.

U.S. Customary Units

In the U.S. customary system, units of length include inches (in.), feet (ft), yards (yd) and miles (mi). Length in the U.S. customary system is usually measured with a yardstick, the inch side of a ruler, or a tape measure.

U.S. Customary Units of Length
1 mile (mi) = 1,760 yards (yd)
1 mile = 5,280 feet (ft)
1 yard = 3 feet
1 foot = 12 inches (in.)

Personal References for U.S. Customary System Units of Length	
About 1 inch	About 1 foot
• Width (diameter) of a quarter • Width of a man's thumb	• Distance from elbow to wrist (for an adult) • Length of a piece of notebook paper
About 1 yard	About 1 mile
• One big step (for an adult) • Width of a front door	• Length of 15 football fields (including the end zones) • 2,000 average-size steps (for an adult)

The distance across a quarter is about 1 inch.

Geometry

Perimeter

The distance around a polygon is called its **perimeter**. To find the perimeter of any polygon, add the lengths of all its sides.

Example

Find the perimeter of polygon ABCDE.

2 cm + 2 cm + 1.5 cm + 2 cm + 2.5 cm = 10 cm

The perimeter is 10 centimeters.

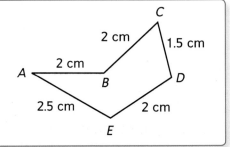

Perimeter Formulas

Rectangles	Squares	Regular Polygons
$p = 2 * (l + w)$	$p = 4 * s$	$p = n * s$
p is the perimeter	p is the perimeter	p is the perimeter
l is the length of the rectangle	s is the length of one side of the square	n is the number of sides
w is the width of the rectangle		s is the length of a side

Examples

Find the perimeter of each polygon.

4 cm / 3 cm

Rectangle

Use the formula
$p = 2 * (l + w)$

- length (l) = 4 cm
- width (w) = 3 cm
- perimeter $p = 2 * (4\ cm + 3\ cm)$
  $= 2 * 7\ cm = 14\ cm$

The perimeter is 14 cm.

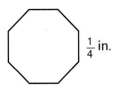

9 ft

Square

Use the formula
$p = 4 * s$

- length of side (s) = 9 ft
- perimeter $p = 4 * 9$ ft
  $= 36$ ft

The perimeter is 36 ft.

$\frac{1}{4}$ in.

Regular Octagon

Use the formula
$p = n * s$

- number of sides (n) = 8
- length of side (s) = $\frac{1}{4}$ in.
- perimeter $p = 8 * \frac{1}{4}$ in.
  $= \frac{8}{4}$ in. = 2 in.

The perimeter is 2 in.

Check Your Understanding

Solve. Include the unit in each answer.

1. Find the perimeter of a rectangle whose dimensions are 9 feet and 3 feet.
2. Find the perimeter of a regular hexagon whose sides are 15 yards long.

Check your answers in the Answer Key.

Geometry

Area

Area is a measure of the amount of surface inside a closed boundary. You can find the area by counting the number of squares of a certain size that cover the region inside the boundary. The squares must cover the entire region. They must not overlap, have any gaps, or cover any surface outside the boundary.

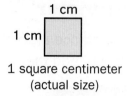
1 square centimeter (actual size)

Sometimes a region cannot be covered by an exact number of squares. In that case, first count the number of whole squares, then the fraction of squares that cover the region.

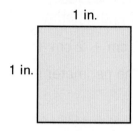
1 square inch (actual size)

Area is reported in square units. Units of area for small regions are square inches (in.²), square feet (ft²), square yards (yd²), square centimeters (cm²), and square meters (m²). For large areas, square miles (mi²) are used in the United States, while square kilometers (km²) are used in most other countries.

You may report area using any of the square units, but you should choose a square unit that makes sense for the region being measured.

Examples

The area of a field-hockey field is reported below in three different ways.

The area of the field is 6,000 square yards.	The area of the field is 54,000 square feet.	The area of the field is 7,776,000 square inches.
Area = 6,000 yd²	Area = 54,000 ft²	Area = 7,776,000 in²

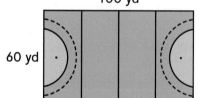

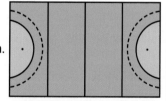

Although each of these measurements is correct, reporting the area in square inches doesn't give a good idea about the size of the field. It is hard to imagine 7,776,000 of anything.

Did You Know?

Tropical rain forests, where more than half the plant and animal species in the world live, once covered more than 6,000,000 square miles of Earth's surface. Because of destruction by people, fewer than 2,400,000 square miles remained in 2014.

SRB
252 two hundred fifty-two

Geometry

Area of a Rectangle

When you cover a rectangular shape with unit squares, the squares can be arranged in rows. Each row will contain the same number of squares and fractions of squares.

Examples

Find the area of the rectangle.

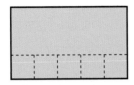
5 squares in a row

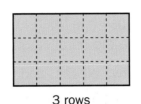
3 rows

3 rows with 5 squares in each row for a total of 15 squares

Area = 15 square units

Note

The distance along one side of a rectangle is its length and the adjacent side is its width. Either dimension of a rectangle can be considered its **length** or **base**. The adjacent side of the rectangle is the **width** or **height**.

To find the area of a rectangle, use any of these formulas:

Area Formulas for Rectangles
Area = (the number of squares in one row) * (the number of rows)

$A = b * h$	$A = l * w$
A is the area, b is the length of the base, and h is the height of the rectangle.	A is the area, l is the length of the rectangle, and w is the width of the rectangle.

Examples

Find the area of the rectangle.
Use the formula $A = b * h$.
- length of base (b) = 4 in.
- height (h) = 3 in.
- area (A) = 4 in. * 3 in.
  = 12 in.²

3 in.
4 in.

The area of the rectangle is 12 in.².

Find the area of the rectangle.
Use the formula $A = b * h$.
- length of base (b) = $3\frac{1}{3}$ ft
- height (h) = 2 ft
- area (A) = $3\frac{1}{3}$ ft * 2 ft

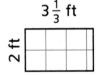
$3\frac{1}{3}$ ft
2 ft

The area of the rectangle is $6\frac{2}{3}$ ft².

Check Your Understanding

Find the area of the following figures. Include the unit in your answers.

1. 4 in.
   10 in.

2. 3 yd
   $8\frac{2}{3}$ yd

Check your answers in the Answer Key.

two hundred fifty-three

Geometry

Area of a Parallelogram

In a parallelogram, either pair of opposite sides can be chosen as its **bases**. The **height** of the parallelogram is the shortest distance between the two bases.

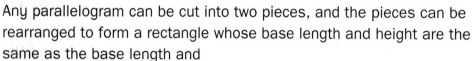

In the parallelograms to the right, the height is shown by a dashed line that is **perpendicular** (at a right angle) to the base. In the second parallelogram, the base has been extended, and the dashed height line falls outside the parallelogram.

Any parallelogram can be cut into two pieces, and the pieces can be rearranged to form a rectangle whose base length and height are the same as the base length and height of the parallelogram. The rectangle has the same area as the parallelogram. So you can find the area of the parallelogram in the same way you find the area of the rectangle—by multiplying the length of the base by the height.

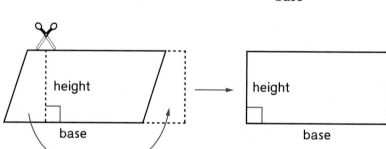

Formula for the Area of a Parallelogram
$A = b * h$

A is the area, b is the length of the base, and h is the height of the parallelogram.

Example

Find the area of the parallelogram.

Use the formula $A = b * h$.
- length of base (b) = 6 cm
- height (h) = 3.8 cm
- area (A) = 6 cm * 3.8 cm = 22.8 cm²

The area of the parallelogram is 22.8 cm².

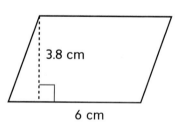

Geometry

Area of a Triangle

Any of the sides of a triangle can be chosen as a **base**. The **height** of the triangle (for the chosen base) is the shortest distance between the line containing that base and the **vertex** opposite that base.

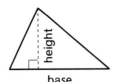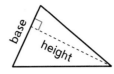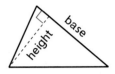

Every triangle has three bases and three heights.

In the triangles below, the height is shown by a dashed line that is **perpendicular** (at a right angle) to the line containing the base. On the right triangle, the height is one of the sides. In the other triangle, the base has been extended and the dashed line showing the height falls outside the triangle.

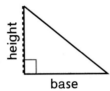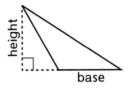

Any triangle can be combined with a second triangle of the same size and shape to form a parallelogram. Each triangle below shares a base with a parallelogram, and the height to that base is the same for both the triangle and the parallelogram. The area of each triangle is half of the area of the parallelogram. Therefore, the area of the triangle is half of the product of the base length and the height.

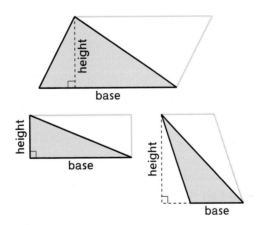

Formula for the Area of a Triangle
$A = \frac{1}{2} * b * h$

A is the area, b is the length of the base, and h is the height of the triangle.

two hundred fifty-five

Geometry

Example

The three pictures at the right show different measurements for the same triangle. What is the area of the triangle?

One way:

Use the formula $A = \frac{1}{2} * b * h$.
One base of the triangle is 10.5 cm.
$b = 10.5$ cm
The height for that base is 4.0 cm.
$h = 4.0$ cm

$A = \frac{1}{2} * (10.5 \text{ cm} * 4.0 \text{ cm}) = \frac{1}{2} * 42 \text{ cm}^2 = \frac{42}{2} \text{ cm}^2$
$= 21 \text{ cm}^2$

The area of the triangle is 21 cm².

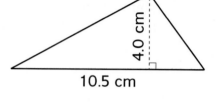

Another way:

Another base of the triangle is 5.0 cm.
$b = 5.0$ cm
The height for that base is 8.4 cm.
$h = 8.4$ cm

$A = \frac{1}{2} * (5.0 \text{ cm} * 8.4 \text{ cm}) = \frac{1}{2} * 42 \text{ cm}^2 = \frac{42}{2} \text{ cm}^2$
$= 21 \text{ cm}^2$

The area of the triangle is 21 cm².

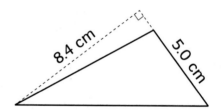

Another way:

Another base of the triangle is 8.5 cm.
$b = 8.5$ cm
The height for that base is 4.9 cm.
$h = 4.9$ cm

$A = \frac{1}{2} * (8.5 \text{ cm} * 4.9 \text{ cm}) = 0.5 * 41.65 \text{ cm}^2$
$= 20.825 \text{ cm}^2$

The area of the triangle rounded to the nearest square centimeter is 21 cm².

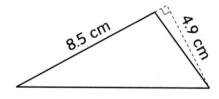

Geometry

Finding Area by Composing and Decomposing Shapes

You can find the area of complex shapes by decomposing the shape into non-overlapping polygons whose areas you know how to calculate. Then you can add the areas of those shapes to find the total area of the shape.

Example

Find the area of this shape.

Divide the shape into non-overlapping polygons.

Calculate side lengths you need.
- The sides of the green square are 5 in. That means the height of the pink right triangle is also 5 in.
- Since the length of the base of the blue rectangle is 15 in., and the side of the square is 5 in., the base of the pink triangle will be the leftover portion, or 10 in. (15 in. − 5 in. = 10 in.)

Find the area of each shape.

Use $A = l * w$ to find the area of the rectangle and the square.
- Area of blue rectangle = 15 in. * 8 in. = 120 in.²
- Area of green square = 5 in. * 5 in. = 25 in.²

Use $A = \frac{1}{2} * b * h$ to find the area of the triangle.
- Area of pink triangle: $\frac{1}{2}$ * (10 in. * 5 in.) = $\frac{1}{2}$ * 50 in. = 25 in.²

Add the areas of the three shapes to find the area of the whole shape.

Area of whole shape = Area of blue rectangle + Area of green square + Area of pink triangle
= 120 in.² + 25 in.² + 25 in.² = 170 in.²

The area of the whole shape is 170 square inches.

Check Your Understanding

Decompose the following shape into non-overlapping shapes to find its area.

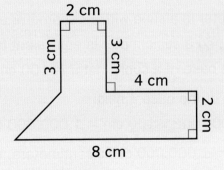

Check your answers in the Answer Key.

two hundred fifty-seven **257**

Geometry

Volume

The **volume** of a solid object such as a box or a ball is the measure of how much space it takes up.

The volume of 3-dimensional objects is measured in cubic units, such as cubic centimeters (cm^3), cubic inches ($in.^3$), cubic feet (ft^3), or cubic meters (m^3).

Each cubic unit is named for the length of its edges. For example, a cube with 1-centimeter edges is a cubic centimeter and a cube with 1-inch edges is a cubic inch. A cubic foot has 1-foot edges; a cubic yard has 1-yard edges; and a cubic meter has 1-meter edges.

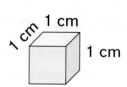
1 cubic centimeter (actual size)

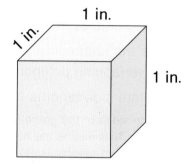
1 cubic inch (actual size)

Examples

How does a cubic yard compare to a cubic foot?

A cubic yard is a cube with 1-yard edges.

A cubic foot is a cube with 1-foot edges.

1 yard = 3 feet, so a cubic yard is equivalent to a 3 ft by 3 ft by 3 ft cube.

The area of the base of a cubic yard is 3 feet * 3 feet = 9 square feet.

The height of a cubic yard is 3 feet.

Volume = area of the base * height
= 9 square feet * 3 feet = 27 cubic feet

1 cubic yard = 27 cubic feet, so a cubic yard is 27 times as large as 1 cubic foot.

How does a cubic meter compare to a cubic centimeter?

A cubic meter is a cube with 1-meter edges.

A cubic centimeter is a cube with 1-centimeter edges.

1 m = 100 cm, so a cubic meter is equivalent to a 100 cm by 100 cm by 100 cm cube.

The area of the base of a cubic meter is 100 cm * 100 cm = 10,000 cm^2.

Volume = area of the base * height
= 10,000 cm^2 * 100 cm = 1,000,000 cm^3

1 cubic meter = 1,000,000 cubic centimeters, so a cubic meter is 1,000,000 times as large as a cubic centimeter.

Geometry

Volume of Geometric Solids

You can think of the volume of a geometric solid as the total number of whole unit cubes and fractions of unit cubes that are used to fill the interior of the solid without gaps or overlaps.

Prisms

In a prism, the cubes can be arranged in layers, with each layer containing the same number of cubes.

> **Example**
>
> Find the volume of the prism.
>
>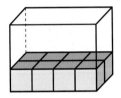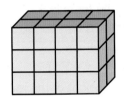
>
> 8 cubes in 1 layer   3 layers
>
> 3 layers with 8 cubes in each layer makes a total of 24 cubes.
>
> Volume = 24 cubic units

Did You Know?

The area of the Atlantic Ocean is about 41 million square miles with its dependent seas. The average depth of the ocean is about 2.1 miles. So the volume of the Atlantic Ocean is about 41 million square miles * 2.1 miles, or about 86 million cubic miles.

The **height** of a prism is the shortest distance between its **bases**. The volume of a prism is the product of the area of its base times its height.

Pyramids

The height of a pyramid is the shortest distance between its base and the vertex opposite its base.

If a prism and a pyramid have the same base and height, the volume of the pyramid is one-third the volume of the prism.

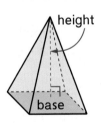

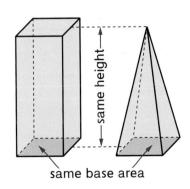

same height
same base area

two hundred fifty-nine **SRB 259**

Geometry

Volume of a Rectangular or Triangular Prism

Volume of a Prism	Area of a Rectangle	Area of a Triangle
$V = B * h$	$A = b * h$	$A = \frac{1}{2} * b * h$
V is the volume, B is the area of the base, h is the height of the prism.	A is the area, b is the length of the base, h is the height of the rectangle.	A is the area, b is the length of the base, h is the height of the triangle.

Example

Find the volume of the rectangular prism.

Step 1 Find the area of the base (B). Use the formula $A = b * h$.
- length of the rectangular base (b) = 8 cm
- height of the rectangular base (h) = 5 cm
- area of the base (B) = 8 cm * 5 cm = 40 cm²

Step 2 Multiply the area of the base by the height of the rectangular prism. Use the formula $V = B * h$.
- area of the base (B) = 40 cm²
- height of prism (h) = 6 cm
- volume (V) = 40 cm² * 6 cm = 240 cm³

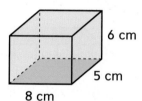

The volume of the rectangular prism is 240 cm³.

Example

Find the volume of the triangular prism.

Step 1 Find the area of the base (B). Use the formula $A = \frac{1}{2} * b * h$.
- length of the triangular base (b) = 5 in.
- height of the triangular base (h) = 4 in.
- area of the base (B) = $\frac{1}{2}$ * (5 in. * 4 in.) = 10 in.²

Step 2 Multiply the area of the base by the height of the triangular prism. Use the formula $V = B * h$.
- area of the base (B) = 10 in.²
- height of prism (h) = 6 in.
- volume (V) = 10 in.² * 6 in. = 60 in.³

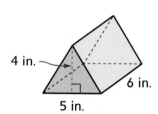

The volume of the triangular prism is 60 in.³

Geometry

Volume of a Rectangular or Triangular Pyramid

Volume of a Pyramid	Area of a Rectangle	Area of a Triangle
$V = \frac{1}{3} * (B * h)$	$A = b * h$	$A = \frac{1}{2} * b * h$
V is the volume, B is the area of the base, h is the height of the pyramid.	A is the area, b is the length of the base, h is the height of the rectangle.	A is the area, b is the length of the base, h is the height of the triangle.

Example

Find the volume of the rectangular pyramid.

Step 1 Find the area of the base (B). Use the formula $A = b * h$.
- length of the rectangular base (b) = 4 cm
- height of the rectangular base (h) = 2.5 cm
- area of the base (B) = 4 cm * 2.5 cm = 10 cm²

Step 2 Find $\frac{1}{3}$ of the product of the area of the base multiplied by the height of the pyramid. Use the formula $V = \frac{1}{3} * (B * h)$.
- area of the base (B) = 10 cm²
- height of pyramid (h) = 9 cm
- volume (V) = $\frac{1}{3} * (10$ cm² $* 9$ cm$) = 30$ cm³

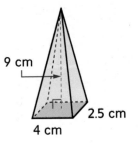

The volume of the rectangular pyramid is 30 cm³.

Example

Find the volume of the triangular pyramid.

Step 1 Find the area of the base (B). Use the formula $A = \frac{1}{2} * b * h$.
- length of the triangular base (b) = 10 in.
- height of the triangular base (h) = 6 in.
- area of the base (B) = $\frac{1}{2} * (10$ in. $* 6$ in.$) = 30$ in.²

Step 2 Find $\frac{1}{3}$ of the product of the area of the base multiplied by the height of the pyramid. Use the formula $V = \frac{1}{3} * (B * h)$.
- area of the base (B) = 30 in.²
- height of pyramid (h) = $4\frac{1}{2}$ in.
- volume (V) = $\frac{1}{3} * (30$ in.² $* 4\frac{1}{2}$ in.$) = 45$ in.³

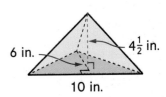

The volume of the triangular pyramid is 45 in.³.

Geometry

Volume of a Rectangular Prism with Fractional Edge Lengths

A cubic unit is a cube with edges that are 1 unit long. For example, one cubic inch is a cube with edges that are each 1 inch long. You could fill a container with cubic inches. You could also fill a container with a smaller unit such as cubes with $\frac{1}{2}$-inch edges.

Example

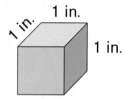

1 cubic inch

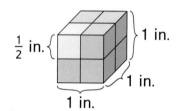

The yellow cube is a $\frac{1}{2}$ inch by $\frac{1}{2}$ inch by $\frac{1}{2}$ inch cube. Since 8 of these smaller cubes make 1 cubic inch, the area of each smaller cube is $\frac{1}{8}$ cubic inch.

Example

Alyssa is filling a box with cubes with $\frac{1}{4}$-foot edge lengths. How many of these cubes will fit in the box?

4 cubes fit along the length of the box, 4 rows of cubes fit along the width of the box, and 4 cubes fit along the box's height.

One layer has 4 * 4 = 16 cubes.

There are 4 layers.

4 * 16 = 64 cubes

64 cubes with edge lengths of $\frac{1}{4}$ foot will fit in a 1 cubic foot box.

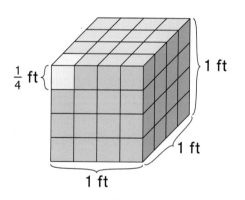

The yellow cube is $\frac{1}{64}$ cubic foot. The box is filled with 64 cubes with edge lengths of $\frac{1}{4}$ foot.

Geometry

Introduction to Surface Area

Surface area is the combined area of all surfaces of a geometric solid. The surface area of a prism or pyramid is the sum of the areas of its faces.

One way to think about and calculate the surface area of a solid is by making a net. A **net** is a 2-dimensional figure that can be folded to form a closed, 3-dimensional shape. When folded, the net forms a **geometric solid**—the surface or collection of surfaces that surround a 3-dimensional shape.

When unfolded, the net shows each of the faces of the solid as 2-dimensional polygons.

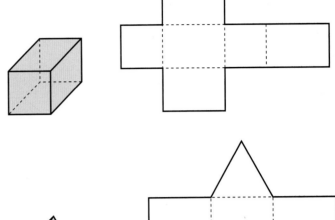

Each net can be folded to form the geometric solid next to it.

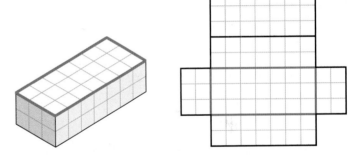

The rectangular face outlined in blue on the rectangular prism on the right is represented by the rectangle outlined on the net.

Check Your Understanding

Match each geometric solid with its net.

1   2   3

A   B   C

Check your answers in the Answer Key.

two hundred sixty-three **263**

Geometry

Calculating Surface Area

To calculate the surface area of a prism or pyramid, it is usually possible to add the areas of each of its faces. You can find this area using a diagram of the geometric solid.

Example

Find the surface area of the cube.

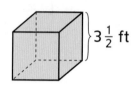

Think: What kind(s) of faces does a cube have?

A cube has 6 square faces. Calculate the area of each face.

Area of square face = s^2 = $(3\frac{1}{2} \text{ ft})^2$ = $3\frac{1}{2}$ ft ∗ $3\frac{1}{2}$ ft = $12\frac{1}{4}$ ft²

Surface area of the cube = $(6 * 12\frac{1}{4} \text{ ft}^2)$ = $73\frac{1}{2}$ ft²

The surface area of the cube is $73\frac{1}{2}$ ft².

Note Nets can be drawn in different ways and can still be accurate. As long as they can be folded to create a geometric solid, the net accurately represents the surface area of that solid.

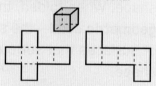

Both of the these nets can be folded to form a cube.

You can also draw a net to find the surface area of a geometric solid.

Example

Find the surface area of the square pyramid.

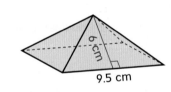

Draw a net of the solid.

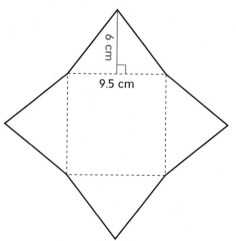

Decompose the net into the faces of the solid.
Calculate the area of each type of face.

Area of triangular face = $\frac{1}{2} * b * h$ = $\frac{1}{2} *$ (9.5 cm ∗ 6 cm)
$= \frac{1}{2} *$ (57 cm²) = 28.5 cm²

Area of square face = s^2 = (9.5 cm)² = 9.5 cm ∗ 9.5 cm
= 90.25 cm²

There are four triangular faces and one square face in the net.
Surface area of the pyramid = (4 ∗ 28.5 cm²) + 90.25 cm²
= 114 cm² + 90.25 cm²
= 204.25 cm².

The surface area of the pyramid is 204.25 cm².

Geometry

Coordinate Grid

A coordinate grid is used to name the location of points on a flat surface. It is made up of two number lines, called **axes,** that are perpendicular, or at right angles to each other. The horizontal number line is the *x*-axis and the vertical number line is the *y*-axis. The axes intersect at their 0 points to form four quadrants. The point where the axes meet is called the **origin.**

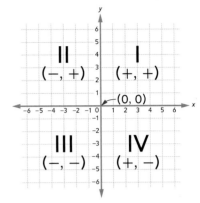

The ordered pair (0, 0) names the origin. The four quadrants are shown with Roman numerals and the signs for their quadrants: I (+, +), II (−, +), III (−, −), IV (+, −).

Every point on a coordinate grid can be named with an **ordered pair** giving the coordinates of the point. The first coordinate, the *x*-coordinate, describes the distance and direction (positive or negative) from 0, where the point will be located on the horizontal axis, or *x*-axis. The second coordinate, the *y*-coordinate, describes the distance and direction (positive or negative) from 0, where the point will be located on the vertical axis, or *y*-axis. On the grid at the right, the ordered pair (7, 5) names point *A*. The numbers 7 and 5 are the coordinates of point *A*. The first number, 7, is the *x*-coordinate and the second number, 5, is the *y*-coordinate.

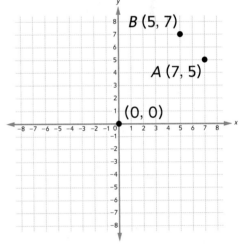

The order of the numbers in an ordered pair is important. The coordinates (5, 7) do not name the same point as the coordinates (7, 5). Point (5, 7) is located 5 units from the origin in the horizontal direction and 7 units from the origin in the vertical direction. Point (5, 7) appears to the left of point (7, 5) and higher on the grid.

Positive numbers are located to the right of the origin (0, 0) on the horizontal axis and above the origin on the vertical axis. Negative numbers are located to the left of the origin on the horizontal axis and below the origin on the vertical axis. If both coordinates are positive, the point will be in Quadrant I. If the *x*-coordinate is negative, and the *y*-coordinate is positive, the corresponding point will be in Quadrant II. If both values are negative, the point will be in Quadrant III. If the *x*-coordinate is positive, and the *y*-coordinate is negative, the point will be in Quadrant IV.

Note For more information on how the quadrants are numbered, see page 95.

two hundred sixty-five

Geometry

A point with 5 as its x-coordinate will be the same distance from 0 along the x-axis as a point with −5 as its x-coordinate, but the points will be on opposite sides of the origin. For example, point Q (5, 0) is the same distance from the y-axis as point P (−5, 0).

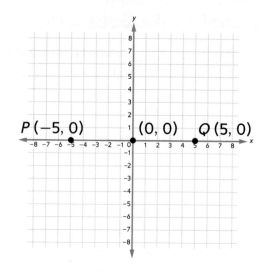

You can plot fractions and decimals on a coordinate grid.

Example

Locate (−2, 3), (−4, −1.75), (3$\frac{1}{2}$, 0), and (4, −2$\frac{1}{4}$).

For each ordered pair:

Locate the first coordinate on the horizontal axis and draw a vertical line. This is the point's distance from the vertical axis.

Locate the second coordinate on the vertical axis and draw a horizontal line. This is the point's distance from the horizontal axis.

The two lines intersect at the point named by the ordered pair.

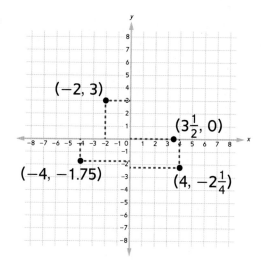

Check Your Understanding

Draw a coordinate grid on graph paper and plot the following points.

1. (3, 1)
2. (4, 2$\frac{1}{2}$)
3. (−2, 4)
4. (−3, −7)
5. (0, 5.25)

Check your answers in the Answer Key.

Plotting Polygons on a Coordinate Grid

You can plot polygons on a coordinate grid by connecting points on the grid with line segments. Each point will be a vertex of the polygon. Remember, in order to be a polygon, the shape must be closed.

> **Example**
>
> Plot and label the following coordinates. What shape do you get when you connect the points?
>
> A: (−2, 2)   B: (2, 2)
> C: (2, 5½)   D: (−2, 5½)
>
> Find the location of each of the coordinate points. Then connect the four points with line segments.
>
> The shape is a rectangle: ABCD.

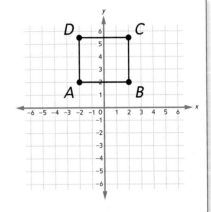

> **Example**
>
> Find the lengths of $\overline{TU}$ and $\overline{VU}$ in the quadrilateral on the right.
>
> **Same Quadrant**
>
> Both endpoints of $\overline{TU}$ are in the same quadrant.
>
> Find the coordinates of both endpoints of $\overline{TU}$:
>
> T: (2.5, −1)   U: (2.5, −4)
>
> The x-coordinate is the same for both points, but the y-coordinates differ.
>
> Subtract the absolute values of the y-coordinates:
> |−4| − |−1| = 4 − 1 = 3.
>
> $\overline{TU}$ is 3 units long.
>
> **Different Quadrants**
>
> The endpoints of $\overline{VU}$ are in different quadrants.
>
> Find the coordinates of both endpoints of $\overline{VU}$:
>
> V: (−3, −4)   U: (2.5, −4)
>
> The y-coordinate is the same for both points, but the x-coordinates differ.
>
> Add the absolute values of the x-coordinates: |−3| + |2.5| = 3 + 2.5 = 5.5.
>
> $\overline{VU}$ is 5.5 units long.

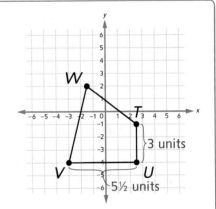

Note For more information on finding distances on a coordinate grid, see page 96.

two hundred sixty-seven

Geometry

Congruent Figures

Figures that have the same shape and size are **congruent figures.** Figures are congruent if one can replace the other as a perfect fit. You can sometimes use a coordinate grid to determine whether figures are congruent.

Example

Line segments are congruent if they have the same length.

$\overline{EF}$ and $\overline{CD}$ are both 3 units long. They have the same shape and the same length. These line segments are congruent.

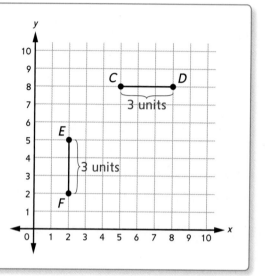

Example

Polygons are congruent if all of their sides have the same length and all of their angles have the same measure.

Quadrilaterals WXYZ and KLMN both have two sides that are 3 units long and two sides that are 4 units long. All of the angles of both quadrilaterals measure 90°. You can replace one with the other and they would be a perfect fit. These figures are congruent.

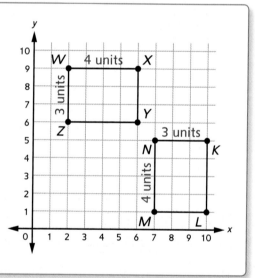

The matching sides of congruent polygons are called **corresponding sides.** Each pair of corresponding sides of congruent polygons is the same length. The polygons on the right are congruent. Slash marks identify pairs of corresponding sides. Sides with the same number of slash marks are corresponding sides. (The number of slashes has nothing to do with the length of the sides.)

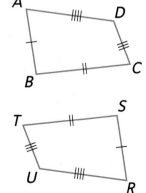

SRB
268 two hundred sixty-eight

Geometry

Similar Figures

Figures that have exactly the same shape are called **similar figures.** Usually, one figure is an enlargement or a reduction of the other. The **scale factor** tells the amount of enlargement or reduction. Congruent figures are similar because they have the same shape. The scale factor for congruent figures is 1:1 because they have the same size.

Example

The triangles CAT and DOG are similar. The larger triangle is an enlargement of the smaller triangle.

Each side and its enlargement form a pair of sides called **corresponding sides.** These sides are marked with the same number of slash marks.

The scale factor is 2:1. Each side in the larger triangle is twice the size of the corresponding side in the smaller triangle. The measure of the angles is the same for both triangles. For example, ∠T and ∠G have the same measure.

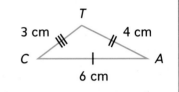

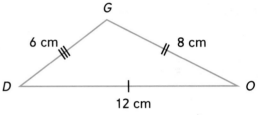

Example

Similar figures can be shown on a coordinate grid by applying the same rule to all of the x- and y-values of each coordinate in the original figure. For example, a scale factor of 1:3 means that you divide the x- and y-values of each coordinate in the original figure by 3. The side lengths of the new image will be one-third as large as the side lengths of the original figure.

Quadrilaterals ABCD and MNOP are similar. You can create the smaller quadrilateral (new figure) by applying a rule:

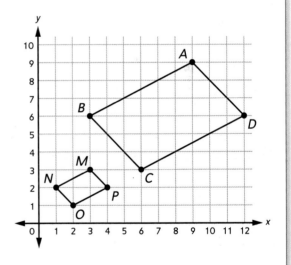

Quadrilateral ABCD (original figure)	Quadrilateral MNOP (new figure) Rule: Divide each coordinate in the original ordered pair by 3.
A (9, 9)	M (3, 3)
B (3, 6)	N (1, 2)
C (6, 3)	O (2, 1)
D (12, 6)	P (4, 2)

two hundred sixty-nine **SRB 269**

Geometry

Line Symmetry

A figure is *symmetric about a line* if the line divides the figure into two parts so that both parts look alike but are facing in opposite directions. In a symmetric figure, each point on one of the halves of the figure is the same distance from the *line of symmetry* as the corresponding point in the other half.

Example

The figure shown here is symmetric about the dashed line. The dashed line is its line of symmetry.

Points A and A' (read as "A prime") are corresponding points. The shortest distance from point A to the line of symmetry is equal to the shortest distance from point A' to the line of symmetry. The same is true of points B and B', C and C', and any other pair of corresponding points.

The line of symmetry is the perpendicular bisector of the line segments that connect corresponding points such as B and B'. It bisects each line segment and is perpendicular to each line segment.

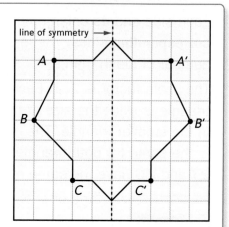

An easy way to check whether a figure has **line symmetry** is to fold it in half. If the two halves match exactly, the figure is symmetric about the fold.

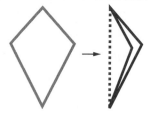

Some figures have more than one line of symmetry.

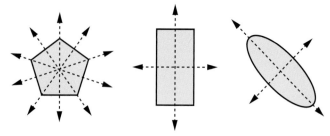

Some figures have no lines of symmetry.

Reflections

A reflection image of a figure appears as a "flipping" motion of the figure. The line that the figure is flipped over is called the **line of reflection.** The preimage and the image are on opposite sides of the line of reflection.

A reflection of a polygon can be created on a coordinate grid by reversing the signs of *either* the *x*-coordinates or the *y*-coordinates in the ordered pair for each vertex. Changing the signs of the *x*-coordinates will create a reflection along the vertical axis, and changing the *y*-coordinates will create a reflection along the horizontal axis. Each point's distance to the reflection axis will remain the same, but it will appear on the opposite side of that axis.

Note A reflection is a *transformation* or an operation on a figure that produces another figure called an *image*. The image has the same size and shape as the original figure, called the *preimage*.

Example

Create a reflection along the vertical axis of the preimage on the right.

Find the coordinates of each vertex in the preimage.

A: $(1\frac{1}{2}, 4)$ B: $(5, 2\frac{1}{2})$ C: $(3, -1)$ D: $(1, 1)$

Reverse the signs of *only* the *x*-coordinates in each ordered pair, and plot the new points. Connect the points with line segments to create the new image.

W: $(-1\frac{1}{2}, 4)$ X: $(-5, 2\frac{1}{2})$ Y: $(-3, -1)$ Z: $(-1, 1)$

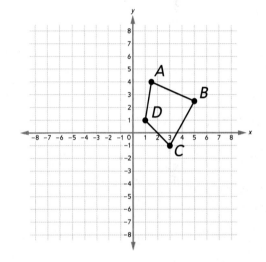

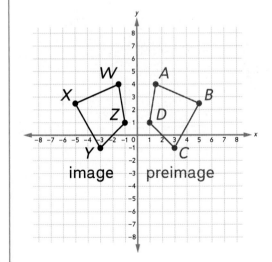

Geometry

When you create reflection on a coordinate grid along one of the axes:

- The preimage and the image have the same size and shape.
- The preimage and the image are reversed.
- The absolute values of the coordinates of each point in the preimage and the absolute values of the coordinates of the corresponding point in the image are the same.

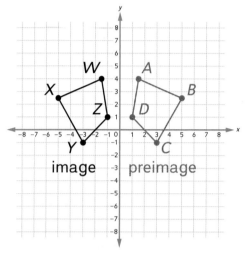

An approximate reflection is shown here. The line of reflection is the water's edge, along the bank.

Paper Folding

From cranes to airplanes, many people around the world fold paper to create art and have fun. In mathematics, paper folding has been used to solve problems and prove theorems.

A current popular paper-folding activity is folding dollar bills into unusual shapes such as this dollar-bill shirt.

Paper folding is an ancient art. It is believed that paper was first invented in China around 100 CE and the first paper folding may have occurred there. However, it was in Japan where paper folding evolved into the art form known as *origami*. Because paper objects tend to break down over time, the exact origin of origami is not known. The first clear reference to origami is found in a poem published in 1680. The poem describes origami butterflies like those on the right, which are used to represent the bride and groom in some traditional Japanese wedding ceremonies.

Traditional origami uses square pieces of paper that are folded into different shapes without cutting or pasting. One of the first books with written instructions for origami was published in 1797. It illustrates how to fold paper cranes. In Japanese legend, it is believed that a person who folds 1,000 cranes, or *senbazuru*, will be granted a wish such as long life or recovery from illness.

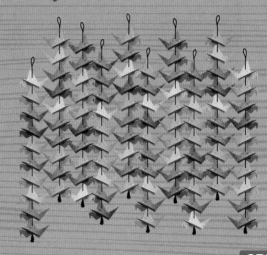

two hundred seventy-three

Modular Origami: Folding a Cube

Modular origami consists of putting together a set number of identical pieces called modules, or units, to form a complete model. An amazing variety of shapes can be made, such as this origami **polyhedron** made of red, yellow, and green pieces.

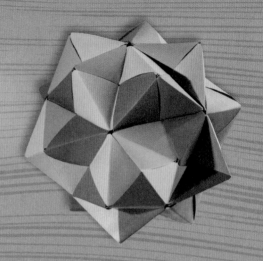

But modular origami doesn't have to be so elaborate. It can be as simple as this cube.

To fold a cube, you need six square pieces of paper.

Fold each square into a shape with flaps and pockets, so the pieces can be connected.

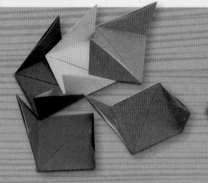

Connect the pieces of folded paper by inserting the triangular end of one piece into the square face of another piece. The connections between the pieces hold the cube together.

Möbius Strips

A mathematician confided
That a Möbius band is one-sided.
And you'll get quite a laugh,
If you cut one in half,
For it stays in one piece when divided.

(Anonymous)

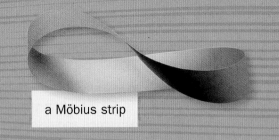
a Möbius strip

The limerick above was inspired by the work of August Ferdinand Möbius (1790–1868), a German mathematician and astronomer. Möbius examined the properties of one-sided surfaces. One such surface, easily made from a strip of paper, became known as a *Möbius strip*. Möbius strips are studied in the branch of mathematics known as topology. While no actual folding is done in making a Möbius strip, a twist of the paper strip is required.

To construct a Möbius strip, cut a strip of a paper about 2 inches wide and as long as possible. Turn over one end of the strip (give one end a half twist), and tape the two ends together to form a loop.

Möbius strips are recognized for their artistic properties. The artist M. C. Escher was intrigued not only by tessellations but also by Möbius strips. In the work above, an artist depicts ants endlessly crawling along a Möbius strip.

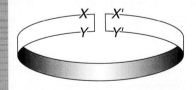 

Once you have assembled the strip, try poking scissors through the middle of the strip and cut it lengthwise all the way around. What happens? Is the limerick at the top of the page correct? Does it stay in one piece when divided?

Flexagons

While studying mathematics at Princeton University in 1939, Arthur Stone, a graduate student from England, found that American notebook pages did not fit into English notebooks. To make them fit, he cut a thin strip off of each page. He began folding these strips and discovered that he had made something special: a hexagon that could show three distinct faces when properly *flexed*, or squeezed. When Arthur shared his discovery with friends, they were so intrigued they formed a "flexagon committee" to investigate the properties of this mysterious face-shifting object.

Martin Gardner

Flexagons are folded-paper polygons that can change their faces when flexed. Since their introduction, many different kinds of flexagons have been developed. A flexagon's name tells both its shape and the number of faces that can be made by flexing it. For example, Arthur Stone's creation shown above is called a *trihexaflexagon* because it is a six-sided flexagon with three different faces (blue, green, red).

A recent addition to paper folding, flexagons were introduced to the public by Martin Gardner in his first "Mathematical Games" column for the *Scientific American* magazine in 1956.

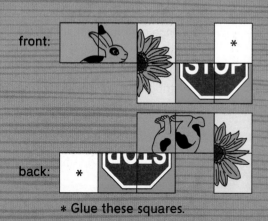

* Glue these squares.

A *tritetraflexagon* (tri-tetra-flexagon) is a four-sided flexagon with three different faces. The tritetraflexagon shown above was made using the template at the left. The three faces of this tritetraflexagon show a rabbit, a sunflower, and a stop sign. If you start with the face showing the rabbit, you can flex to reveal the second face with the sunflower. Then you can turn it over to see the stop sign on the third face.

Hexaflexagons

The most well-known flexagon is the hexaflexagon. The simplest of these is a trihexaflexagon like the one first constructed by Arthur Stone. A trihexaflexagon is made by folding a strip of 10 equilateral triangles into a hexagon. The strips at the right show a template for making a trihexaflexagon similar to the one on page 276.

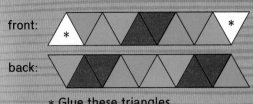

* Glue these triangles.

One of the most interesting hexaflexagons is the hexahexaflexagon. This flexagon is made by folding a strip of 19 equilateral triangles into a hexagon and gluing the triangles at either end together. The template to the right can be used to make a hexahexaflexagon.

* Glue these triangles.

Once this flexagon is assembled, the different faces are found by "flexing," or squeezing, the corners of the flexagon.

When flexed properly, these six faces can be made.

There are many types of hexaflexagons. While trihexaflexagons and hexahexaflexagons are the ones most commonly folded, hexaflexagons with more sides can be made. Bryant Tuckerman, one of the original members of the "flexagon committee," was able to make a workable hexaflexagon with 48 faces.

What connections can you find between the faces of a flexagon and its template?

Paper Airplanes

In 1967, *Scientific American* held the First International Paper Airplane Competition. There were 11,851 original designs for paper airplanes submitted from 28 different countries. (About 5,000 children submitted entries.)

Note More information about the First International Paper Airplane Competition, as well as templates and directions for making each of the winning designs, can be found in *The Great International Paper Airplane Book* by Jerry Mander, George Dippel, and Howard Gossage: Galahad Books, 1998.

Two of the categories in the competition were duration aloft (the time the paper airplane stays in the air) and distance flown. The winning time in the duration aloft category was 10.2 seconds, and the winning distance was 91 feet and 6 inches. Each winner received a trophy called *The Leonardo*, named after Leonardo da Vinci (1452–1519), whom *Scientific American* refers to as the "Patron Saint of Paper Airplanes."

The Leonardo

Over the years new records have been set for time aloft and distance flown. In 2012, Takuo Toda's *Sky King* paper airplane stayed aloft for 27.9 seconds, setting a new world record.

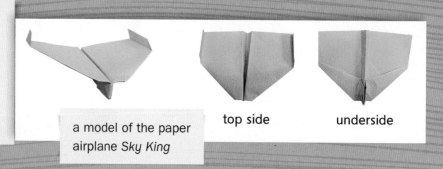

a model of the paper airplane *Sky King* | top side | underside

a model of the paper airplane *Suzanne*

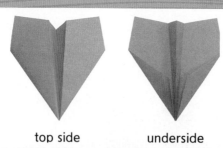

top side | underside

Suzanne, a paper airplane designed by John Collins, set a distance record of 226 feet and 10 inches, or 69.14 meters, in February 2012. John Collins didn't have the arm strength to set the record, so he asked former professional quarterback Joe Ayoob to make the winning throw.

Plan your own paper plane competition. You need a tape measure, a stopwatch, and an open space to fly the planes. Will you fly the planes inside or outside? How long do you think your plane will stay in the air? How far will it fly?

Statistics

Statistics

Statistical Questions

Statistics involves collecting, organizing, describing, representing, and analyzing information. We often collect and analyze information in order to answer a **statistical question.** When you ask a statistical question, you expect that you will need a variety of information to answer it. Consider the following questions:

- How many students attend your school?
- How many students typically attend a middle school in New York City?

The first question has just *one* answer, for *your* school. It is *not* a statistical question. The second question *is* a statistical question because in order to answer the question for *typical* attendance at middle schools in New York City, you need to consider the many different numbers of students that attend individual middle schools in New York City. The information gathered to answer the question is called **data.**

For the statistical question of how many students typically attend a middle school in New York City, the data are counts from each school. Since each school in New York City does not have the same number of students, the numbers vary. The data are **variable.** You can analyze the data you collect about numbers of students to make a reasonable estimate of how many students typically attend a school in the city.

Answering a statistical question involves collecting and organizing data. Data may be numerical, involving numbers such as measures or counts. Data may also be categorical, or fit into categories such as color or favorite food. The type of data often determines the best collection methods.

Check Your Understanding

Consider the following questions.
Label each one a *statistical question* or *not a statistical question*.

1. Do sixth graders generally run faster than fifth graders?
2. What was the highest score on the last test?
3. What was a typical score on the last test?
4. How many minutes does a typical sixth grader exercise each day?

Check your answers in the Answer Key.

Statistics

Collecting Data

There are several ways that data can be collected to answer statistical questions. Some data are collected through measurement, observation, or research.

Example

Highway engineers might ask the statistical question, "What is the typical speed that vehicles travel this stretch of road?" To answer this question, they may use video data to observe and analyze vehicle speeds and driving patterns.

Example

Birdwatchers might ask, "How many species of birds typically visit Chicago during December and January?" To answer this question, they observe and count the number of different bird species in Chicago during those months. The counts are combined to create a final data set.

From recent Chicago bird counts:

Year	Number of Species Seen
2003	78
2004	71
2005	71
2006	68
2007	65
2008	68
2009	73
2010	66
2011	75
2012	71

Surveys

Another method of collecting data is a **survey.** Surveys ask people to answer questions about themselves, what they think, or what they do. Survey data can be collected in several ways, including interviews, questionnaires, and group discussions.

Information from surveys is often used to help make decisions. Stores survey their customers to find out which products they should carry. Television stations survey viewers to learn which programs are popular. Politicians survey people to learn how they plan to vote in elections or what issues are important to them.

A face-to-face interview

Statistics

Samples

The **population** for a survey is the group of people or things that is being studied. Because the population may be very large, it may not be possible to collect data from every member of the population. Therefore, sometimes data are collected only from a sample group to provide information for the population. A **sample** is a part of the population that is chosen to represent the whole population.

Large samples usually give more dependable estimates than small ones. For example, if you want to know what percent of adults typically drive to work, a sample of 100 adults provides a better estimate than a sample of 10.

Example

Researchers use a survey to collect data about the population of people aged 8 to 18. There are over 40 million people between these ages in the United States, so it is not possible to collect data from everyone in that group. Instead, researchers collect data from a sample of young people. The table below shows results from a 2010 survey of about 2,000 youth.

The Way a Typical Young Person Divides His or Her Media Time	
Watching TV/movies	48%
Listening to music	22%
Using computers	14%
Playing video games	11%
Reading print media	4%

A **random sample** is a sample that gives all members of the population the same chance of being selected. Random samples give more dependable information than those that are not random.

Example

Suppose you want to know how a typical voter will vote in an election and whether most voters are likely to support Mr. Jenkins.

- If you use a sample of Mr. Jenkins's 100 close friends, the sample is *not* a random sample. People who do not know Mr. Jenkins have no chance of being selected.

- A sample of close friends will not fairly represent the entire population. It will not provide a dependable estimate of how the entire population will vote. If you were to use Mr. Jenkins's friends as your sample, you might find that 100% of them will vote for Mr. Jenkins. This probably wouldn't happen if you use a random sample of the population.

Organizing Data

Once the data have been collected, it helps to organize the information in order to make the data easier to understand. **Tally charts** and **dot plots** are two methods of organizing data. Dot plots are similar to **line plots** but use dots instead of Xs.

> **Example**
>
> Ms. Halko's class got the following scores on a 20-word vocabulary test. Make a dot plot and tally chart to show the data below.
>
> 20   15   18   17   20   12   15   17   19   18   20   16   16
> 17   14   15   19   18   18   15   9   20   19   18   15   18
>
>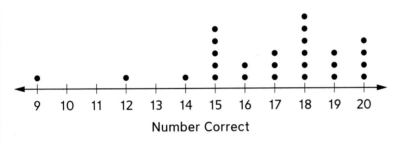
>
> Scores on a 20-Word Vocabulary Test
>
> | Number Correct | Number of Students |
> |---|---|
> | 9 | / |
> | 10 | |
> | 11 | |
> | 12 | / |
> | 13 | |
> | 14 | / |
> | 15 | //// |
> | 16 | // |
> | 17 | /// |
> | 18 | //// / |
> | 19 | /// |
> | 20 | //// |
>
> The dot plot and tally chart make it easier to describe the data. For example:
> - 4 students have 20 correct (a perfect score).
> - 18 correct is the score that came up most often.
> - 9, 12, and 14 correct are the scores that came up least often.
> - 0 to 8, 10, 11, and 13 correct are scores that did not occur at all.

> **Check Your Understanding**
>
> Below are the numbers of hits made by 18 players in a baseball game:
>
> 3   1   0   4   0   2   1   0   0   2   3   0   2   2   1   2   0   0
>
> 1. Organize the data by making a dot plot.
> 2. Which number of hits came up most often?
> 3. How many players had 2 hits in the game?
>
> Check your answers in the Answer Key.

two hundred eighty-three   **283**

Statistics

Statistical Landmarks: Measures of Center

The landmarks for a set of data are used to describe the data. A **measure of center** is a value near the center of a data set. The mean, median, and mode are landmarks that are often used as measures of center. A measure of center can be used to represent a typical value of that set.

Measures of Center

The **mode** is the value (or values) that occurs most often.

The **median** is the middle value. One way to find the median is to list the numbers in order. Cross out one number from each end of the list. Continue crossing out one number from each end of the list.

- If there are an odd number of data values, the median is the number that remains after all others have been crossed out.
- If there are an even number of data values, the median is the number halfway between the two middle values. If the two middle numbers are the same, then that number is the median.

The **mean** is another measure of center, often referred to as the *average* value. See pages 285–288 for more information about the mean.

> **Example**
>
> The dot plot shows students' scores on a 20-word vocabulary test. Find the mode and median of this data.
>
> **Scores on a 20-Word Vocabulary Test**
>
>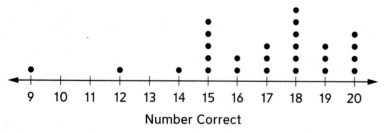
>
> Number Correct
>
> The mode is 18 because more students scored 18 than any other score.
>
> median = 17.5
>
> 9̶ 1̶2̶ 1̶4̶ 1̶5̶ 1̶5̶ 1̶5̶ 1̶5̶ 1̶5̶ 1̶6̶ 1̶6̶ 1̶7̶ 1̶7̶ | 17 18 | 1̶8̶ 1̶8̶ 1̶8̶ 1̶8̶ 1̶8̶ 1̶9̶ 1̶9̶ 1̶9̶ 2̶0̶ 2̶0̶ 2̶0̶ 2̶0̶
>
> middle scores
>
> There are two middle scores, 17 and 18. The median is 17.5, which is the number halfway between 17 and 18.
>
> The mode and the median are measures of center. They can be used to describe a typical score on this vocabulary test.

Statistics

Measures of Center: The Mean (or Average)

The **mean** of a set of numbers is often called the average. There are two ways to think of the mean: as a balance point for the values in the data set and as the value that a data point would have if all data points were redistributed evenly.

Finding the Mean as a Balance Point

The balance point, or mean, is the point on the number line where:

| the sum of the distances from each data point *below* the balance point to the balance point | = | the sum of the distances from each data point *above* the balance point to the balance point |

Note The distance between a balance point at 5 and a data point at 2 is their difference, $5 - 2 = 3$ units.

To find a balance point of a data set:
First, make a good guess about what a typical value might be. You can use symmetry or other reasoning to make a good guess. Often it makes sense that the balance point would be near the center of the data.

Then check whether your guess is the balance point.

- For every data point in the set, find its distance from the chosen balance point.
- Then compare the sums of the distances of the points above and below the balance point.
- If the sums are equal, you guessed correctly. If not, adjust your guess and check again.

Example

Seven of Jason's classmates walk to school. This dot plot shows the number of blocks each of those students walks to school. Jason wants to know the typical number of blocks walked by his classmates. He finds the balance point, or mean, of this data set.

Jason guesses that a typical student walks 5 blocks. Then he finds the distance from each data point to 5. He finds the sum of the distances *below* his chosen balance point and the sum of the distances *above* it.

Number of Blocks Walked to School

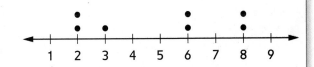

Number of Blocks Walked to School

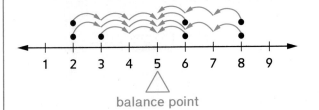
balance point

Sum of the distances from 5

Distances from the data points *below* 5 to 5:
$3 + 3 + 2 = 8$

Distances from the data points *above* 5 to 5:
$1 + 1 + 3 + 3 = 8$

5 is the balance point, or mean, of this set of data because these sums are equal.

two hundred eighty-five 285

Statistics

Example

If the mean, or average, high temperature in Nairobi during a week in December was 76 degrees Fahrenheit, what could have been the high temperatures on Friday and Saturday?

Day of the Week	High Temperature
Sunday	72
Monday	79
Tuesday	75
Wednesday	77
Thursday	78
Friday	
Saturday	

For 76 to be the balance point, the next two dots need to make the sums of the distances to the balance point on the two sides equal.

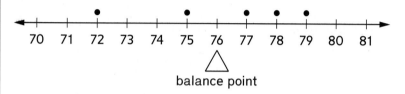

For example, if Friday's temperature was 80 degrees (a distance of 4 above the balance point), the sum of distances to the right would be $1 + 2 + 3 + 4 = 10$.

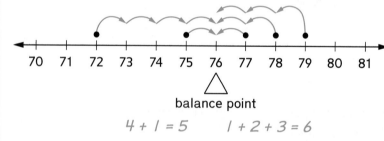

To make the total distance on the left of the balance point equal to the right, the distance of Saturday's temperature from 76 needs to be 5, because $4 + 1 + 5 = 10$.

Saturday's high temperature would be $76 - 5 = 71$ degrees.

Check Your Understanding

Use the information above to determine a Saturday temperature that would make the balance point 76 degrees.

1. If Friday's high temperature was 73 degrees, what was the high temperature on Saturday?
2. If Friday's high temperature was 70 degrees, what was the high temperature on Saturday?

Check your answers in the Answer Key.

Statistics

Finding the Mean as a Fair Share

Another way to find the mean is to think of it as a fair sharing situation. You can "redistribute" the data to identify the value representing an equal distribution of the data points. To find the mean by redistribution:

Step 1: Add the numbers.

Step 2: Divide the sum by the number of addends.

> **Example**
>
> Find the mean of this set of numbers: 2, 6, 7.
>
> The bars at the right represent these three numbers.
>
> **Step 1:** Add the numbers: 2 + 6 + 7 = 15
>
> **Step 2:** Divide the sum by the number of addends:
> 15 / 3 = 5
>
> The mean of 2, 6, and 7 is 5.
>
> Replace the original bar picture by one that has three bars of length 5.
>
> If you have a set of numbers, the mean can be used to divide the total into equal shares.
>
>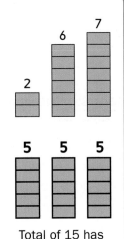
>
> Total of 15 has been divided into 3 equal shares.

> **Example**
>
> On a four-day trip, Kenny's family drove 200, 120, 160, and 260 miles. What is the mean number of miles they drove per day?
>
> **Step 1:** Add the numbers: 200 + 120 + 160 + 260 = 740
>
> **Step 2:** Divide by the number of addends: 740 / 4 = 185
>
> The mean is 185.
> They drove an average of 185 miles per day.

Statistics

You can organize data in a dot plot or tally chart to help find the mean of a large data set.

Example

Find the mean score on this vocabulary test.

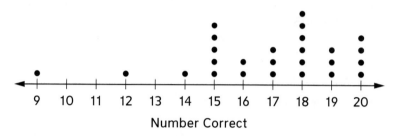

Number Correct	Number of Students
9	/
10	
11	
12	/
13	
14	/
15	ＨＨ
16	//
17	///
18	ＨＨ /
19	///
20	////

Scores on a 20-Word Vocabulary Test

The number of tallies or the number of dots for each value tells you how many times that value needs to be added into the total. For example, the 5 tallies next to 15 in the tally chart (or the 5 dots above 15 in the dot plot) tell you to add the value 15 five times. This number model shows the calculations for this data set:

9 + 12 + 14 + 15 + 15 + 15 + 15 + 15 + 16 + 16 + 17 + 17 + 17 + 18 + 18 + 18 + 18 + 18 + 18 + 19 + 19 + 19 + 20 + 20 + 20 + 20 = 438

With a large data set, you can multiply to simplify the number model. Multiply each score by the number of students with that score and add the products.

(1 * 9) + (1 * 12) + (1 * 14) + (5 * 15) + (2 * 16) + (3 * 17) + (6 * 18) + (3 * 19) + (4 * 20) = 438

It makes sense to use a calculator. After you have found the sum, divide it by the number of students. You can count tally marks or dots to find the number of students, 26.

Division with a calclulator shows this result: 438 / 26 ≈ 16.84615385

The mean, rounded to the nearest whole number, is 17 points.

Check Your Understanding

The dot plot shows Meghan's scores on math tests.

1. Find Meghan's mean score.
2. Describe the method you used to find the mean.

Meghan's Math Test Scores

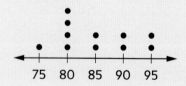

Check your answers in the Answer Key.

Choosing a Measure of Center

Sometimes you need to decide which landmark makes the most sense as a typical value for a data set. The mean, median, and mode are landmarks that can make sense as measures of center.

Example

Jalen gets $5.00 each week for an allowance. He collects data from classmates about their allowances:

Student	Allowance Amount
Taylor	$10.00
Caleb	$4.00
Noah	$5.00
Paige	$10.00
Lily	$8.00
Nate	$0.00
Maya	$25.00
Ryan	$4.00
Sydney	$2.00
Peyton	$4.00

Jalen figured out the mean, median, and mode of this data set.

Mean: $7.20

Median: $4.50

Mode: $4.00

- If Jalen wants to convince his parents to increase his allowance, which landmark might he choose to use in his argument?

- Which landmark best represents a typical value for a weekly allowance for these students?

The mean value of $7.20 for the allowances is higher than the mode or median. Maya's allowance of $25.00 per week is much greater than any of the other students' and has a greater effect on the balance point, so $7.20 is a less typical value.

The mode, $4.00, is very close to the median value of $4.50. Half of the students receive less than the median of $4.50, and half of the students receive more. Three out of the ten students receive exactly $4.00, the value of the mode, each week. Both the mode and the median can be considered typical values for this data set.

If Jalen uses the mean value of $7.20 to persuade his parents to increase his allowance because it is higher than his current weekly allowance, will his argument convince his parents?

Statistics

Example

Kyla and Nadia run the 100-meter dash for the school track team. They recorded their times for the last five races and are trying to decide who is a faster runner. The 100-meter dash is often won or lost by tenths of a second.

100-Meter Dash Race Times in Seconds

Runner	Race #1	Race #2	Race #3	Race #4	Race #5
Kyla	16.67	17.99	19.83	16.84	17.04
Nadia	17.41	18.22	17.35	17.26	17.58

Which landmark would you use to determine who is a faster runner, the mean or median?

Kyla's mean time: 17.67 seconds   Nadia's mean time: 17.56 seconds

If you use the mean as a typical time, you would say that Nadia is the faster runner. Her mean time is more than a tenth of a second less than, or faster than, Kyla's mean time. However, if you look at each race time, Kyla beat Nadia in four out of five races because her times were less than Nadia's in Races 1, 2, 4, and 5. In Race 3, Kyla's time was almost two seconds greater than any other race. This slower time made her mean time higher.

Kyla's median time: 17.04 seconds   Nadia's median time: 17.41 seconds

If you use the median as a typical time, it seems that Kyla is the faster runner. In fact, Kyla's median time is faster than every one of Nadia's times. In this situation, median is a better measure of center for this data.

Check Your Understanding

A sixth-grade class made boats from aluminum foil in their science lab and tested them to see how many pennies they could hold before sinking. So far, Maribel has tested her boat four times. These are her results:

1. Find the mean and median of this data set without Trial 5.
2. Maribel decided to test her boat one more time. In Trial 5, her boat only held 4 pennies before sinking. Find the mean and median of this data set if Trial 5 is 4 pennies.
3. Which landmark (mean or median) is affected the most by this new data point?
4. After testing five times, which landmark would be the best to use as a typical number of pennies held by Maribel's boat? Why?

Pennies Held Before Sinking

Trial Number	Number of Pennies
1	22
2	20
3	16
4	18
5	

Check your answers in the Answer Key.

Statistical Landmarks: Measures of Spread and Variability

Some landmarks describe how data values vary or how they are spread out across the distribution of the data points. These landmarks describe the **variability** of the data.

The **minimum** is the smallest value. The **maximum** is the largest value.
The **range** is the difference between the maximum and the minimum.
An **outlier** is a data value that is much smaller or much larger than the others in the data set. Outliers can have a great effect on landmarks of a data set, especially the mean and range.

Example

Find the minimum, maximum, and range for this dot plot. What is the outlier?

Scores on a 20-Word Vocabulary Test

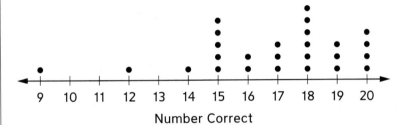

Minimum: 9
Maximum: 20
Range: 20 − 9 = 11

The score of 9 is an outlier. It is much smaller than most of the other test scores.

Example

The range tells you how spread out the values in a data set are. Look at the scores for a science quiz in the dot plot below. The same number of students took this science quiz and the vocabulary test in the example above. The median, mean, and mode are the same.

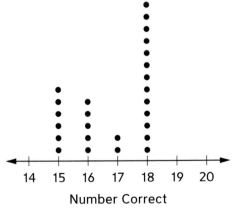

Scores on a 20-Question Science Quiz

Find the minimum, maximum, and range for this dot plot. Is there an outlier?

Minimum: 15
Maximum: 18
Range: 3

There is not an outlier in this data set.

Even though the measures of center are the same for both quizzes, the scores on the vocabulary test are more spread out. The scores on the science quiz vary by only 3 points while the scores on the vocabulary test vary by 11 points.

two hundred ninety-one

Statistics

Quartiles

The **lower quartile (Q1)** is the middle value of the data below the median.
The **upper quartile (Q3)** is the middle value of the data above the median.
The data value at the median is *not* included when finding the upper and lower quartiles.

> **Example**
>
> Find the median, lower quartile, and upper quartile of this set of numbers:
>
> 7, 3, 6, 4, 5, 8, 5, 3, 6, 2, 5
>
> First, put the numbers in order. Because there are an odd number of data points, the middle number is the median.
>
> 5 is the median in this data set:
>
>   *median*
> 2, 3, 3, 4, 5, **5**, 5, 6, 6, 7, 8
>
> You should exclude just that one data point of the median, 5, when finding the upper and lower quartiles:
>
>   Q1 *median* Q3
> 2, 3, **3**, 4, 5, **5**, 5, 6, **6**, 7, 8

Note The median is sometimes called the *middle quartile (Q2)*, so that for a data set you can find three quartiles: Q1, the median, and Q3. They are called quartiles because these three landmarks divide the data set into four useful parts, two parts below the median and two parts above the median.

> **Example**
>
> Find the median, lower quartile, and upper quartile of the scores on Ms. Halko's 20-word vocabulary test:
>
> 9, 12, 14, 15, 15, 15, 15, 15, 16, 16, 17, 17, 17, 18, 18, 18, 18, 18, 18, 19, 19, 19, 20, 20, 20, 20
>
> Because there are an even number of data points, the median will be between the two middle numbers.
>
>   *median = 17.5*
> 9, 12, 14, 15, 15, 15, 15, 15, 16, 16, 17, 17, 17, | 18, 18, 18, 18, 18, 18, 19, 19, 19, 20, 20, 20, 20
>
> In this case, the median is not one of the data points in the set, so you don't exclude any data points when finding the upper and lower quartiles.
>
>   Q1   *median = 17.5*   Q3
> 9, 12, 14, 15, 15, 15, |15, 15, 16, 16, 17, 17, 17,| 18, 18, 18, 18, 18, 18, |19, 19, 19, 20, 20, 20, 20

Interquartile Range

The length of the interval between the lower and upper quartiles is called the **interquartile range (IQR)**. You can think of the IQR as the distance between the upper and lower quartiles.

Note We also say that the middle half of the data is *in* the interquartile range.

> **Example**
> Find the interquartile range of the vocabulary test scores.
>
> Q1 median = 17.5 Q3
> 9, 12, 14, 15, 15, 15, |5, 15, 16, 16, 17, 17, 17,|18, 18, 18, 18, 18, 18, 1|9, 19, 19, 20, 20, 20, 20
>
> Interquartile range = Q3 − Q1 = 19 − 15 = 4 points
> Remember that for all the scores, the range is 20 − 9 = 11 points.
> For the middle half of the scores, the range is only 4 points.

Mean Absolute Deviation

The *range* of data in a set is a rough measure of the set's variability, or spread of the data, based only on the minimum and maximum values. The *interquartile range* is a more detailed measure based on the data in quartile groupings. An even more detailed statistic for variability uses *each* data value in the set. The **mean absolute deviation (m.a.d.)** in a data set is the average (mean) distance between individual data values and the mean of those values.

There are three steps to calculating the mean absolute deviation.

Step 1: Calculate the mean of the data.

Step 2: Find the distance of each data value from the mean found in Step 1.

Step 3: Calculate the mean of the distances found in Step 2.

See page 294 for an example showing how to find mean absolute deviation.

Statistics

Example

Calculate the mean absolute deviation of the calories in popular beef burgers as shown in the table to the right:

Company	Beef Burger	Total Calories
Brand A	Regular Burger	290
Brand B	Large Burger	470
Brand C	Large Burger	350
Brand D	Regular Burger	310
Brand E	Regular Burger	390
Brand F	Regular Burger	310
Brand G	Regular Burger	250
Brand H	Junior Burger	310
Brand I	Junior Burger	230
Brand J	Regular Burger	140

The table to the right shows how to calculate the mean absolute deviation of the beef burger data.

Step 1: The first column shows the total calories of each burger. The bottom row shows the mean of the data: 305 calories.

Step 2: The second column shows the deviations (differences) of the calorie data from the mean. Values below the mean have a negative deviation, and values above the mean have a positive deviation. The third column shows the absolute values of those deviations or differences.

Step 3: The mean of the absolute deviations is shown at the bottom of the third column. The m.a.d. of the beef burger data is 62 calories. This means that the typical difference between a data point and the mean is 62 calories.

Total Calories	Deviation	Absolute Deviation
290	−15	15
470	165	165
350	45	45
310	5	5
390	85	85
310	5	5
250	−55	55
310	5	5
230	−75	75
140	−165	165
mean = 305		**mean = 62**

Statistics

Data Representations: Bar Graphs

A **bar graph** is a drawing that uses bars to represent numbers. Bar graphs display information in a way that makes it easy to show comparisons. The title of a bar graph describes the information in the graph. Each bar has a label. Units are given to show how something was counted or measured. When possible, the graph gives the source of the information.

Example

This is a vertical bar graph.

- Each bar represents the number of pet cats in the country that is named below the bar.
- It is easy to compare cat populations by comparing the bars. China has about 4 or 5 times as many cats as Russia, Brazil, and France. The ratio of cats in the United States to cats in China is about 75 to 50, or about 3 to 2.

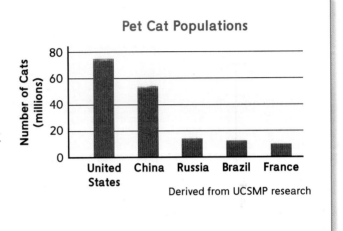

Sometimes two or more bar graphs are related to the same situation. Related bar graphs are often combined into a single graph. The combined graph saves space and makes it easier to compare the data.

Example

The bar graphs below show the number of sports teams that boys and girls joined during a one-year period.

The bars within each graph can be stacked on top of one another. The **stacked bar graph** to the right includes each of the stacked bars.

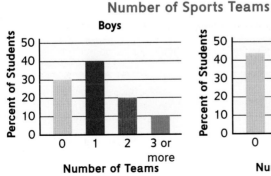
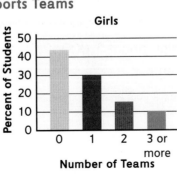
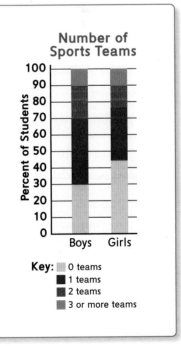

two hundred ninety-five **295**

Statistics

Data Representations: Histograms

A **histogram** is a bar graph of numerical data grouped into **bins,** or **intervals,** along one axis. While histograms don't display individual data points, they can be particularly useful for analyzing and describing the shape of a data distribution.

Histograms often display **continuous** data, such as measurements, that can have any numerical value within the interval.

Note When making statements about the data represented in histograms, you compare the area of the bars. The number of data points that fall within each interval determines the area of each bar. The histograms in Grade 6 have bins of equal width, so it is only necessary to compare the height of the bars to compare their areas.

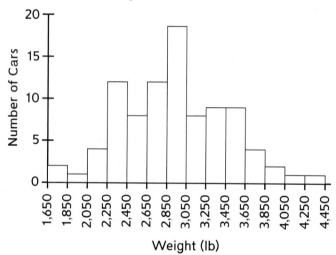

The horizontal axis of this histogram is a number line with bins of 200 pounds each. The first bin is the interval between 1,650 and 1,850 pounds. The graph shows that about 12 types of cars weigh between 2,250 and 2,450 pounds. This data is continuous because the actual weight of a car could be between whole pounds.

Check Your Understanding

Refer to the histogram above to answer the following questions:

1. About how many cars weigh between 2,850 and 3,050 pounds?
2. Are sedans more likely to weigh about 1,000, 2,000, 3,000, or 4,000 pounds? How do you know?

Check your answers in the Answer Key.

How to Make a Histogram

Example

For a health project, the students in Mr. Preston's class took each other's pulse rates. (A *pulse rate* is the number of heartbeats per minute.) These were the results:

92	72	90	86	102	78	88	75	72	82
90	94	70	94	78	75	90	102	65	94
70	94	85	88	105	86	78	75	86	108
94	75	88	86	99	78	86			

Follow these steps to make a histogram of this data.

Step 1: Organize the data. It may help to put the data in order or to find the minimum and maximum of the data set.

Minimum: 65 heartbeats per minute

Maximum: 108 heartbeats per minute

Step 2: Consider how to divide the data into intervals or bins. The data ranges from 65 to 108 heartbeats per minute, so a bin width of 10 will work: from 60 to 70, 70 to 80, and so on.

Step 3: Count the number of data values within each of the intervals. If a data point falls on the endpoint of a bin interval, count it in the bar to the right. For example, count 90 heartbeats with 90–100 NOT with 80–90.

Pulse Rates of Students

Heartbeats per Minute	Number of Students
60–70	/
70–80	⊦⊦⊦⊦ ⊦⊦⊦⊦ //
80–90	⊦⊦⊦⊦ ⊦⊦⊦⊦
90–100	⊦⊦⊦⊦ ⊦⊦⊦⊦
100–110	////

Statistics

Example (Continued)

Step 4: Draw a line segment for the horizontal axis of the histogram that is long enough to include the minimum and maximum. Divide the line segment into intervals that make sense for the data values. (See Steps 2 and 3 on page xxx.) Label the endpoints of each bin. Give the horizontal axis a label.

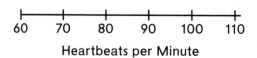

Step 5: Label the vertical axis and choose a convenient scale for it. For this histogram, the greatest frequency in any bin is 12, so a scale of 1 will work.

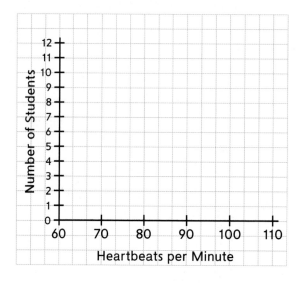

Step 6: Shade the bars to show the frequency of data points in each bin (the number of data points in each bin). Finish the histogram with a title.

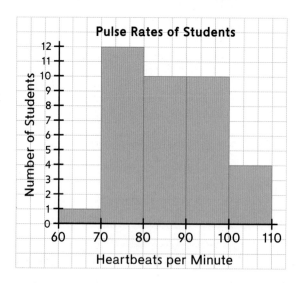

Statistics

Example

A histogram's in width can have a large impact on how the histogram looks and how it is analyzed. These histograms display the same data as the histogram on page 298.

This histogram is made with bins of just 5 beats per minute. In this histogram, you might notice that there is only one student with a heart rate between 80 and 85 beats per minute. This information isn't apparent in histograms with larger bin intervals.

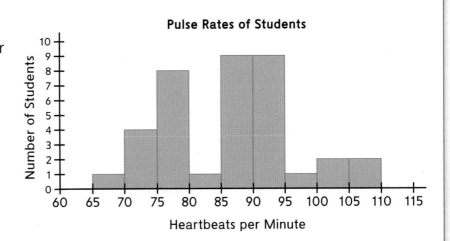

With a bin width of 25 beats per minute, the histogram looks like this.

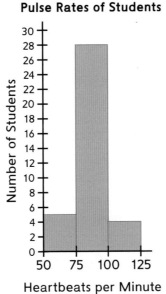

How would you interpret the data differently looking at the three histograms?

Check Your Understanding

The first manned spaceflight occurred during the early 1960s.

The ages of the first ten space travelers were:

27, 25, 43, 37, 32, 31, 39, 36, 28, 26.

Make a histogram of this data.

Check your answers in the Answer Key.

two hundred ninety-nine 299

Statistics

Data Representations: Box Plots

Box plots display the spread of a data set using five landmarks that measure center and variability, sometimes referred to as the five-number summary. Box plots are constructed using the minimum, the lower quartile (Q1), the median, the upper quartile (Q3), and the maximum.

Note Box plots are also commonly called box-and-whiskers plots.

Example

A sixth-grade class measured the height of the different varieties of tomato plants they were planting in their school garden.

The landmarks for this data are:

Minimum: 14 inches
Lower Quartile: 16 inches
Median: 20 inches
Upper Quartile: 25 inches
Maximum: 32 inches

The box plot below shows the spread of the tomato plant data around these landmarks.

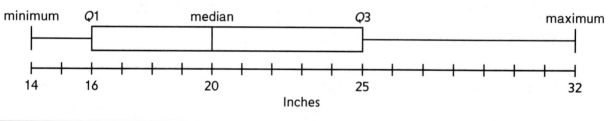

Here is information about the tomato plants in the garden according to the box plot:

- About one quarter of the plants are 14 to 16 inches tall.
- About one quarter are 16 to 20 inches tall.
- About one quarter are 20 to 25 inches tall.
- About one quarter are 25 to 32 inches tall.

Overall, the range of the plant heights is $32 - 14 = 18$ inches. The length of the interval between the lower and upper quartiles, the *interquartile range*, for the tomato plant data is $Q3 - Q1 = 25 - 16 = 9$ inches. So for the middle half of the tomato plant heights, the range is only 9 inches.

Note In a box plot, you can find the median, minimum, maximum, and range. Box plots can also be useful for analyzing data that are close together or spread out. However, you can't see individual data values and you don't know how many data points there are. Therefore, you can't use a box plot to find the mode or mean.

Statistics

How to Draw a Box Plot

The calories in some popular chicken snacks are given in the table below.

Company	Chicken Snack	Total Calories
Brand A	Chicken Nuggets	250
Brand A	Chicken Strips	630
Brand B	Chicken Tenders	250
Brand B	Chicken Fries	260
Brand C	Chicken Nuggets	230
Brand D	Chicken Tenders	630
Brand D	Popcorn Chicken	531
Brand E	Chicken Strips	630
Brand F	Chicken Strips	1,270
Brand G	Popcorn Chicken	550
Brand H	Chicken Strips	710
Brand I	Chicken Rings	340

Five summary numbers determine the length of each part in the box plot. These summary numbers are the minimum, maximum, median, and Q1 and Q3.

Follow these four steps to make a box plot of the calorie data.

Step 1: Order the data and find the summary numbers: minimum, maximum, median, lower quartile (Q1), and upper quartile (Q3).

230, 250, 250, 260, 340, 531, 550, 630, 630, 630, 710, 1,270

minimum   Q1 = 255   median = 540.5   Q3 = 630   maximum

Step 2: Draw a number line long enough to include the five summary numbers you found in Step 1. Choose a convenient scale. The range of data from the 230 minimum to the 1,270 maximum is 1,270 − 230 = 1,040, or about 1,000. So a scale of 100 will work.

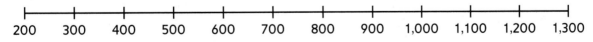

Step 3: Draw tick marks above the number line to mark the five landmarks.

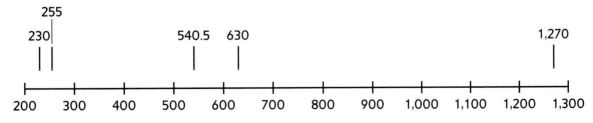

three hundred one   **301**

Statistics

Step 4: Draw the box and whiskers of the box plot. Connect the tick marks at Q1 and Q3 with horizontal lines to form the box. Show the median inside the box. Draw one whisker from the minimum tick mark to the Q1 box end. Draw the other whisker from the Q3 box end to the maximum tick mark. Finish the box plot with a title.

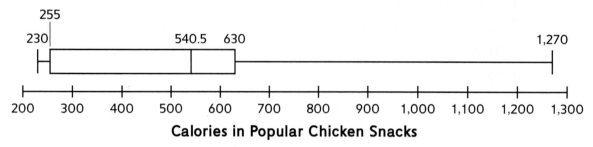

Calories in Popular Chicken Snacks

Comparing Box Plots

Box plots can be useful to compare different sets of data that measure the same thing with the same units. The calories in popular beef burgers are shown in the table below.

Company	Beef Burger	Total Calories
Brand A	Regular Burger	290
Brand B	Large Burger	470
Brand C	Large Burger	350
Brand D	Regular Burger	310
Brand E	Regular Burger	390
Brand F	Regular Burger	310
Brand G	Regular Burger	250
Brand H	Junior Burger	310
Brand I	Junior Burger	230
Brand J	Regular Burger	140

Below, a box plot for the beef burger calorie data is drawn above the chicken snack box plot. By using the same scale on the number line, you can compare the landmarks and spreads of the different types of foods. When the boxes or whiskers are longer, there is more variation in the data in those quartiles. When boxes or whiskers are shorter, the data is more concentrated in those quartiles.

Which type of food has more variation in the number of calories, chicken snacks or beef burgers?

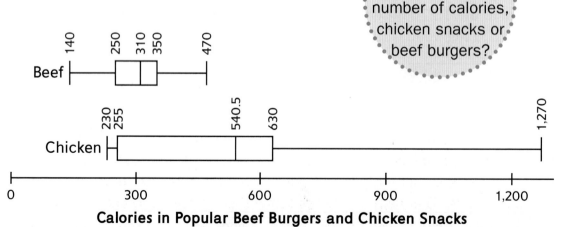

Calories in Popular Beef Burgers and Chicken Snacks

Reading and Analyzing Graphs

Data can be represented in a variety of ways. Different representations have advantages and disadvantages, depending on the data and what questions you are trying to answer. When you read a graph, there are many important features to notice and questions to consider:

- What type of representation is it? Bar graph? Histogram? Dot plot? Box plot?
- What information can you get from this type of representation?
- What is the scale on the graph? Is the scale consistent? Does it start at zero?
- Is the bin width reasonable? Are the bars equal widths?
- How spread out is the data? Are there outliers?
- Is it symmetrical? Where is the center of the data? Are there clusters, peaks, or gaps? What measure of center might best represent the data?

Think about the box plot showing calories in popular chicken snacks from page 302.

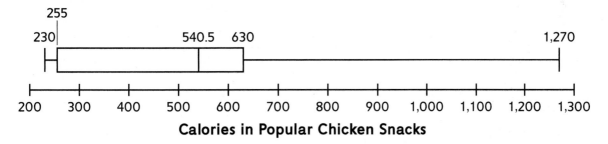

A box plot shows the minimum, maximum, median, and upper and lower quartiles. You can see the spread of the data and the relative concentration of data within each quartile. However, in a box plot you can't see any gaps in the data.

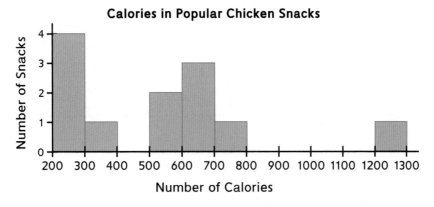

This histogram has a bin width of 100 calories. While you can't find the exact maximum, minimum, or median, you can see peaks, clusters, and gaps in the data. Here, you can see that the greatest number of chicken snacks have between 200 and 300 calories. There are no snacks containing between 400 and 500 or between 800 and 1,200 calories.

Statistics

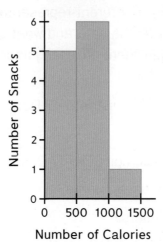

This histogram has a bin width of 500 calories. Can you see the same peaks, clusters, or gaps? What information gets lost when you make a histogram with wider bins?

When you read and analyze a graph, it is also important to consider how the data were collected. For example, here are some questions you might ask yourself about the chicken snack and beef burger data:

- How many different types of chicken snacks were used in collecting this data?
- How many types of snacks were left out?
- How were "popular" chicken snacks determined? Is this a random sample?
- Who collected the data for these graphs? Are they trying to prove something about chicken snacks or beef burgers?

The Shape of the Data Distribution

Look at the graphs on pages 305–307. Pay attention to the shape or **distribution** of the data. Do you notice any patterns? Do you see bumps, holes, outliers, or clusters of data? Sometimes it helps to picture the line or curve that would fit the data. This curve is drawn in red on the graphs that follow. Observing and describing the shape can help you ask and answer meaningful questions about the data.

A symmetrical curve that looks like a hill or a bump approximately fits the data in the graph below. It shows that the data are clustered around the mean and spread out evenly in both directions. A curve like this is called a bell curve and represents an approximately **normal distribution** of data. The single peak in the center of the curve shows the value of three measures of center—the mean, median, and mode—which are approximately equal.

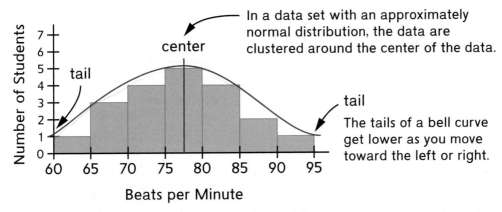

In a box plot of the same data, you can see how the data still appears symmetrical. The short boxes show how the middle half of the data is concentrated close to the median.

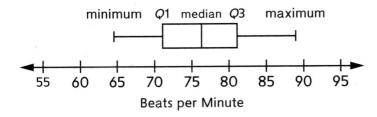

three hundred five 305

Statistics

Sometimes the top of a graph looks relatively flat. This data set has close to a **uniform distribution,** or **plateau distribution.**

Because of the way the data is spread across the range, measures of the center (mode, median, and mean) are not useful for describing the data set.

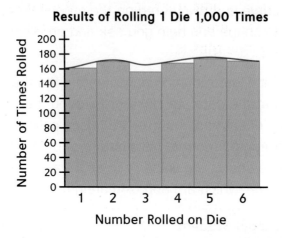

Remember that the **mode** of a data set is the value that occurs most often. Some graphs have two peaks or hills representing values that occur more often than the values around them. When there are two values that occur more often relative to the nearby data, even if one occurs more often than the other, the data set has two modes and the graph has a **bimodal distribution.**

Chicago has two professional basketball teams. The Chicago Bulls are all men, and their tallest player has a height of 6 feet, 11 inches or 83 inches. The Chicago Sky are all women, and their tallest player has a height of 6 feet, 6 inches or 78 inches. Why do you think this graph has two separate peaks?

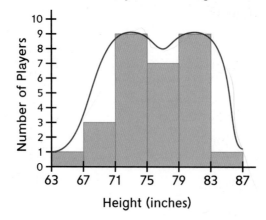

This data has a **skewed distribution.** The mode, median, and mean do not share the same value. The curve that fits this graph is not symmetrical. This distribution is right-skewed because the longer, flatter tail goes toward the right.

Skewed distribution often happens when there is a restriction on the lowest or highest values in the data set. The driving time from Chicago to St. Louis cannot be much shorter than 4 hours because there is a limit to how fast a car can go. What do you think could have made some trips last up to 8 hours?

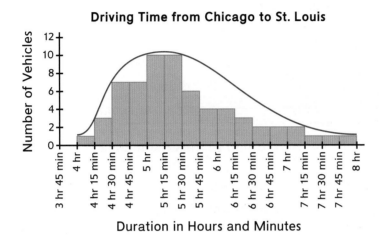

This box plot shows the same data as the histogram above. The long whisker on the right side shows the right-skewed distribution.

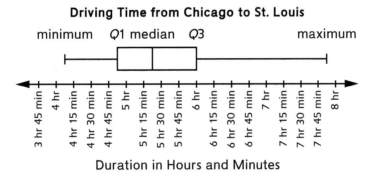

Check Your Understanding

Use the dot plot to answer the following questions.

1. What shape is this data?
2. Why isn't there a tail on the right side of the graph?
3. Find the mean, median, and mode of these test scores. Are all of these measures of center in the same location on the graph?

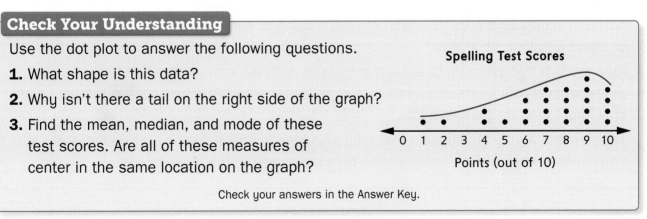

Check your answers in the Answer Key.

Statistics

Persuasive Graphs

There are many ways to make a graph, so it is important to understand how graphs can be used to persuade people, intentionally or unintentionally, into believing something about the data. One way that persuasive graphs are made is by adjusting the scale. As you make and interpret graphs, consider the starting point and scale of numerical data.

Example

All of the sixth graders at a school voted on where to go for a field trip. The table shows the results of their votes.

Kyle and Janise each made a bar graph of the results.

Why do you think their graphs are so different?
What do you think Kyle and Janise were trying to make someone believe?

Field Trip Location	Number of Votes
Zoo	18
Arboretum	22
History Museum	20
Space Museum	24

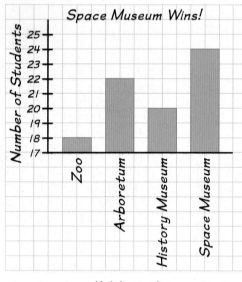

Kyle's graph

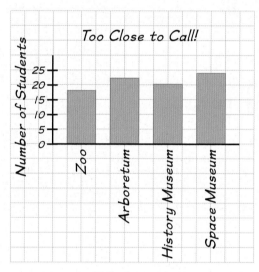

Janise's graph

Both graphs show the same data, but they are different because Kyle and Janise used different scales on the vertical axis. They are trying to persuade the readers of their graphs to agree with their opinions.

Kyle made his graph to try to show that they should plan a trip to the Space Museum. Kyle's vertical axis starts at 17 rather than 0. This makes the differences in the bar height look larger than they actually are.

Janise made a graph to suggest that they should vote again. Janise started her scale at 0, and chose a scale of 5 to make all four bars look close to the same height.

The U.S. Census

Every ten years since 1790, the U.S. government has undertaken the monumental task of counting and gathering information about every person living in the United States. With the growth of the U.S. population and advances in technology, this process, called the decennial census, has changed a great deal.

A Constitutional Mandate

The authors of the Constitution wanted the number of members in the House of Representatives from each state to be based on that state's total population. For this reason, a *census*, or count of the people, is required by the Constitution.

Article I, Section 2 of the Constitution calls for an "enumeration" of the entire U.S. population every ten years.

The 1790 census was conducted by 17 U.S. Marshals and hundreds of assistants. They recorded information by hand on paper. It took 18 months to gather data and make this report, which shows a total of about 3.9 million people.

Dealing with Data

From the beginning, the census counted people and gathered basic information, such as names, ages, and genders. In 1820, the government added questions in order to learn more about manufacturing, agriculture, and social issues, such as literacy and crime. The task of processing the data became much more complex.

As in this 1870 scene, census-takers, or *enumerators*, handwrote answers to the census questions during home visits. This face-to-face method of data collection was not changed until 1960.

In 1880, the population was over 50 million people, and processing the census information took seven years. With the invention of a punch-card tabulating system, the 1890 census was processed in just two and a half years.

A punch card machine makes holes in specific locations on a card to represent each person's information.

A tabulating machine pushes metal pins through the holes in each card, which completes an electrical circuit and causes dial counters to move.

In 1940, in addition to the decennial census, the U.S. Census Bureau began tracking unemployment through monthly surveys of a representative part of the population. Statisticians used the data to draw conclusions about the whole country. This method, called **sampling**, saved time and money, and led to the development of surveys on more than 100 topics.

In 1940, a detailed set of questions about housing was permanently added to the census. The government hoped to use this information to improve the standard of living for the population of 132 million Americans.

The Census Bureau helped pay for the development of the first electronic computer for processing data. Starting with the 1950 census, data from punch cards were transferred to magnetic tape, which enabled the room-sized UNIVAC (Universal Automatic Computer) to tabulate 4,000 items per minute.

More Advances

To take advantage of the ability of computers to process data quickly, advances were needed in data collection and the preparation of data for processing.

This equipment scanned reels of microfilmed questionnaires. The responses were transferred to magnetic tape that computers could read.

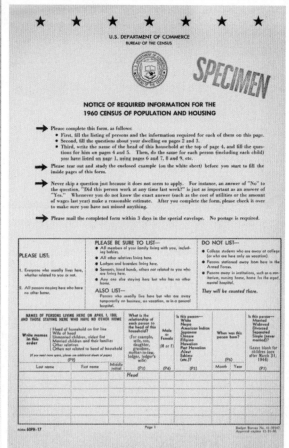

In 1960, the Census Bureau began using self-enumeration forms, which were mailed to households for people to fill out themselves. Enumerators then collected the completed forms.

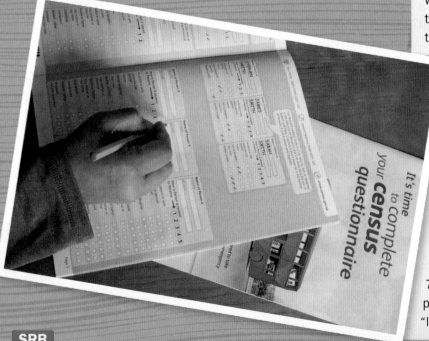

Sampling techniques have improved the efficiency and value of the decennial census. In 2000, most of the population received a "short form" questionnaire with 6 or 7 basic questions. About 17% of the population received the more detailed "long form."

Since 1970, the Census Bureau has asked people to return their questionnaires by mail. In 2010, 74% of American households returned their forms by mail.

Enumerators visit households that don't return their questionnaires to help the Bureau reach as close to 100% of the population as possible. Enumerators use handheld computers with GPS (the Global Positioning System) to locate addresses of households that need to be counted, and they use laptops to enter census data.

In 2000, the census was easier to fill out and faster to process due to the use of computers capable of reading handwriting. In 2010, the Census Bureau used better technology to locate each household address, counting a total population of 308,745,538. As the population grows, the Census Bureau continues to research ways to make their data collection more efficient and less expensive.

Putting the Data to Use

The U.S. Census Bureau publishes thousands of reports every year, which are used by governments, businesses, journalists, social scientists, non-profit organizations, and ordinary people. Because of the huge amount of information it collects and shares, the Bureau is often called "The Fact Finder for the Nation."

The number of representatives each state sends to Congress is still determined by census results. For example, the population of Texas grew from 20,903,994 people in 2000 to 25,268,418 people in 2010. Therefore, they gained 4 seats in the House of Representatives.

Government leaders use census data to plan how to distribute money to schools, libraries, healthcare clinics, and senior centers.

Business planners use census data to decide which communities are good markets for things such as stores, movie theaters, and restaurants.

You can access most of the Census Bureau data on the Internet or at your library. What do you want to know about your community?

Games

Games

Games

Throughout the year, you will play games that help you practice important math skills. Playing mathematics games gives you a chance to practice math skills in a different way. We hope that you will play often and have fun.

In this section of your *Student Reference Book*, you will find the directions for many games. The numbers in most games are generated randomly. This means that when you play games over and over, it is unlikely that you will repeat the same problems.

Many students have created their own variations of these games to make them more interesting. We encourage you to do this too.

Materials

The materials for each game are different and may include cards, dice, coins, counters, and calculators. Many of these materials are also available in your eToolkit. For some games you will have to make a gameboard, a score sheet, or a set of cards that are not number cards. These instructions are included with the game directions. More complicated gameboards and card decks are available from your teacher.

Number Cards. You need a deck of number cards for many of the games. You can use an Everything Math Deck, a deck of regular playing cards, or make your own deck out of index cards. An Everything Math Deck is part of the eToolkit.

An Everything Math Deck includes 54 cards. There are 4 cards each for the numbers 0–10. And there is 1 card for each of the numbers 11–20.

Number Cards

You can also use a deck of regular playing cards after making a few changes. A deck of playing cards includes 54 cards (52 regular cards, plus 2 jokers). To create a deck of number cards, use a permanent marker to mark the cards in the following ways:

- Mark each of the 4 aces with the number 1.
- Mark each of the 4 queens with the number 0.
- Mark the 4 jacks and 4 kings with the numbers 11 through 18.
- Mark the 2 jokers with the numbers 19 and 20.

Games

Game	Skill	Page
Absolute Value Sprint	Determining the absolute value of a number	318
Algebra Election	Substituting variables; solving equations	320
Build-it	Comparing and ordering fractions	322
Daring Division	Estimating quotients for whole numbers and decimals	323
Divisibility Dash	Recognizing multiples; using divisibility tests	324
Doggone Decimal	Estimating products of whole numbers and decimals	325
Factor Captor	Finding factors of a number	326
First to 100	Substituting variables; solving equations	327
Fraction Action, Fraction Friction	Estimating sums of fractions	328
Fraction Capture	Adding fractions; finding equivalent fractions	329
Fraction Top-It	Multiplying, dividing, and comparing fractions	330
Fraction/Whole Number Top-It	Multiplying whole numbers and fractions	331
Getting to One	Estimating	332
Hidden Treasure	Plotting ordered pairs; calculating distances on a coordinate grid	333
High-Number Toss (Decimal Version)	Understanding decimal place value, subtraction, and addition	334
Landmark Shark	Finding the range, mode, median, and mean	335
Mixed-Number Spin	Adding and subtracting fractions and mixed numbers; solving inequalities	337
Multiplication Bull's Eye	Estimating products of 2- and 3-digit numbers	338
Multiplication Wrestling (Mixed-Number Version)	Multiplying mixed numbers using partial products	339
Name That Number	Naming numbers with expressions; using order of operations	340
Percent Spin	Reasoning with percents; calculating percents	341
Polygon Capture	Identifying properties of polygons	343
Ratio Comparison	Comparing ratios	344
Ratio Dominoes	Recognizing equivalent ratios	345
Ratio Memory Match	Matching visual representations of ratios with ratio notation	346
Solution Search	Solving inequalities	347
Spoon Scramble	Multiplying fractions, decimals, and percents	348
Top-It Games	Multiplying, dividing, adding, or subtracting	349

three hundred seventeen

Games

Absolute Value Sprint

Materials
- ☐ 1 *Absolute Value Sprint* Gameboard (*Math Masters*, p. G18)
- ☐ 1 *Absolute Value Sprint* Record Sheet 1 for each player (*Math Masters*, p. G19)
- ☐ 32 counters (16 each of two different colors)
- ☐ number cards 0–10 (4 of each)

Players 2

Skill Determining the absolute value of a number

Object of the Game To win the most counters based on comparing distances from the centerline.

Directions

1. Each player chooses a side of the gameboard and a color of counters to use, then places one counter on each letter on his or her side of the board.

2. Shuffle the cards and place them in a pile next to the gameboard.

3. Players take turns. When it is your turn:

    - Draw a card. This number indicates how many steps to move toward 0.
    - Choose one counter and move it that number of steps.

    Players alternate turns until each player has drawn 4 cards and moved each counter once.

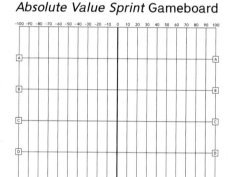

Absolute Value Sprint Gameboard

4. Each player takes one more turn, drawing a fifth card. For this card, players may choose any one of their counters to move, or they may choose not to move a counter. A counter may continue past 0 to land on the other side of the centerline.

5. Complete the game record sheet to determine the winner of the round.

    - List the position of each counter on the board.
    - Record each counter's distance from 0 (the center line).
    - Compare pairs of counters (compare your Counter A with your partner's Counter A). Circle the distance of the counter in each pair that is closer to 0.

6 The player with more counters closer to 0 wins all of the counters.
If each player has 2 counters closer to 0, split the counters so that each player wins 2.

7 Switch sides of the gameboard and repeat Steps 1–6 to play another round.

8 The game ends after 4 rounds. The player with the most counters wins. In the case of a tie, each player adds up his or her distances for all four rounds. The player with the smaller total wins the tie.

> **Example**
>
> Becca and Shania are playing *Absolute Value Sprint*.
>
> **End of the Round**
>
>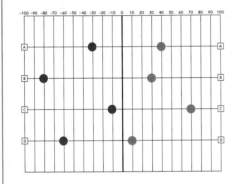
>
> **Shania's Record Sheet for the Round**
>
> | | My Counters | | My Partner's Counters | | |
> |---|---|---|---|---|---|
> | | Position | Distance | Position | Distance | |
> | A | −30 | (30) | 40 | 40 | A |
> | B | −80 | 80 | 30 | (30) | B |
> | C | −10 | (10) | 70 | 70 | C |
> | D | −60 | 60 | 10 | (10) | D |
>
> Shania's Counter A and Counter C are closer to 0. Becca's Counter B and Counter D are closer to 0. They each win 2 counters for this round.

Games

Algebra Election

Materials
- ☐ 1 set of *Algebra Election* Cards (*Math Journal 2*, Activity Sheets AS7–AS8)
- ☐ 1 Electoral Vote Map (*Math Masters*, pp. 261–262)
- ☐ 1 six-sided die
- ☐ 4 counters (different colors)
- ☐ 1 calculator

Players 2 teams of 2 players each

Skill Substituting variables; solving equations

Object of the Game To collect 270 or more votes to win the election. Winning team members become President and Vice President.

Directions

1. Shuffle the *Algebra Election* Cards. Place them writing-side down on the table.

2. Each player chooses a counter and puts it on Iowa to begin the game.

3. Alternate turns between teams and partners: Team 1, Player 1; Team 2, Player 1; Team 1, Player 2; Team 2, Player 2. When it is your turn:

   - Roll the die. The result is the number of moves you must make from your current state. Each new state counts as one move. Moves can be in any direction as long as you move between states that share a common border. *Exceptions:* Players can move to and from Alaska by way of Washington State and to and from Hawaii by way of California. Once a player has been in or through a state, the player may not return to that state on the same turn.

   - Move your counter and take the top card from the deck. Substitute the state's number of electoral votes for the variable x in the problem on the card. Solve the problem and say the answer. If the card contains several problems, you must answer all of the questions correctly. The other team checks your answer or answers with a calculator.

Algebra Election Cards

Tell whether each is true or false. $10 * x > 100$ $\frac{1}{2} * x * 100 < 10^3$ $x^3 * 1{,}000 > 4 * 10^4$	$T = B - (2 * \frac{H}{1{,}000})$ If $B = 80$ and $H = 100x$, what does T equal?	Insert parentheses in $\frac{1}{10} * x - 2$ so that its value is greater than 0 and less than 4.	Find: x squared x to the fourth power $\frac{1}{x}$
Evaluate. $x * 10^2$ $x * 10^5$	Find n. $n = \frac{(2 * x)}{10}$ $n + 1 = 2 * x$	Find n. $n + 10 = 4x$ $n - 10 = 4x$	Find n. $1{,}000 - n = x$ $1{,}000 \div x = n$
A boulder dropped off a cliff and fell approximately $16 * x^2$ feet in x seconds. How many feet is that?	Suppose you earn x dollars per hour. Complete the table. **Time** **Earnings** 1 hr _____ 2 hr _____ 4 hr _____	What is the value of n? $n = \frac{(5x - 4)}{2}$	Complete. $x * 10^6 =$ _____ million $x * 10^9 =$ _____ billion $x * 10^{12} =$ _____
Which is less, $\frac{x^2}{10}$ or $(x + 10)^2$?	Which is greater ... x^2 or 10^3? x^3 or 10^4?	What is the value of n? $20 + \|x\| = n$ $\|x\| + n = 200$	What is the value of n? $\|-20\| + x = n$ $\|-100\| - x = n$

Note In these directions, "state" means either a state or the District of Columbia (D.C.).

SRB

320 three hundred twenty

- If you answer correctly, your team wins that state's electoral votes. Record the state's name and its electoral votes on a piece of scratch paper. Write your initials in pencil on the state. Once a state is won, the opposing team may land on the state, but they cannot capture its votes.

- If you do *not* answer correctly, the state remains open. Players may still try to win its votes.

4 When all the cards have been used, shuffle the deck and continue playing.

5 The game ends when one team gets at least 270 votes and wins the election.

Variation

For a shorter version of the game, play through all 32 *Algebra Election* Cards just once. The team with more votes after all of the cards have been used is the winning team.

Electoral Vote Map

NOTE: Alaska and Hawaii are not drawn to scale.

Games

Build-It

Materials	☐ 1 set of *Build-It* Cards (*Math Masters*, p. G4)
	☐ 1 *Build-It* Gameboard for each player (*Math Masters*, p. G5)
Players	2
Skill	Comparing and ordering fractions
Object of the Game	To be the first player to arrange 5 fraction cards in order from least to greatest.

Directions

1 Shuffle the cards. Deal one card number-side up on each of the 5 spaces on each *Build-It* gameboard. Do not change the order of the cards.

2 Put the remaining cards number-side down in a draw pile. Turn the top card over and place it number-side up to start a discard pile.

3 Players take turns. When it is your turn:

- Take either the top card from the draw pile or the top card from the discard pile.

- Decide whether to keep this card or put it on the discard pile.

- If you keep the card, it must replace 1 of the 5 cards on your gameboard. Put the replaced card on the discard pile.

4 If all of the cards in the draw pile are used, shuffle the discard pile. Place the cards number-side down in a new draw pile. Turn over the top card to start a new discard pile.

5 When you think your 5 cards are in order from least to greatest, say, "Built it!" and have the other player check the order of your cards. If the cards are in order, you are the winner. If not, continue playing until someone wins.

Build-It Cards

$\frac{5}{9}$	$\frac{1}{3}$	$\frac{11}{12}$	$\frac{1}{12}$
$\frac{7}{12}$	$\frac{3}{8}$	$\frac{1}{4}$	$\frac{1}{5}$
$\frac{2}{3}$	$\frac{3}{7}$	$\frac{4}{7}$	$\frac{3}{4}$
$\frac{3}{5}$	$\frac{4}{5}$	$\frac{7}{9}$	$\frac{5}{6}$

Build-It Gameboard

Closest to 0 — Closest to 1

three hundred twenty-two

Daring Division

Materials
- ☐ number cards 0–9 (4 of each)
- ☐ 4 index cards labeled 0.1, 1, 10, 100
- ☐ 2 counters per player (to use as decimal points)
- ☐ 1 calculator for each player

Players 2

Skill Estimating quotients for whole numbers and decimals

Object of the Game To collect more number cards.

Directions

1. One player shuffles the number cards and deals 4 cards to each player.

2. The other player shuffles the index cards, places them number-side down, and turns over the top card. The number that appears (0.1, 1, 10, or 100) is the *target number.*

3. Using 4 number cards, each player forms a division problem.
   - Players try to form 2 numbers whose quotient is as close as possible to the target number. The number cards may be used to make two 2-digit numbers or one 3-digit number and one 1-digit number.
   - The decimal point can go anywhere in a number. See examples on the right.

4. Each player computes the quotient of his or her numbers using a calculator.

5. The player whose quotient is closer to the target number takes all 8 number cards.

6. Deal 4 new number cards to each player and turn over a new target number. Repeat Steps 3–5 using the new target number.

7. The game ends when all of the target numbers have been used. The player with more number cards wins the game. In the case of a tie, reshuffle the index cards and turn over a new target number. Play one tie-breaking round.

Variation

Each player receives 4 number cards. Players must make at least one fraction when forming their numbers.

three hundred twenty-three **323**

Games

Divisibility Dash

Materials	☐ number cards 0–9 (4 of each)
	☐ number cards 2, 3, 5, 6, 9, and 10 (2 of each)
	☐ 1 *Divisibility Dash* Record Sheet for each player (*Math Masters*, p. G13)
Players	2
Skill	Recognizing multiples; using divisibility tests
Object of the Game	To earn the most points.

Note The number cards 2, 3, 5, 6, 9, and 10 (2 of each) are the *divisor cards*.

The number cards 0–9 (4 of each) are the *draw cards*. This set of cards is also called the *draw pile*.

Directions

1. Shuffle the divisor cards and place them number-side down on the table. Shuffle the draw cards and deal 8 cards to each player. Place the remaining draw cards number-side down next to the divisor cards.

2. For each round, turn the top divisor card number-side up. When it is your turn:
   - Use the cards in your hand to make as many 2-digit numbers that are multiples of the divisor card as you can. A card used to make a 2-digit number may not be used again to make another number. Both players record the divisor and multiples played.
   - If a player disagrees that a 2-digit number is a multiple of the divisor card, that player may challenge. Use the divisibility test for the divisor to check. Any numbers that are not multiples of the divisor card must be returned to the player's hand.
   - Place all the cards you used to make 2-digit numbers in a discard pile.
   - If you cannot make a 2-digit number that is a multiple of the divisor card, you must take a card from the draw pile. Your turn is over. You score 0 points.

3. At the end of each round, each player finds the sum of his or her multiples. Circle the set of multiples on the record sheet with the greatest sum. Record each player's points. Score 1 point for every multiple made, and 1 point for having the greatest sum.

4. If the draw pile or divisor cards have all been used, shuffle and put them back into play.

5. The game ends when one player runs out of cards or at the end of 5 rounds. The player with the most points is the winner.

Example

The divisor card is 3.
Andrew uses his cards to make 2 numbers that are multiples of 3: 15 and 75.
Jayla uses her cards to make 1 number that is a multiple of 3: 96.
Andrew scores 2 points: one for each multiple.
Jayla scores 2 points: 1 point for her multiple and 1 point for having the greater sum since $96 > 15 + 75$.

Doggone Decimal

Materials
- ☐ number cards 0–9 (4 of each)
- ☐ 4 index cards labeled 0.1, 1, 10, and 100
- ☐ 2 counters per player (to use as decimal points)
- ☐ 1 calculator for each player

Players 2

Skill Estimating products of whole numbers and decimals

Object of the Game To collect more number cards.

Directions

1. One player shuffles the number cards and deals 4 cards to each player.

2. The other player shuffles the index cards, places them number-side down, and turns over the top card. The number that appears (0.1, 1, 10, or 100) is the *target number*.

3. Using 4 number cards and 2 counters as decimal points, each player forms 2 numbers. Each number must have 2 digits and a decimal point.

   - Players form numbers whose product is as close as possible to the target number.
   - The decimal point can go anywhere in a number. For example:

   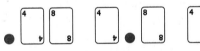

4. Each player computes the product of his or her numbers, and then uses a calculator to check the other player's product.

5. The player whose product is closer to the target number takes all 8 number cards if he or she computed correctly. Otherwise, shuffle all 8 cards back into the deck and play again with the same target number.

6. Deal 4 new number cards to each player and turn over a new target number. Repeat Steps 3–5 using the new target number.

7. The game ends when all of the target numbers have been used. The player with more number cards wins the game. In the case of a tie, reshuffle the index cards and turn over a new target number. Play one tie-breaking round.

Example

The target number is 10.
Brian is dealt 1, 4, 8, and 8. He forms the numbers 8.8 and 1.4. His product is 12.32.
Eve is dealt 2, 3, 6, and 9. She forms the numbers 2.6 and 3.9. Her product is 10.14.
Eve's product is closer to 10. She wins the round and takes all 8 cards.

Games

Factor Captor

Materials
- ☐ 1 *Factor Captor* Grid—either Grid 1 or Grid 2 (*Math Masters*, pp. G2 or G3)
- ☐ coin-size counters (48 for Grid 1; 70 for Grid 2)
- ☐ 1 calculator for each player

Players 2

Skill Finding factors of a number

Object of the Game To have the higher total score.

Directions

1. Player 1 chooses a 2-digit number on the number grid, covers it with a counter, and records the number on scratch paper. This is Player 1's score for the first round.

2. Player 2 covers all of the factors of Player 1's number. Player 2 finds the sum of the factors and records it on scratch paper. This is Player 2's score for the first round.

A factor may be covered only once during a round.

3. If Player 2 missed any factors, Player 1 can cover them with counters and add them to his or her score.

4. In the next round, players switch roles. Player 2 chooses a number that is not covered by a counter. Player 1 covers all factors of that number.

5. Any number that is covered by a counter is no longer available and may not be used again.

6. The first player in a round may not cover a number that is less than 10, unless no other numbers are available.

7. Play continues with players trading roles until all numbers on the grid are covered. Players use calculators to find their total scores. The player with the higher score wins.

1	2	2	2	2	2
2	3	3	3	3	3
3	4	4	4	4	5
5	5	5	6	6	7
7	8	8	9	9	10
10	11	12	13	14	15
16	18	20	21	22	24
25	26	27	28	30	32

Grid 1
(Beginning Level)

1	2	2	2	2	3
3	3	3	3	4	4
4	5	5	5	5	6	6
6	7	7	8	8	9	9
10	10	11	12	13	14	15
16	17	18	19	20	21	22
23	24	25	26	27	28	30
32	33	34	35	36	38	39
40	42	44	45	46	48	49
50	51	52	54	55	56	60

Grid 2
(Advanced Level)

Example

Round 1: James covers 27 and scores 27 points. Emma covers 1, 3, and 9, and scores $1 + 3 + 9 = 13$ points.

Round 2: Emma covers 18 and scores 18 points. James covers 2, 3, and 6, and scores $2 + 3 + 6 = 11$ points. Emma covers 9 with a counter, because 9 is also a factor of 18. Emma adds 9 points to her score.

First to 100

Materials	☐ one set of *First to 100* Problem Cards (*Math Journal 2*, Activity Sheets AS11 and AS12)
	☐ 2 six-sided dice
	☐ 1 calculator
Players	2 to 4
Skill	Substituting variables; solving equations
Object of the Game	To be the first player to collect at least 100 points by solving problems.

Directions

1 Shuffle the problem cards and place them writing-side down on the table.

2 Players take turns. When it is your turn:

- Roll 2 dice and find the product of the numbers.
- Turn over the top problem card and substitute the product for the variable x in the problem on the card.
- Solve the problem mentally or use paper and pencil. Give the answer. (You have 3 chances to use a calculator to solve difficult problems during a game.) Other players check your answer with a calculator.
- If your answer is correct, you get the number of points equal to the product that was substituted for the variable x. Some problem cards require two or more answers. In order to win any points, you must answer all parts of the problem correctly.
- Put the used problem card at the bottom of the deck.

3 The first player to get at least 100 points wins the game.

Example

Alice rolls a 5 and a 6. The product is 30.
She turns over a problem card: $20 * x = ?$
She substitutes 30 for x and answers 600.
Her answer is correct. Alice gets 30 points.

three hundred twenty-seven

Games

Fraction Action, Fraction Friction

Materials ☐ 1 *Fraction Action, Fraction Friction* Card Deck (Math Masters, p. G14)

☐ 1 or more calculators

Players 2 or 3

Skill Estimating sums of fractions

Object of the Game To collect a set of fraction cards with a sum as close as possible to 2, without going over 2.

Directions

1. Shuffle the deck and place it number-side down on the table.

2. Players take turns.

   - On each player's first turn, he or she takes a card from the top of the pile and places it number-side up on the table.

   - On each of the player's following turns, he or she announces one of the following:

Fraction Action, Fraction Friction Card Deck

$\frac{1}{2}$	$\frac{1}{3}$	$\frac{2}{3}$	$\frac{1}{4}$
$\frac{3}{4}$	$\frac{1}{6}$	$\frac{1}{6}$	$\frac{5}{6}$
$\frac{1}{12}$	$\frac{1}{12}$	$\frac{5}{12}$	$\frac{5}{12}$
$\frac{7}{12}$	$\frac{7}{12}$	$\frac{11}{12}$	$\frac{11}{12}$

"Action" This means that the player wants an additional card. The player believes that the sum of the fraction cards he or she already has is not close enough to 2 to win the hand. The player thinks that another card will bring the sum of the fractions closer to 2, without going over 2.

"Friction" This means that the player does not want an additional card. The player believes that the sum of the fraction cards he or she already has is close enough to 2 to win the hand. The player thinks there is a good chance that taking another card will make the sum of the fractions greater than 2.

Once a player says "Friction," he or she cannot say "Action" on any turn after that.

3. Play continues until all players have announced "Friction" or have a set of cards whose sum is greater than 2. The player whose sum is closest to 2 without going over 2 is the winner of that round. Players may check each other's sums on their calculators.

4. Reshuffle the cards and begin again. The winner of the game is the first player to win 5 rounds.

Fraction Capture

Materials ☐ 1 *Fraction Capture* Gameboard (*Math Masters*, p. G6)
☐ 2 six-sided dice

Players 2

Skill Adding fractions; finding equivalent fractions

Object of the Game To capture more squares on the *Fraction Capture* Gameboard.

Directions

1. Player 1 rolls the 2 dice and makes a fraction with both numbers. A fraction equal to a whole number is NOT allowed. For example, if a player rolls a 3 and a 6, the fraction cannot be $\frac{6}{3}$, because $\frac{6}{3}$ equals 2. If the 2 dice show the same number, the player rolls again.

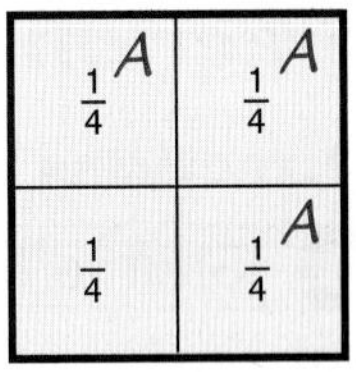

2. Player 1 initials sections of one or more gameboard squares to show the fraction formed. This *claims* the sections for the player.

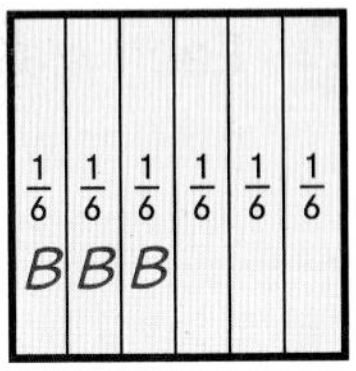

Figure 1

Examples

- A player rolls a 4 and a 3 and makes $\frac{3}{4}$. The player claims three $\frac{1}{4}$ sections by initialing them. (Figure 1)
- Equivalent fractions can be claimed. If a player rolls a 1 and a 2 and makes $\frac{1}{2}$, the player can initial one $\frac{1}{2}$ section of a square, or two $\frac{1}{4}$ sections, or three $\frac{1}{6}$ sections. (Figure 2)
- The fraction may be split between squares. A player can show $\frac{5}{4}$ by claiming $\frac{1}{2}$ and $\frac{1}{2}$ on one square and $\frac{1}{4}$ on another square. (Figure 3)

Figure 2

3. Players take turns. If a player forms a fraction, but cannot claim enough sections to show that fraction, no sections can be claimed and the player's turn is over.

4. A player *captures* a square when that player has claimed sections making up *more than* $\frac{1}{2}$ of the square. If each player has initialed $\frac{1}{2}$ of a square, no one has captured that square.

5. Blocking is allowed. For example, if Player 1 initials $\frac{1}{2}$ of a square, Player 2 may initial the other half, so that no one can capture the square.

6. Play ends when all of the squares have either been captured or blocked. The winner is the player who has captured more squares.

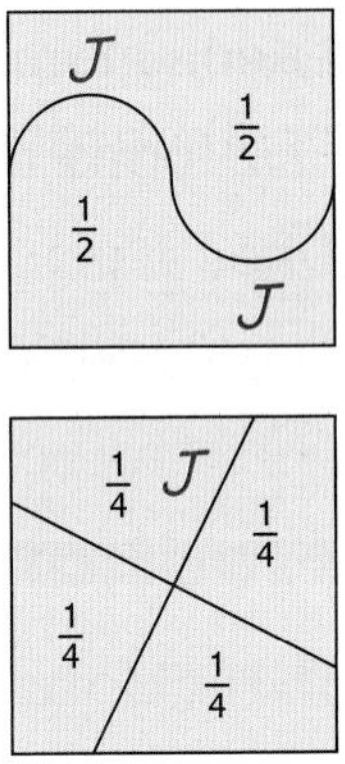

Figure 3

three hundred twenty-nine

Games

Fraction Top-It

Materials ☐ number cards 1–10 (4 of each)
             ☐ calculator (optional)
Players   2 to 4
Skill   Multiplying, dividing, and comparing fractions
Object of the Game   To collect the most cards.

Fraction Top-It (Multiplication)

Directions

1. Shuffle the cards and place them number-side down on the table.

2. Each player turns over 4 cards and forms one fraction from the first 2 cards and a second fraction from the last 2 cards. All fractions must be less than or equal to 1.

3. Each player calculates the product of his or her fractions and calls out the product. The player with the largest product takes all the cards. Players may use a calculator to compare their products.

4. In the case of a tie, each tied player repeats Steps 2 and 3. The player with the largest product takes all the cards from both plays.

5. The game ends when there are not enough cards left for each player to have another turn. The player with the most cards wins.

> **Example**
>
> Kenny turns over the cards 2, 1, 4, and 8, in that order.
> Liz turns over the cards 2, 3, 5, and 5, in that order.
> Kenny forms the fractions $\frac{1}{2}$ and $\frac{4}{8}$. His product is $\frac{1}{2} * \frac{4}{8} = \frac{4}{16} = \frac{1}{4}$.
> Liz forms the fractions $\frac{2}{3}$ and $\frac{5}{5}$. Her product is $\frac{2}{3} * \frac{5}{5} = \frac{2}{3} * 1 = \frac{2}{3}$.
> $\frac{2}{3} > \frac{1}{4}$
> Liz's product is larger, so she takes all the cards.

Fraction Top-It (Division)

Directions

This game is similar to *Fraction Top-It* (Multiplication). Instead of multiplying, each player divides one fraction by the other and calls out the quotient. Carefully consider the order of your division problem, as different orders can result in different quotients. For example, $\frac{1}{2} \div \frac{2}{3}$ is different from $\frac{2}{3} \div \frac{1}{2}$. The player with the largest quotient takes all the cards.

Fraction/Whole Number Top-It

Materials ☐ number cards 1–10 (4 of each)
              ☐ calculator (optional)

Players   2 to 4

Skill   Multiplying whole numbers and fractions

Object of the Game   To collect the most cards.

Directions

1 Shuffle the cards and place them number-side down on the table.

2 Each player turns over 3 cards. The card numbers are used to form one whole number and one fraction.

- The first card drawn is placed number-side up on the table. This card number is the whole number.

- The second and third cards drawn are used to form a fraction and are placed number-side up next to the first card. The fraction that these cards form must be less than or equal to 1.

3 Each player calculates the product of his or her whole number and fraction and calls it out. The player with the largest product takes all the cards. Players may use a calculator to compare their products.

4 In case of a tie, each tied player repeats Steps 2 and 3. The player with the largest product takes all the cards from both plays.

5 The game ends when there are not enough cards left for each player to have another turn. The player with the most cards wins.

Example

Amy turns over a 3, a 9, and a 5, in that order.
Roger turns over a 7, a 2, and an 8, in that order.
Amy's product is $3 * \frac{5}{9} = \frac{15}{9}$.
Roger's product is $7 * \frac{2}{8} = \frac{14}{8}$.
$\frac{15}{9} = \frac{5}{3} = 1\frac{2}{3}$ $\frac{14}{8} = \frac{7}{4} = 1\frac{3}{4}$

Roger's product is larger, so he takes all the cards.

three hundred thirty-one

Games

Getting to One

Materials ☐ 1 calculator

Players 2

Skill Estimating

Object of the Game To correctly guess a mystery number in as few tries as possible.

Directions

1. Player 1 chooses a mystery number that is between 1 and 100.

2. Player 2 guesses the mystery number.

3. Player 1 uses a calculator to divide Player 2's guess by the mystery number. Player 1 then reads the answer in the calculator display. If the answer has more than 2 decimal places, only the first 2 decimal places are read.

4. Player 2 continues to guess until the calculator result is 1. Player 2 keeps track of the number of guesses.

5. When Player 2 has guessed the mystery number, players trade roles and repeat Steps 1–4. The player who guesses the mystery number in the fewest number of guesses wins the round. The first player to win 3 rounds wins the game.

Note For a decimal number, the places to the right of the decimal point with digits in them are called decimal places. For example, 4.06 has 2 decimal places, and 0.780 has 3 decimal places.

Example

Player 1 chooses the mystery number 65.

Player 2 guesses 45. Player 1 keys in 45 ÷ 65 =.
Answer: 0.69   Too small.

Player 2 guesses 73. Player 1 keys in 73 ÷ 65 =.
Answer: 1.12   Too big.

Player 2 guesses 65. Player 1 keys in 65 ÷ 65 =.
Answer: 1   Just right!

Variation Allow mystery numbers up to 1,000.

SRB
332   three hundred thirty-two

Games

Hidden Treasure

Materials	☐ 1 sheet of *Hidden Treasure* Gameboards for each player (*Math Masters*, p. G11)
	☐ 2 pencils
	☐ 1 red pen or crayon
Players	2
Skill	Plotting ordered pairs; calculating distances on a coordinate grid
Object of the Game	To find the other player's hidden point on a coordinate grid.

Directions

1. Each player uses 2 grids. Players sit so they cannot see what the other is writing.

2. Each player secretly marks a point on his or her Grid 1. Use a red pen or crayon. These are the "hidden" points.

3. Player 1 guesses the location of Player 2's hidden point by naming an ordered pair. Player 1 records the guess on his or her Grid 2.

4. If Player 2's hidden point is at that location, Player 1 wins.

5. If the hidden point is not at that location, Player 2 marks the guess in pencil on his or her Grid 1. Player 2 tells Player 1 the fewest number of "square sides" needed to travel from the hidden point to the guessed point. Repeat Steps 3–5 with Player 2 guessing and Player 1 answering.

6. Play continues until one player finds the other's hidden point.

Example

Player 1 marks a hidden point at (2, 5). | Player 2 marks a hidden point at (2, −4).

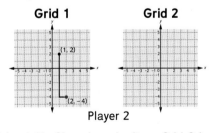

- Player 1 guesses that Player 2's hidden point is at (1, 2) and marks it on Grid 2 in pencil.
- Player 2 marks the point (1, 2) in pencil on Grid 1 and tells Player 1 that (1, 2) is 7 units (square sides) away from the hidden point.
- Player 1 writes "7" next to the point (1, 2) on his or her Grid 2. Player 1's turn is over. Player 2 makes a guess.

three hundred thirty-three **SRB 333**

Games

High-Number Toss (Decimal Version)

Materials ☐ number cards 0–9 (4 of each)
☐ 1 scorecard for each player

Players 2

Skill Understanding decimal place value, subtraction, and addition

Object of the Game To make the largest decimal numbers possible.

Directions

1. Each player makes a scorecard like the one at the right. Players fill out their own scorecards at each turn.

2. Shuffle the cards and place them number-side down on the table.

3. In each round:
   - Player 1 draws the top card from the deck and writes the number on any of the 3 blanks on the scorecard. It need not be the first blank—it can be any of them.
   - Player 2 draws the next card from the deck and writes the number on one of his or her blanks.
   - Players take turns doing this 2 more times. The player with the larger number wins the round.

4. The winner's score for a round is the difference between the two players' numbers. (Subtract the smaller number from the larger number.) The other player scores 0 points for the round.

Scorecard

Game 1

Round 1   Score
0. __ __ __   _____
Round 2
0. __ __ __   _____
Round 3
0. __ __ __   _____
Round 4
0. __ __ __   _____
        Total:
                 _____

Example

Player 1: 0 . _7_ _6_ _3_
Player 2: 0 . _9_ _2_ _1_

Since 0.921 − 0.763 = 0.158, Player 2 scores 0.158 point for the round. Player 1 scores 0 points.

5. Players take turns starting a round. At the end of 4 rounds, they find their total scores. The player with the larger total score wins the game.

SRB
334 three hundred thirty-four

Landmark Shark

Materials
- ☐ 1 set of number cards
- ☐ 1 each of the range, median, and mode *Landmark Shark* Cards for each player (*Math Masters*, p. G9)
- ☐ 1 *Landmark Shark* Score Sheet (*Math Masters*, p. G10)

Players 2 or 3

Skill Finding the range, mode, median, and mean

Object of the Game To score the most points by finding data landmarks.

Directions

1 To play a round:
- The dealer shuffles the number cards and deals 5 cards number-side down to each player.
- Players put their cards in order from the smallest number to the largest.
- There are 3 ways a player may score points using his or her five cards:

Range: The player's score is the range of the 5 numbers.

Example
Brian's hand:
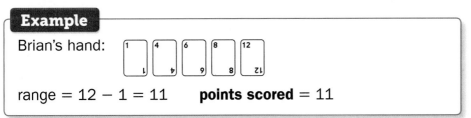
range = 12 − 1 = 11   **points scored** = 11

Median: The player's score is the median of the 5 numbers.

Example
Liz's hand:
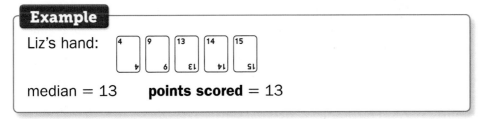
median = 13   **points scored** = 13

Mode: The player must have at least 2 cards with the same number. The player's score is found by multiplying the mode of the 5 numbers by the number of modal cards. If there is more than one mode, the player uses the mode that will produce the most points.

Example
Caroline's hand:
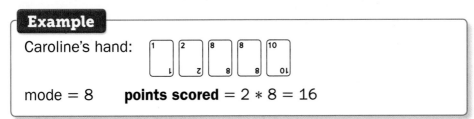
mode = 8   **points scored** = 2 * 8 = 16

Games

Landmark Shark (continued)

② Each player decides which landmark will yield the highest score for the hand. A player indicates his or her choice by placing 1 of the 3 *Landmark Shark* Cards (range, median, or mode) on the table.

③ Players can try to improve their scores by exchanging up to 3 of their cards for new cards from the deck. However, the *Landmark Shark* Card they chose stays the same.

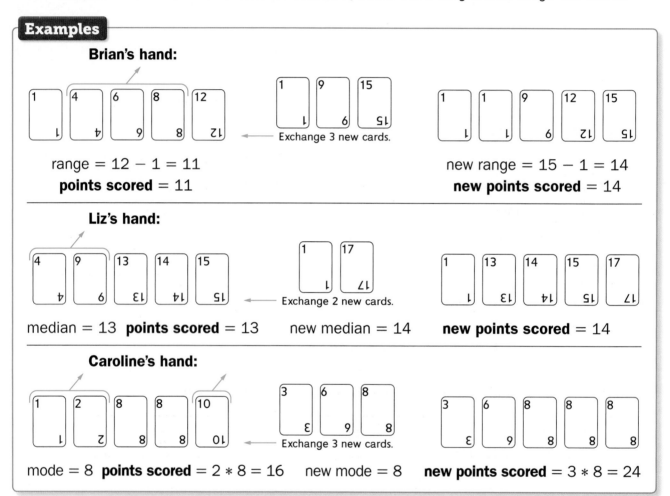

Examples

Brian's hand:

range = 12 − 1 = 11
points scored = 11

Exchange 3 new cards.

new range = 15 − 1 = 14
new points scored = 14

Liz's hand:

median = 13 **points scored = 13**   new median = 14   **new points scored = 14**

Exchange 2 new cards.

Caroline's hand:

mode = 8 **points scored = 2 * 8 = 16**   new mode = 8   **new points scored = 3 * 8 = 24**

Exchange 3 new cards.

④ Players lay down their cards and record their points scored on the score sheet.

Round 1:

	Player 1	Player 2	Player 3
Points Scored			
Bonus Points			
Round 1 Score			

⑤ **Bonus Points:** Each player calculates the *mean* of his or her card numbers to the nearest tenth. Each player's score for the round is the sum of his or her points scored plus any bonus points.

⑥ Repeat Steps 1–5 for each round. The winner is the player with the highest total after 5 rounds.

SRB
336 three hundred thirty-six

Mixed-Number Spin

Materials	☐ 1 *Mixed-Number Spin* Record Sheet (*Math Masters*, p. G7)
	☐ 1 *Mixed-Number Spin* Spinner (*Math Masters*, p. G8)
	☐ 1 large paper clip
Players	2
Skill	Adding and subtracting fractions and mixed numbers; solving inequalities
Object of the Game	To complete 10 number sentences that are true.

Directions

1 Each player writes his or her name in one of the boxes on the record sheet.

2 Players take turns. When it is your turn:
- Anchor the paper clip to the spinner with the point of your pencil. Spin.
- Write the fraction or mixed number you spin on any one of the blanks below your name.

3 The first player to complete 10 true number sentences is the winner.

Example

Ella spun 2 and $1\frac{1}{8}$ on her first two turns and filled in 2 blank spaces.

On her next turn, Ella spins $1\frac{7}{8}$. She has two choices:
- She can write $1\frac{7}{8}$ in a sentence where there are 2 blanks.
- She can use $1\frac{7}{8}$ to form the true number sentence $1\frac{7}{8} - 1\frac{1}{8} > \frac{1}{2}$.

Ella cannot use $1\frac{7}{8}$ in the first sentence because $2 + 1\frac{7}{8}$ is not < 3.

Ella	Ella	Ella
$\underline{2} + \underline{} < 3$	$\underline{2} + \underline{} < 3$	$\underline{2} + \underline{} < 3$
$\underline{} + \underline{} > 3$	$\underline{} + \underline{} > 3$	$\underline{} + \underline{} > 3$
$\underline{} - \underline{} < 1$	$\underline{1\frac{7}{8}} - \underline{} < 1$	$\underline{} - \underline{} < 1$
$\underline{} - \underline{1\frac{1}{8}} > \frac{1}{2}$	$\underline{} - \underline{1\frac{1}{8}} > \frac{1}{2}$	$\underline{1\frac{7}{8}} - \underline{1\frac{1}{8}} > \frac{1}{2}$

Games

Multiplication Bull's Eye

Materials ☐ number cards 0–9 (4 of each)
  ☐ 1 six-sided die

Players 2

Skill Estimating products of 2- and 3-digit numbers

Object of the Game To score more points.

Directions

1. Shuffle the deck and place it number-side down on the table.

2. Players take turns. When it is your turn:
   - Roll the die. Look up the target range of the product in the table at the right.
   - Take 4 cards from the top of the deck.
   - Use the cards to try to form 2 numbers whose product falls within the target range. Do not use a calculator or pencil and paper.
   - Multiply the 2 numbers on paper using U.S. traditional multiplication to determine whether the product falls within the target range. If it does, you have hit the bull's eye and score 1 point. If it doesn't, you score 0 points.
   - Sometimes it is impossible to form 2 numbers whose product falls within the target range. If this happens, you score 0 points for that turn.

3. The game ends when each player has had 5 turns. The player with more points wins.

Number on Die	Target Range of Product
1	500 or less
2	501–1,000
3	1,001–3,000
4	3,001–5,000
5	5,001–7,000
6	more than 7,000

Example

Tom rolls a 3, so the target range of the product is from 1,001 to 3,000. He turns over cards 5, 7, 9, and 2.

Tom uses estimation to try to form 2 numbers whose product falls within the target range—for example, 97 and 25.

He then finds the product using U.S. traditional multiplication.

Since the product is between 1,001 and 3,000, Tom has hit the bull's eye and scores 1 point.

Some other possible winning products from the 5, 7, 9, and 2 cards are: 25 * 79, 27 * 59, 9 * 257, and 2 * 579.

```
       1
       3
       9 7
  *    2 5
  -------
       4 8 5
  + 1 9 4 0
  ---------
    2 4 2 5
```

Multiplication Wrestling (Mixed-Number Version)

Materials ☐ number cards 0–9 (4 of each)

Players 2

Skill Multiplying mixed numbers using partial products

Object of the Game To get the larger product of two mixed numbers.

Directions

1. Shuffle the deck and place it number-side down on the table.

2. Each player draws 6 cards and forms 2 mixed numbers. Each player creates two "wrestling teams" by writing each of their numbers as a sum of the whole part and the fractional part.

3. Each player's teams wrestle. Multiply each member of the first team (for example, 3 and $\frac{4}{5}$) by each member of the second team (for example, 6 and $\frac{2}{8}$). Then add the 4 partial products. The player with the larger product wins the round and receives 1 point.

4. To begin a new round, each player draws 6 new cards to form 2 new mixed numbers. Play 3 rounds. The player with more points wins.

Example

Player 1: Draws 4, 3, 8, 2, 5, 6.
Forms $3\frac{4}{5}$ and $6\frac{2}{8}$.

$$3\frac{4}{5} * 6\frac{2}{8}$$

Team 1   **Team 2**
$(3 + \frac{4}{5})$ * $(6 + \frac{2}{8})$

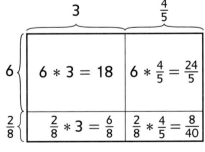

$18 + \frac{24}{5} + \frac{6}{8} + \frac{8}{40} = 23\frac{3}{4}$
$3\frac{4}{5} * 6\frac{2}{8} = 23\frac{3}{4}$

Player 2: Draws 8, 5, 9, 5, 6, 3.
Forms $8\frac{5}{6}$ and $9\frac{3}{5}$.

$$8\frac{5}{6} * 9\frac{3}{5}$$

Team 1   **Team 2**
$(8 + \frac{5}{6})$ * $(9 + \frac{3}{5})$

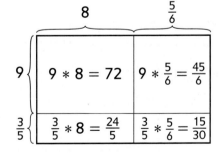

$72 + \frac{45}{6} + \frac{24}{5} + \frac{15}{30} = 84\frac{8}{10}$
$8\frac{5}{6} * 9\frac{3}{5} = 84\frac{8}{10}$

three hundred thirty-nine

Games

Name That Number

Materials	☐ 1 set of number cards
Players	2 or 3
Skill	Naming numbers with expressions; using order of operations
Object of the Game	To collect the most cards.

Directions

1. Shuffle the deck and deal 5 cards to each player. Place the remaining cards number-side down on the table. Turn over the top card and place it beside the deck. This is the *target number* for the round.

2. Players try to match the target number by adding, subtracting, multiplying, or dividing the numbers on as many of their cards as possible. Number cards can also be used to make fractions or as exponents in an expression. A card may only be used once.

3. Players write their solutions on a sheet of paper using grouping symbols as needed. When players have written and shared their best solutions:

   - Each player sets aside the cards he or she used to match the target number.
   - Each player replaces the cards he or she set aside by drawing new cards from the top of the deck.
   - The old target number is placed on the bottom of the deck.
   - A new target number is turned over, and another round is played.

4. Play continues until there are not enough cards left to replace all of the players' cards. The player who has set aside the most cards wins the game.

Example

Target number: 16

Player 1's cards: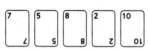

Some possible solutions:

$10 + 8 - 2 = 16$ (3 cards used)

$10 + (7 * 2) - 8 = 16$ (4 cards used)

$10 / (5 * 2) + 8 + 7 = 16$ (all 5 cards used)

$5^2 - (10 - 8) - 7 = 16$ (all 5 cards used)

Games

Percent Spin

Materials
- ☐ 1 *Percent Spin* Record Sheet for each player (*Math Masters*, p. G24)
- ☐ 1 *Percent Spin* Spinner for each player (*Math Masters*, p. G23)
- ☐ 1 straightedge
- ☐ 10 counters per player
- ☐ 1 large paper clip

Players 2

Skill Reasoning with percents; calculating percents

Object of the Game To capture as many counters as possible in 10 spins.

Directions

1. Each player makes a spinner:
   - Divide the spinner into three sections of different sizes.
   - Use a straightedge to draw lines connecting the center of the spinner to the tick marks on the circle.
   - Label the sections of your circle A, B, and C.

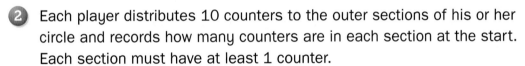

2. Each player distributes 10 counters to the outer sections of his or her circle and records how many counters are in each section at the start. Each section must have at least 1 counter.

> **Example**
>
> Sara started with 5 counters in Section A, 2 counters in Section B, and 3 counters in Section C.
>
> **Start**
>
> | Section Letter | Number of Counters |
> |---|---|
> | A | 5 |
> | B | 2 |
> | C | 3 |
>
> Sara's Record Sheet
>
>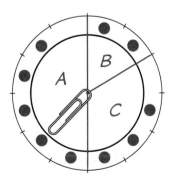
>
> Sara's spinner at the start

SRB
three hundred forty-one 341

Games

3 Players take turns. On your turn:
  - Spin your spinner. Capture 1 counter from the section where the spinner lands.
  - On your record sheet, keep track of the total number of *landings,* or times the spinner lands in each section. Be sure to record every landing, even if you are not able to capture a counter on that spin.

4 After each player has had 10 turns, each player uses the total number of landings to calculate the percent of landings for each section.

5 Each player's score is the total number of counters that he or she removed in 10 spins. The player with the higher score wins.

> **Example**
>
> Sara's spinner landed 6 times on Section A, 2 times on Section B, and 2 times on Section C. She was able to remove all 5 counters from Section A, the 2 counters from Section B, and 2 out of the 3 counters from Section C.
>
>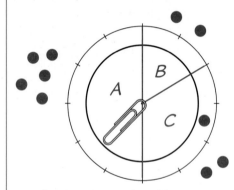
>
> Finish
>
> | Section Letter | Number of Landings | Percent of __10__ Spins |
> |---|---|---|
> | A | ℍℍℍ / | 60% |
> | B | // | 20% |
> | C | // | 20% |
>
> Sara's spinner after 10 spins   Sara's Record Sheet
>
> Score: __9__
>
> Sara's score is 9 because she was able to remove 9 counters.

6 Play again with the same spinner. Use your experience to try to improve your score.

Games

Polygon Capture

Materials	☐ 1 set of *Polygon Capture* Pieces (*Math Masters*, p. G25)
	☐ 1 set of *Polygon Capture* Property Cards (*Math Masters*, p. G26)
Players	2, or two teams of 2
Skill	Identifying properties of polygons
Object of the Game	To collect more polygons.

Polygon Capture Pieces

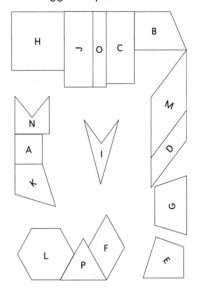

Polygon Capture Property Cards (property-side up)

There is only one right angle.	There is at least one right angle.	All angles are right angles.	There are no right angles.
There is at least one acute angle.	At least one angle is greater than 90°.	All angles are right angles.	There are no right angles.
All opposite sides are parallel.	Only one pair of sides is parallel.	There are no parallel sides.	All sides are the same length.
All opposite sides are parallel.	Some sides have the same length.	All opposite sides have the same length.	**Wild Card:** Pick any property about sides.

Directions

1 Write the letter A on the back of each property card that mentions "angle." Write the letter S on the back of each property card that mentions "side."

2 Spread out the *Polygon Capture* Pieces on the table. Shuffle the *Polygon Capture* Property Cards and sort them property-side down into "A" and "S" piles.

3 Players take turns. When it is your turn:

- Draw the top card from each pile of property cards.
- Take all of the polygons that have *both* of the properties shown on the property cards you drew.
- If there are no polygons with both properties, draw one additional property card from either pile. Look for polygons that have this new property *and* one of the properties already drawn. Take these polygons.
- If you did not capture a polygon that you could have taken, the other player or team may name it, capture it, and add it to their score for the round.

4 When all the property cards in either pile have been drawn, shuffle all of the cards. Sort them into "A" and "S" piles. Continue play.

5 The game ends after 5 rounds, or when there are fewer than 3 polygons left. Players add the number of polygons captured in each round to find their total score. The winner is the player or team who has captured more polygons.

Example

Liz has these property cards: "All angles are right angles" and "All sides are the same length." She can "capture" all the squares (Polygons A and H).

three hundred forty-three **SRB 343**

Games

Ratio Comparison

Materials
- ☐ 1 set of *Ratio Comparison* Situation Cards (*Math Masters*, p. G28)
- ☐ 1 set of *Ratio Comparison* Ratio Cards (*Math Masters*, p. G27)

Players 2 or 3

Skill Comparing ratios

Object of the Game To collect the most cards.

Directions

1. Shuffle each set of cards separately. Place the situation cards facedown in one pile and the ratio cards facedown in a second pile. Turn over the top situation card.

2. Each player selects a ratio card from the pile.

3. All players display and compare their ratios. Use a ratio/rate table or tape diagram to make your comparisons.

4. The player whose ratio best meets the criteria on the situation card wins all of the cards used in that round. If there is a tie because the ratios are equivalent, each player secretly writes down a new ratio, and the player with the ratio that best fits the situation on the card wins all of the cards.

5. Repeat Steps 1–4 until there are no situation cards left. The player with the most cards wins the game.

Example

Sean draws the ratio card showing 2 : 3. He knows this means **2** pints red paint : **3** pints white paint. Jada draws the ratio card showing 5 : 6. She knows this means **5** pints red paint : **6** pints white paint.

Sean's Ratio Table

Red	2	8
White	3	12

Jada's Ratio Table

Red	5	10
White	6	12

Situation Card

4
Paint Mixture

Pints of red paint : Pints of white paint

Which ratio represents a darker shade of pink?

With 12 pints of white paint, Sean's ratio uses 8 pints of red paint while Jada's ratio uses 10 pints of red paint. Jada's ratio represents a darker shade of pink, so she wins all of the cards.

Games

Ratio Dominoes

Materials	☐ 1 set of Ratio Dominoes (*Math Journal 1*, Activity Sheets AS4–AS5)
Players	2 or 3
Skill	Recognizing equivalent ratios
Object of the Game	To use equivalent ratios to play all of your dominoes.

Directions

1 Remove the blank dominoes from the set. Place the rest of the dominoes facedown on the table and mix them up.

2 Each player draws 5 dominoes.

3 The player with the double domino (a domino with equivalent ratios) that shows the largest ratio (closest to $\frac{1}{2}$) goes first and plays it in the middle of the table as shown in the example below. If no one has a double, all players continue to draw, one domino at a time, until someone gets a double to begin the game.

4 Play continues to the left. When it is your turn:

- If you have a domino with a ratio that matches one of the ratios on an open end, play it. Dominoes are played in a horizontal line from either end, with doubles played vertically, perpendicular to the other dominoes. After dominoes have been played horizontally from each side of a double, then you may also play dominoes vertically from either end of the double.

- If you can't play, draw a domino. If you draw one with a match, play it. If not, you lose that turn.

5 The game ends when one player has played all of his or her dominoes. That player is the winner.

Example

This game began with the double domino showing $\frac{5}{50}$ and 10 : 100.
Here is what the game looks like after 8 turns:

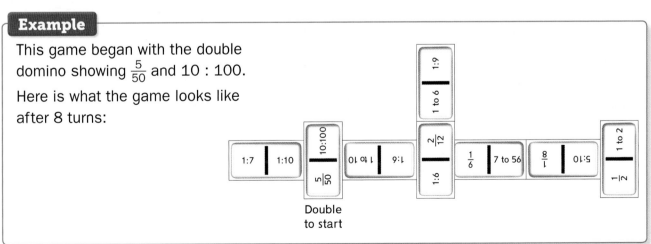

three hundred forty-five **345**

Games

Ratio Memory Match

Materials ☐ 1 set of *Ratio Memory Match* Cards (*Math Journal 1*, Activity Sheet AS6)

Players 2

Skill Matching visual representations of ratios with ratio notation

Object of the Game To make the most ratio matches.

Directions

1. Shuffle the *Ratio Memory Match* Cards and place them facedown on the table, arranged in an array.

2. Players take turns. When it is your turn:

    - Turn over two cards. A *match* consists of one card with a ratio picture of white squares to blue squares and one card with the same ratio of white squares to blue squares using ratio notation.

    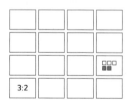

    - If the two cards match, remove the cards and keep them in your match pile. You may continue with your turn until you miss a match.

    - If the two cards do not match, turn them facedown. It is now the other player's turn.

3. The game ends when all of the matches have been made. The player with the most cards in his or her match pile wins the game.

Solution Search

Materials ☐ 1 set of Solution Search Cards (Math Masters, p. G17)
☐ 1 set of number cards

Players 3 or 4

Skill Solving inequalities

Object of the Game To be the first player to discard all cards.

Solution Search Cards

$x \geq 9$	$y < 5$	$m \leq 3.5$	$q * 2 \geq 20$
$x \geq 9$	$y < 5$	$m \leq 3.5$	$q * 2 \geq 20$
$(6 * z) + 2 \geq 65$	$100 / k \geq 25$	$s \neq 5$	$b \leq 6$
$(9 * z) + 2 \geq 65$	$100 / k \geq 25$	$s \neq 5$	$b \leq 6$
$a - 2 \geq 5$	$w - 3 \leq 2$	$r / 2 \geq 5$	$4 \leq p$
$a - 2 \geq 5$	$w - 3 \leq 2$	$r / 2 \geq 5$	$4 \leq p$
$8 \geq f$	$c * 7 \leq 14$	$10 \leq 50 / d$	$2\frac{1}{4} \leq t$
$8 \geq f$	$c * 7 \leq 14$	$10 \leq 50 / d$	$2\frac{1}{4} \leq t$

Directions

1. Shuffle the Solution Search Cards and place the deck facedown in the center of the table.

2. Shuffle the number cards and deal 8 cards to each player. Place the remainder of the deck number-side down in the center of the table. Any player may begin the first round.

3. The player who begins a round turns over the top Solution Search Card and takes the first turn in that round. Play continues in a clockwise direction.

4. When it is your turn:
   - Discard any one of your number cards that is a solution to the inequality on the Solution Search Card.
   - If you do not have a number card that is a solution, continue to draw from the deck of number cards until you draw a solution. Then discard that number card.

5. The next player tries to discard a solution to the same inequality on the Solution Search Card for that round. The round is over when each player has had a turn. The player who takes the last turn in a round begins the next round. Follow Steps 3–5 to play another round.

6. When all of the Solution Search Cards have been used, turn the used pile facedown. Without shuffling, take the next card. When no more number cards remain, shuffle the discard pile and place it number-side down. Continue play.

7. The winner is the first player to discard all of his or her cards.

Variation

Play with 2s and 7s as special cards. Use 2s as WILD cards. A player may choose to play a 2 card with its given value of 2, or a player may assign any value he or she wishes to the 2 card. The value of the 7 card is always 7. However, if a player plays the 7 card, the next player loses his or her turn.

three hundred forty-seven

Games

Spoon Scramble

Materials
- ☐ 1 set of *Spoon Scramble* Cards (*Math Journal 2*, Activity Sheet AS9 or AS10)
- ☐ 3 spoons

Players 4

Skill Multiplying fractions, decimals, and percents

Object of the Game To avoid getting all of the letters in the word SPOONS.

Directions

1. Place the spoons in the center of the table.

2. The dealer shuffles the deck and deals 4 cards number-side down to each player.

3. Players look at their cards. If a player has 4 cards of equal value, proceed to Step 5. Otherwise, each player chooses a card to discard and passes it, number-side down, to the player on the left.

4. Each player picks up the new card and repeats Step 3. The passing of the cards should proceed quickly.

5. As soon as a player has 4 cards of equal value, the player places the cards number-side up on the table and grabs a spoon.

6. The other players then try to grab 1 of the 2 remaining spoons. The player left without a spoon is assigned a letter from the word SPOONS, starting with the first letter. If a player incorrectly claims to have 4 cards of equal value, that player receives a letter instead of the player left without a spoon.

7. Players put the spoons back in the center of the table. The dealer shuffles and deals cards to begin a new round.

8. Play continues until 3 players each get all the letters in the word SPOONS. The remaining player is the winner.

Variations

- For 3 players: Eliminate one set of 4 equivalent *Spoon Scramble* Cards. Use only 2 spoons.

- For a more challenging version of the game, players can make their own deck of *Spoon Scramble* Cards. Each player writes 4 computation problems that have equivalent answers on 4 index cards. Check to be sure the players have all chosen different values.

Top-It Games

Materials	☐ number cards 1–9 (4 of each)
	☐ calculator (optional)
Players	2 to 4
Skill	Multiplying, dividing, adding, or subtracting
Object of the Game	To collect the most cards.

Multiplication Top-It (Extended Facts)

Directions

1 Shuffle the deck and place it number-side down on the table.

2 Each player turns over 2 cards. Multiply the first card drawn by 10. Multiply the product by the second card. For example, if 7 is the first card drawn and 5 is the second card drawn, compute 70 * 5 = 350.

3 Compare products. The player with the largest product takes all the cards.

4 The game ends when there are not enough cards left for each player to have another turn. The player with the most cards wins.

Multiplication Top-It

Directions

1 Shuffle the deck and place it number-side down on the table.

2 Each player turns over 4 cards. Choose 3 of them to form a 3-digit number, then multiply by the remaining number. Carefully consider how to form your 3-digit numbers. For example, 462 * 5 = 2,310 while 256 * 4 = 1,024.

3 Compare products. The player with the largest product takes all the cards.

4 The game ends when there are not enough cards left for each player to have another turn. The player with the most cards wins.

Variation

Use the 4 cards to make two 2-digit numbers to multiply.

three hundred forty-nine

Games

Division Top-It

Directions

1. Shuffle the deck and place it number-side down on the table.

2. Each player turns over 4 cards. Choose 3 of them to form a 3-digit number, then divide the 3-digit number by the remaining number. Ignore the remainder. Carefully consider how to form your 3-digit numbers. For example, 462 / 5 is greater than 256 / 4.

3. Compare quotients. The player with the largest quotient takes all the cards.

4. The game ends when there are not enough cards left for each player to have another turn. The player with the most cards wins.

Addition Top-It

Directions

Addition Top-It is played like the other *Top-It* games, except each player turns over 6 cards, forms two 3-digit numbers, and finds the sum of the numbers. Players should carefully consider how they form their numbers, since different arrangements have different sums. For example, 741 + 652 has a greater sum than 147 + 256. The player with the largest sum takes all the cards.

Subtraction Top-It

Directions

Subtraction Top-It is played like the other *Top-It* games, except each player turns over 6 cards, forms two 3-digit numbers, and finds the difference of the numbers. Players should carefully consider how they form their numbers. For example, 751 − 234 has a greater difference than 517 − 342. The player with the largest difference takes all the cards.

Decimal Top-It

Directions

Decimal Top-It can be played with any of the *Top-It* games listed above. Each player uses 2 counters to make decimal numbers with the cards they turn over. Players may choose where to place the decimal points, but each player must make *at least one* number with *at least one* digit to the right of the decimal point.

three hundred fifty

Real-World Data

Real-World Data

Introduction

We use data as a part of everyday life. Scientists, historians, mathematicians, and others collect data. They collect data by counting, measuring, or observing and then organize their data to share with others. Analyzing data helps answer questions about real-life problems or situations. This section provides a collection of interesting data from the world around you.

Some of the information in this section could be easy to find on your own because it comes from a single source. For example, you can find the locations of the national parks in Utah by looking at an atlas or finding a map online.

Sometimes the information comes from many different sources. In those cases, it would take you a long time to collect the data on your own. For example, to find different types of data about the United States, such

as each state's population, area, and number of electoral votes, you might need to consult several different sources. The information in this section has been collected and organized into tables and maps for you, so you can focus on using the data.

You might be curious about some of the material in the Real-World Data section. You can collect more information about these things through measurement, observation, or research. You might use tools such as a computer, the Internet, or reference books to do more research. Pages 371–372 provide a list of sources used to compile the data in this section.

Real-World Data

United States: Utah National Parks

There are five national parks in Utah. Here is a map showing their locations, and a mileage chart showing the distance in road miles between them. Salt Lake City, the capital of Utah, is also included.

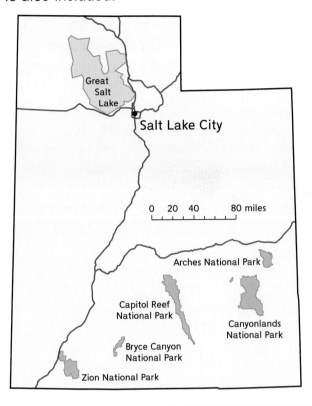

Arches National Park

Utah National Parks Mileage Chart						
	Arches National Park	Bryce Canyon National Park	Canyonlands National Park	Capitol Reef National Park	Salt Lake City	Zion National Park
Arches National Park		272	80	142	229	341
Bryce Canyon National Park	272		310	91	270	85
Canyonlands National Park	80	310		221	308	420
Capitol Reef National Park	142	91	221		222	168
Salt Lake City	229	270	308	222		309
Zion National Park	341	85	420	168	309	

three hundred fifty-three

Real-World Data

United States: State Facts

In 2010, the United States had a population of 308,745,538 people. Its land area is 3,531,905 square miles, or 9,147,593 square kilometers.

The following table lists some facts about each of the 50 states (and the District of Columbia), including each state's 2010 population and land area in square miles and square kilometers. It also includes the number of representatives each state has in the United States House of Representatives and the number of electoral votes that has been allocated to each state.

State Facts					
State	2010 Population	Area (in square miles)	Area (in square kilometers)	Seats in the House of Representatives	Electoral Votes
Alabama	4,779,736	50,645	131,171	7	9
Alaska	710,231	570,641	1,477,953	1	3
Arizona	6,392,017	113,594	294,207	9	11
Arkansas	2,915,918	52,035	134,771	4	6
California	37,253,956	155,779	403,466	53	55
Colorado	5,029,196	103,642	268,431	7	9
Connecticut	3,574,097	4,842	12,542	5	7
Delaware	897,934	1,949	5,047	1	3
District of Columbia	601,723	61	158	1*	3
Florida	18,801,310	53,625	138,887	27	29
Georgia	9,687,653	57,513	148,959	14	16
Hawaii	1,360,301	6,423	16,635	2	4
Idaho	1,567,582	82,643	214,045	2	4
Illinois	12,830,632	55,519	143,793	18	20
Indiana	6,483,802	35,826	92,789	9	11
Iowa	3,046,355	55,857	144,669	4	6
Kansas	2,853,118	81,759	211,754	4	6
Kentucky	4,339,367	39,486	102,269	6	8
Louisiana	4,533,372	43,204	111,898	6	8
Maine	1,328,361	30,843	79,883	2	4
Maryland	5,773,552	9,707	25,142	8	10
Massachusetts	6,547,629	7,800	20,202	9	11

Real-World Data

State Facts

State	2010 Population	Area (in square miles)	Area (in square kilometers)	Seats in the House of Representatives	Electoral Votes
Michigan	9,883,640	56,539	146,435	14	16
Minnesota	5,303,925	79,627	206,232	8	10
Mississippi	2,967,297	46,923	121,531	4	6
Missouri	5,988,927	68,742	178,040	8	10
Montana	989,415	145,546	376,962	1	3
Nebraska	1,826,341	76,824	198,974	3	5
Nevada	2,700,551	109,781	284,332	4	6
New Hampshire	1,316,470	8,953	23,187	2	4
New Jersey	8,791,894	7,354	19,047	12	14
New Mexico	2,059,179	121,298	314,161	3	5
New York	19,378,102	47,126	122,057	27	29
North Carolina	9,535,483	48,618	125,920	13	15
North Dakota	672,591	69,001	178,711	1	3
Ohio	11,536,504	40,861	105,829	16	18
Oklahoma	3,751,351	68,595	177,660	5	7
Oregon	3,831,074	95,988	248,608	5	7
Pennsylvania	12,702,379	44,743	115,883	18	20
Rhode Island	1,052,567	1,034	2,678	2	4
South Carolina	4,625,364	30,061	77,857	7	9
South Dakota	814,180	75,811	196,350	1	3
Tennessee	6,346,105	41,235	106,798	9	11
Texas	25,145,561	261,232	676,587	36	38
Utah	2,763,885	82,170	212,818	4	6
Vermont	625,741	9,217	23,871	1	3
Virginia	8,001,024	39,490	102,279	11	13
Washington	6,724,540	66,456	172,119	10	12
West Virginia	1,852,994	24,038	62,259	3	5
Wisconsin	5,686,986	54,158	140,268	8	10
Wyoming	563,626	97,093	251,470	1	3

*The District of Columbia is represented by one non-voting delegate.

Real-World Data

United States: Native American Population in 2010

The following map shows the Native American population of each state and the District of Columbia in the year 2010, according to data from the United States Census. Data are reported in the thousands.

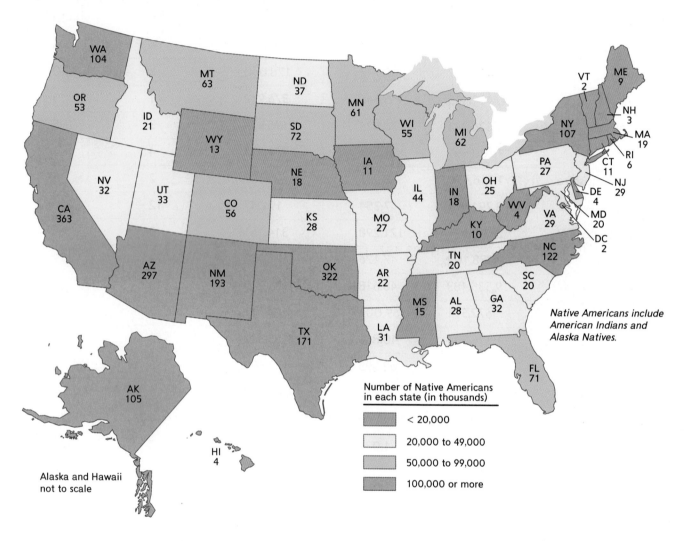

Example

In 2010, approximately 62,000 Native Americans lived in Michigan.
Approximately 122,000 Native Americans lived in North Carolina.

Real-World Data

United States: Foreign-Born Population over Time

Approximately 13% of people living in the United States in 2010 were born in another country. Mexico is the most common country of birth.

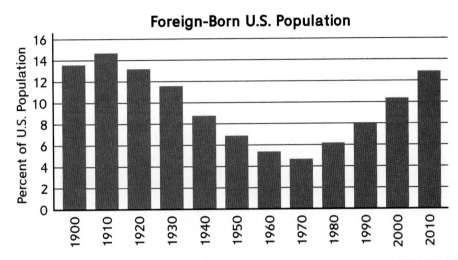

Foreign-Born U.S. Population

	1920			1960			2012	
Country	Number (millions)	%	Country	Number (millions)	%	Country	Number (millions)	%
Germany	1.7	12.1	Italy	1.3	12.9	Mexico	11.6	28.3
Italy	1.6	11.6	Germany	1.0	10.2	China/Hong Kong	2.3	5.6
Soviet Union	1.4	10.1	Canada	1.0	9.8	India	2.0	4.8
Poland	1.1	8.2	Great Britain	0.8	7.9	Philippines	1.9	4.6
Canada	1.1	8.2	Poland	0.7	7.7	El Salvador	1.3	3.1
Great Britain	1.1	8.2	Soviet Union	0.7	7.1	Vietnam	1.3	3.1
Ireland	1.0	7.5	Mexico	0.6	5.9	Cuba	1.1	2.7
Sweden	0.6	4.5	Ireland	0.3	3.5	Korea	1.1	2.7
Austria	0.6	5.1	Austria	0.3	3.1	Dominican Republic	1.0	2.3
Mexico	0.5	3.5	Hungary	0.2	2.5	Guatemala	0.9	2.1

Example

In 1920, 500,000 (0.5 million) people living in the United States were born in Mexico. They accounted for 3.5% of all foreign-born people living in the United States.

By 2012, 11.6 million people living in the United States were born in Mexico. They accounted for more than $\frac{1}{4}$ of all foreign-born people living in the United States.

Real-World Data

World: Largest Countries by Land Area

The table below lists the largest countries in the world by land area. The population of each country in 2013 is also shown.

	Largest Countries by Land Area		
Rank	Country	Area (in square kilometers)	Population in 2013
1	Russia	16,404,742	142,500,482
2	China	9,569,901	1,349,585,838
3	United States	9,158,960	316,668,567
4	Canada	9,093,507	34,568,211
5	Brazil	8,460,415	201,009,622
6	Australia	7,682,300	22,262,501
7	India	2,973,193	1,220,800,359
8	Argentina	2,736,690	42,610,981
9	Kazakhstan	2,699,700	17,736,896
10	Algeria	2,381,741	38,087,812
11	Democratic Republic of the Congo	2,267,048	75,507,308
12	Greenland*	2,166,086	57,714
13	Saudi Arabia	2,149,690	26,939,583
14	Mexico	1,943,945	116,220,947
15	Indonesia	1,811,569	251,160,124
16	Libya	1,759,540	6,002,347
17	Mongolia	1,553,556	3,226,516
18	Iran	1,531,595	79,853,900
19	Peru	1,279,996	29,849,303
20	Niger	1,266,700	16,899,327
21	Chad	1,259,200	11,193,452
22	Angola	1,246,700	18,565,269
23	Mali	1,220,190	15,968,882
24	South Africa	1,214,470	48,601,098
25	Sudan	1,086,788	34,847,910

*Autonomous country within the Kingdom of Denmark

Check Your Understanding

1. Which country in the table above is the most densely populated?
2. Which country in the table above is the least densely populated?

Check your answers in the Answer Key.

Real-World Data

World: Tallest and Largest

The following tables list the tallest buildings in the world and the largest water reservoirs in the world.

World's Tallest Buildings in 2014					
Rank	Name	Location	Height (in meters)	Number of Stories	Year Completed
1	Burj Khalifa	Dubai, United Arab Emirates	828	163	2010
2	Makkah Royal Clock Tower Hotel	Mecca, Saudi Arabia	601	120	2013
3	One World Trade Center	New York City, United States	541	104	2014
4	Taipei 101	Taipei, Taiwan	508	101	2004
5	Shanghai World Financial Center	Shanghai, China	492	101	2008
6	International Commerce Centre	Kowloon, Hong Kong	484	108	2010
7	Petronas Tower I	Kuala Lumpur, Malaysia	452	88	1998
8	Petronas Tower II	Kuala Lumpur, Malaysia	452	88	1998
9	Zifeng Tower	Nanjing, China	450	66	2010
10	Willis Tower	Chicago, United States	442	108	1974

Reservoirs are lakes, often created by humans, used as a water supply. The largest reservoirs in the world are measured by their maximum capacity in cubic meters.

World's Largest Reservoirs in 2013			
Rank	Name	Location	Capacity (in millions of cubic meters)
1	Kariba	Zambia/Zimbabwe	180,600
2	Bratsk	Russia	169,000
3	Aswan High (Lake Nasser)	Egypt	162,000
4	Kpong	Ghana	150,000
5	Akosombo (Lake Volta)	Ghana	150,000
6	Daniel-Johnson	Canada	141,851
7	W. A. C. Bennett	Canada	74,300
8	Krasnoyarsk	Russia	73,300
9	Zeya	Russia	68,400
10	Lajeado	Brazil	64,530

three hundred fifty-nine

Real-World Data

World: Most-Visited Museums

This is a list of the most popular museums in the world in 2013. Some of these museums have free admission, while others require visitors to pay a fee.

World's Most-Visited Museums in 2013			
Rank	Name	Location	Number of Visitors in 2013
1	Louvre Museum	Paris, France	9,334,000
2	National Museum of Natural History	Washington, D.C., United States	8,000,000
3	National Museum of China	Beijing, China	7,450,000
4	National Air and Space Museum	Washington, D.C., United States	6,970,000
5	British Museum	London, United Kingdom	6,701,000
6	Metropolitan Museum of Art	New York City, United States	6,280,000
7	National Gallery	London, United Kingdom	6,031,000
8	Vatican Museums	Vatican City	5,459,000
9	Natural History Museum	London, United Kingdom	5,250,000
10	American Museum of Natural History	New York City, United States	5,000,000

Real-World Data

Sports: Boston Marathon Records

The tables below show the top 10 fastest women's and men's finishing times for the Boston Marathon as of 2014. Marathons are 26.2 miles, or 42.2 kilometers, for both women and men. The Boston Marathon has been held every year since 1897.

Top Women's Finishing Times for the Boston Marathon

Rank	Name	Country	Time (in hr:min:sec)	Year
1	Rita Jeptoo	Kenya	2:18:57	2014
2	Buzunesh Deba	Ethiopia	2:19:59	2014
3	Mare Dibaba	Ethiopia	2:20:35	2014
4	Jemima Jelagat Sumgong	Kenya	2:20:41	2014
5	Margaret Okayo	Kenya	2:20:43	2002
6	Catherine Ndereba	Kenya	2:21:12	2002
7	Uta Pippig	Germany	2:21:45	1994
8	Caroline Kilel	Kenya	2:22:36	2011
9	Desiree Davila	United States	2:22:38	2011
10	Sharon Cherop	Kenya	2:22:42	2011

Top Men's Finishing Times for the Boston Marathon

Rank	Name	Country	Time (in hr:min:sec)	Year
1	Geoffrey Mutai	Kenya	2:03:02	2011
2	Moses Mosop	Kenya	2:03:06	2011
3	Gebregziabher Gebremariam	Ethiopia	2:04:53	2011
4	Ryan Hall	United States	2:04:58	2011
5	Robert Kiprono Cheruiyot	Kenya	2:05:52	2010
6	Abreham Cherkos	Ethiopia	2:06:13	2011
7	Robert Kiprono Cheruiyot	Kenya	2:06:43	2011
8	Philip Kimutai Sanga	Kenya	2:07:10	2011
9	Robert Kipkoech Cheruiyot	Kenya	2:07:14	2006
10	Cosmas Ndeti	Kenya	2:07:15	1994

Check Your Understanding

1. How many of the top 10 fastest men's finishing times occurred in 2011?
2. How much faster was the top men's finishing time than the top women's finishing time?

Check your answers in the Answer Key.

Real-World Data

Sports: Tour de France Records

The table below shows the top 10 fastest finishing times for the Tour de France. The Tour de France is a 21-day bicycle race, in which cyclists from around the world ride across terrain throughout France and sometimes in surrounding countries. The Tour de France started in 1903. The data below were taken from the first 110 years of the race, from 1903 to 2013.

Top Finishing Times for the Tour de France				
Year	Name	Country	Distance (in km)	Time (in hr:min:sec)
2013	Chris Froome	United Kingdom	3,404	83:56:20
1988	Pedro Delgado	Spain	3,286	84:27:53
2009	Alberto Contador	Spain	3,459	85:48:35
2011	Cadel Evans	Australia	3,430	86:12:22
2012	Bradley Wiggins	United Kingdom	3,496	87:34:47
1989	Greg LeMond	United States	3,285	87:38:35
2008	Carlos Sastre	Spain	3,559	87:52:52
1990	Greg LeMond	United States	3,504	90:43:20
1982	Bernard Hinault	France	3,507	92:08:46
1995	Miguel Indurain	Spain	3,635	92:44:59

Example

Chris Froome rode 3,404 kilometers in 21 days. This means he rode his bike an average of 162 kilometers, or 101 miles, every day. Where might you be if you traveled 101 miles from your home or school? Now imagine going through the mountains of the Alps on your way.

Real-World Data

Sports: Olympic Medal Counts

Every four years, countries from around the world send their top athletes to participate in the Olympic Games. There are two Olympics, the Summer Olympics and the Winter Olympics, with different sports in each. The Summer Olympics include sports such as swimming, gymnastics, track and field, tennis, and soccer. The Winter Olympics focus on cold-weather sports, such as skating, ice hockey, skiing, and bobsledding.

Summer Olympics 2012

In 2012, 204 countries sent athletes to London, England to compete in the Summer Olympics. 85 countries received at least one gold, silver, or bronze medal. The table below lists the top 15 countries that participated in the 2012 Summer Olympics. Countries are ranked by the number of gold, then silver, and then bronze medals that they received during the Olympic Games in August 2012.

Winter Olympics 2014

In 2014, the Winter Olympics were held in Sochi, Russia. A total of 88 countries participated in the Olympic Games, and 26 countries won at least one gold, silver, or bronze medal. The table below lists the top 15 countries from the 2014 Winter Olympics, according to the number of gold, then silver, and then bronze medals they received in February 2014.

Medal Counts for the 2012 Summer Olympics					
Rank	Country	Gold	Silver	Bronze	Total
1	United States	46	29	29	104
2	China	38	27	23	88
3	Great Britain	29	17	19	65
4	Russia	24	26	32	82
5	South Korea	13	8	7	28
6	Germany	11	19	14	44
7	France	11	11	12	34
8	Italy	8	9	11	28
9	Hungary	8	4	5	17
10	Australia	7	16	12	35
11	Japan	7	14	17	38
12	Kazakhstan	7	1	5	13
13	Netherlands	6	6	8	20
14	Ukraine	6	5	9	20
15	New Zealand	6	2	5	13

Medal Counts for the 2014 Winter Olympics					
Rank	Country	Gold	Silver	Bronze	Total
1	Russia	13	11	9	33
2	Norway	11	5	10	26
3	Canada	10	10	5	25
4	United States	9	7	12	28
5	Netherlands	8	7	9	24
6	Germany	8	6	5	19
7	Switzerland	6	3	2	11
8	Belarus	5	0	1	6
9	Austria	4	8	5	17
10	France	4	4	7	15
11	Poland	4	1	1	6
12	China	3	4	2	9
13	South Korea	3	3	2	8
14	Sweden	2	7	6	15
15	Czech Republic	2	4	2	8

Real-World Data

Sports: International Athletes in the NBA and MLB

Many athletes who play in American sports leagues grew up in other countries. The tables on this page and the next show the birthplaces of international athletes who play sports in the United States. The first table shows data for the National Basketball Association (NBA) and the second table shows data for Major League Baseball (MLB).

National Basketbal Association Data

At the start of 2013–2014 NBA season, 92 of the 385 players were born outside the United States, representing 39 different countries. That's 23.9% of the total players in the NBA.

Top 10 Birthplaces of International NBA Players		
Rank	Country	Number of Players
1	France	9
2	Canada	8
3	Australia	5
4	Spain	5
5	Argentina	4
6	Brazil	4
7	Italy	4
8	Russia	4
9	Turkey	4
10	Germany	3

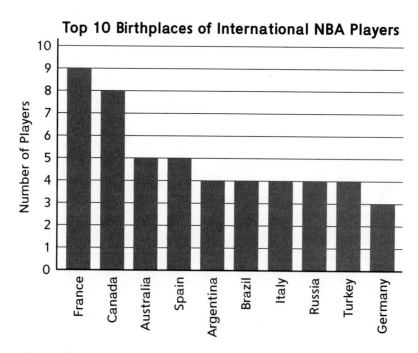

Top 10 Birthplaces of International NBA Players

Real-World Data

Major League Baseball Data

At the start of the 2014 MLB season, 224 of the 853 players were foreign born, from 16 different countries. That's 26.3% of the players in the league.

Top 10 Birthplaces of International MLB Players

Rank	Country	Number of Players
1	Dominican Republic	83
2	Venezuela	59
3	Cuba	19
4	Puerto Rico*	11
5	Canada	10
6	Japan	9
7	Mexico	9
8	Curaçao	5
9	Colombia	4
10	Panama	4

*Puerto Rico is an unincorporated territory of the United States. Puerto Ricans are United States citizens.

Did You Know?

In 1980, less than 2% of the NBA's players were born in countries other than the United States, and only 8.5% of players in the MLB were internationally born.

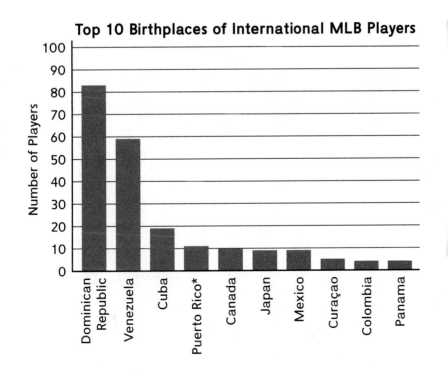

Note Different y-axis scales can make very different data look similar at first glance. Even though the bars showing the number of NBA players from France and the number of MLB players from the Dominican Republic look similar, there are actually almost 10 times as many Dominican players in the MLB than French players in the NBA.

Real-World Data

Science and Nature: Biodiversity

Some places on Earth are home to more species of plants and animals than others. The *biodiversity* in these places is a vital part of each of their ecosystems. The tables below list some of the countries with the greatest number of different types of mammals, birds, reptiles, and plants on Earth. Do you notice anything about the location of these countries? Why do you think there is more diversity in the plants and animals that live there?

Top Countries for Mammals

Country	Continent	Number of Known Species
Indonesia	Asia	670
Brazil	S. America	648
China	Asia	551
Mexico	N. America	523
Peru	S. America	467
Colombia	S. America	442
United States	N. America	440
Dem. Rep. of the Congo	Africa	430
India	Asia	412
Kenya	Africa	376

Top Countries for Birds

Country	Continent	Number of Known Species
Colombia	S. America	1,821
Peru	S. America	1,781
Brazil	S. America	1,712
Indonesia	Asia	1,604
Ecuador	S. America	1,515
Bolivia	S. America	1,414
Venezuela	S. America	1,392
China	Asia	1,221
India	Asia	1,180
Dem. Rep. of the Congo	Africa	1,148

Real-World Data

Top Countries for Reptiles

Country	Continent	Number of Known Species
Australia	Australia	880
Mexico	N. America	837
Indonesia	Asia	749
Brazil	S. America	651
India	Asia	521
Colombia	S. America	518
China	Asia	424
Ecuador	S. America	419
Malaysia	Asia	388
Madagascar	Africa	383

Top Countries for Plants

Country	Continent	Number of Known Species
Brazil	S. America	56,215
Colombia	S. America	51,220
China	Asia	32,200
Indonesia	Asia	29,375
Mexico	N. America	26,071
South Africa	Africa	23,420
Venezuela	S. America	21,073
United States	N. America	19,473
Ecuador	S. America	19,362
India	Asia	18,664

Real-World Data

Science and Nature: Tallest and Deepest

The tables on these pages rank the top 10 tallest mountains, deepest lakes, and tallest trees on Earth.

World's Tallest Mountains

Rank	Mountain	Height (in feet)	Range	Location
1	Everest	29,035	Himalayas	Nepal/Tibet
2	K2 (Godwin-Austen)	28,251	Karakoram	Kashmir
3	Kangchenjunga	28,169	Himalayas	India/Nepal
4	Lhotse I	27,923	Himalayas	Nepal/Tibet
5	Makalu	27,824	Himalayas	Nepal/Tibet
6	Lhotse II	27,560	Himalayas	Nepal/Tibet
7	Dhaulagiri	26,795	Himalayas	Nepal
8	Manaslu I	26,781	Himalayas	Nepal
9	Cho Oyu	26,750	Himalayas	Nepal/Tibet
10	Nanga Parbat	26,660	Himalayas	Kashmir

World's Deepest Lakes

Rank	Lake	Maximum Depth (in feet)	Continent
1	Baikal	5,712	Asia
2	Tanganyika	4,825	Africa
3	Caspian Sea	3,363	Asia
4	Malawi	2,316	Africa
5	Issyk-kul	2,191	Asia
6	Great Slave	2,015	North America
7	Matano	1,936	Asia
8	Crater	1,932	North America
9	Toba	1,736	Asia
10	Hornindals	1,686	Europe

Did You Know?

Lake Baikal in Siberia is $4\frac{1}{2}$ times as deep at its deepest point as the Empire State Building is tall.

Real-World Data

World's Tallest Trees

Rank	Name	Height (in meters)	Diameter (in meters)	Location
1	Hyperion	115.61	4.84	Redwood National Park, California
2	Helios	114.58	4.96	Redwood National Park, California
3	Icarus	113.14	3.78	Redwood National Park, California
4	Stratosphere Giant	113.05	5.18	Humboldt Redwoods State Park, California
5	National Geographic	112.71	4.39	Redwood National Park, California
6	Orion	112.63	4.33	Redwood National Park, California
7	Federation Giant	112.62	4.54	Humboldt Redwoods State Park, California
8	Paradox	112.51	3.90	Humboldt Redwoods State Park, California
9	Mendocino	112.32	4.19	Montgomery Woods State Reserve, California
10	Millennium	111.92	2.71	Humboldt Redwoods State Park, California

Check Your Understanding

1. What is the mean height of the 10 tallest trees in the world?
2. How many trees the height of Hyperion would it take to reach the lake floor at the deepest point in Lake Baikal? Hint: Pay attention to the units of each measurement.
3. World-class climbers can ascend Mount Everest at a rate of around 900–1,000 feet per hour. How long would it take climbers to reach the summit if they climbed it at 900 feet per hour without stopping? Give your answer in days and hours.

Check your answers in the Answer Key.

Real-World Data

Science and Nature: Electricity Consumption

Countries produce electricity in many different ways. Some electricity is produced from non-renewable resources, such as natural gas, coal, or uranium (used for nuclear power). Other means of generating electricity are renewable. Solar, wind, and hydroelectric power (power from water) are examples of renewable sources of electricity. As countries grow more concerned about environmental issues and dwindling resources, many are moving toward producing their electricity from renewable resources. In 2013, 22% of the world's electricity came from renewable resources. The table below lists the 10 countries with the largest percentage of their electricity produced from renewable resources. The electricity usage is reported in terawatt-hours (TWh).

Countries with the Highest Percentage of Renewable Energy in 2012

Rank	Country	Amount of Electricity Produced from Renewable Resources in 2012 (in TWh)	Percent of Electricity Produced from Renewable Sources
1	Iceland	17.5	100%
2	Ethiopia	7.0	99.9%
3	Norway	144.8	98.0%
4	Costa Rica	9.1	91.4%
5	Brazil	462.2	82.7%
6	Colombia	48.3	80.6%
7	Uganda	2.3	78.6%
8	Austria	50.9	74.6%
9	Cameroon	4.6	74.4%
10	Kenya	6.0	73.1%

Note A terawatt-hour (TWh) is equal to the power of 1,000,000,000 watts providing energy for one hour. The energy of one TWh is enough to power approximately 90,000 average homes in the United States for a year.

Although the United States is one of the leading producers of renewable energy in the world, the population consumes so much electricity that the United States actually produces a smaller *percent* of renewable energy than the average country in the world. The circle graphs compare the renewable and non-renewable electricity produced by the United States and Brazil.

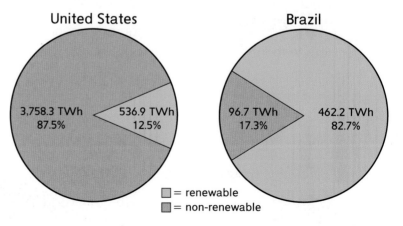

Renewable and Non-Renewable Electricity Produced by the United States and Brazil

United States: 3,758.3 TWh 87.5%; 536.9 TWh 12.5%

Brazil: 96.7 TWh 17.3%; 462.2 TWh 82.7%

☐ = renewable
■ = non-renewable

Real-World Data

Data Sources

The lists below give many of the sources for the data in this section. The information on some pages came from just one or two sources. On other pages the information came from many different sources. Often the information came from one or more sources and then was checked using different sources.

The numbers on these pages are not always what you might find if you looked them up yourself. Data may change as experts discover or collect new information. Or you may find the results of different sets of data from those shown here.

United States: Utah National Parks

www.nps.gov

United States: State Facts

www.census.gov

United States: Native American Population in 2010

Janssen, S. (Ed.). (2013). *The World Almanac and Book of Facts 2014*. New York: World Almanac Books.

United States: Foreign-Born Population over Time

www.census.gov

Janssen, S. (Ed.). (2013). *The World Almanac and Book of Facts 2014*. New York: World Almanac Books.

World: Largest Countries by Land Area

www.fao.org

Janssen, S. (Ed.). (2013). *The World Almanac and Book of Facts 2014*. New York: World Almanac Books.

World: Tallest and Largest

Janssen, S. (Ed.). (2013). *The World Almanac and Book of Facts 2014*. New York: World Almanac Books.

Council on Tall Buildings and Urban Habitat

World: Most Visited Museums

www.teaconnect.org

three hundred seventy-one | SRB 371

Real-World Data

Sports: Boston Marathon Records

Boston Athletic Association

Sports: Tour de France Records

British Broadcasting Corporation

www.letour.fr

Union Cycliste Internationale

Sports: Olympic Medal Counts

British Broadcasting Corporation

NBC Olympics

www.olympic.org

Sports: International Athletes in the NBA and MLB

ESPN

Major League Baseball

National Basketball Association

Science and Nature: Biodiversity

www.iucnredlist.org

World Conservation Monitoring Centre of the United Nations Environment Programme

Science and Nature: Tallest and Deepest

Ash, Caroline. (2012). *Top 10 of Everything 2013*. New York: Sterling.

Janssen, S. (Ed.). (2013). *The World Almanac and Book of Facts 2014*. New York: World Almanac Books.

www.conifers.org

Science and Nature: Electricity Consumption

Wall Street Journal

www.eia.gov

www.energies-renouvelables.org

Tables and Charts

Place-Value Chart

billions	100 millions	10 millions	millions	100 thousands	10 thousands	thousands	hundreds	tens	ones	.	tenths	hundredths	thousandths
1,000 millions	100,000,000s	10,000,000s	1,000,000s	100,000s	10,000s	1,000s	100s	10s	1s	.	0.1s	0.01s	0.001s
10^9	10^8	10^7	10^6	10^5	10^4	10^3	10^2	10^1	10^0	.	10^{-1}	10^{-2}	10^{-3}

Rules for Order of Operations

1. Do operations inside parentheses or other grouping symbols first.
2. Calculate all expressions with exponents (e.g., $10^2 = 10 * 10 = 100$).
3. Multiply and divide in order, from left to right.
4. Add and subtract in order, from left to right.

Prefixes

uni-	one	tera-	trillion (10^{12})
bi-	two	giga-	billion (10^9)
tri-	three	mega-	million (10^6)
quad-	four	kilo-	thousand (10^3)
penta-	five	hecto-	hundred (10^2)
hexa-	six	deca-	ten (10^1)
hepta-	seven	uni-	one (10^0)
octa-	eight	deci-	tenth (10^{-1})
nona-	nine	centi-	hundredth (10^{-2})
deca-	ten	milli-	thousandth (10^{-3})
dodeca-	twelve	micro-	millionth (10^{-6})
icosa-	twenty	nano-	billionth (10^{-9})

Multiplication and Division Table

*, /	1	2	3	4	5	6	7	8	9	10
1	1	2	3	4	5	6	7	8	9	10
2	2	4	6	8	10	12	14	16	18	20
3	3	6	9	12	15	18	21	24	27	30
4	4	8	12	16	20	24	28	32	36	40
5	5	10	15	20	25	30	35	40	45	50
6	6	12	18	24	30	36	42	48	54	60
7	7	14	21	28	35	42	49	56	63	70
8	8	16	24	32	40	48	56	64	72	80
9	9	18	27	36	45	54	63	72	81	90
10	10	20	30	40	50	60	70	80	90	100

The numbers on the diagonal are square numbers.

three hundred seventy-three

Tables and Charts

Metric System

Units of Length

1 kilometer (km) = 1,000 meters (m)
1 meter = 10 decimeters (dm)
= 100 centimeters (cm)
= 1,000 millimeters (mm)
1 decimeter = 10 centimeters
1 centimeter = 10 millimeters

Units of Area

1 square meter (m^2) = 100 square decimeters (dm^2)
= 10,000 square centimeters (cm^2)
1 square decimeter = 100 square centimeters
1 square kilometer (km^2) = 1,000,000 square meters

Units of Volume

1 cubic meter (m^3) = 1,000 cubic decimeters (dm^3)
= 1,000,000 cubic centimeters (cm^3)
1 cubic decimeter = 1,000 cubic centimeters

Units of Capacity and Liquid Volume

1 kiloliter (kL) = 1,000 liters (L)
1 liter = 1,000 milliliters (mL)
1 cubic centimeter = 1 milliliter

Units of Mass and Weight

1 metric ton (t) = 1,000 kilograms (kg)
1 kilogram = 1,000 grams (g)
1 gram = 1,000 milligrams (mg)

U.S. Customary System

Units of Length

1 mile (mi) = 1,760 yards (yd)
= 5,280 feet (ft)
1 yard = 3 feet
= 36 inches (in.)
1 foot = 12 inches

Units of Area

1 square yard (yd^2) = 9 square feet (ft^2)
= 1,296 square inches (in.^{2})
1 square foot = 144 square inches
1 acre (a.) = 43,560 square feet
1 square mile (mi^2) = 640 acres

Units of Volume

1 cubic yard (yd^3) = 27 cubic feet (ft^3)
1 cubic foot = 1,728 cubic inches (in.^{3})

Units of Capacity and Liquid Volume

1 gallon (gal) = 4 quarts (qt)
1 quart = 2 pints (pt)
1 pint = 2 cups (c)
1 cup = 8 fluid ounces (fl oz)
1 fluid ounce = 2 tablespoons (tbs)
1 tablespoon = 3 teaspoons (tsp)

Units of Mass and Weight

1 ton (T) = 2,000 pounds (lb)
1 pound = 16 ounces (oz)

System Equivalents

1 inch is about 2.5 centimeters (2.54)
1 kilometer is about 0.6 mile (0.621)
1 mile is about 1.6 kilometers (1.609)
1 meter is about 39 inches (39.37)
1 liter is about 1.1 quarts (1.057)
1 ounce is about 28 grams (28.350)
1 kilogram is about 2.2 pounds (2.205)

Units of Time

1 century = 100 years
1 decade = 10 years
1 year (yr) = 12 months
= 52 weeks (plus one or two days)
= 365 days (366 days in a leap year)
1 month (mo) = 28, 29, 30, or 31 days
1 week (wk) = 7 days
1 day (d) = 24 hours
1 hour (hr) = 60 minutes
1 minute (min) = 60 seconds (sec)

Tables and Charts

Personal References of Measurement

Metric System

Units of Length

About 1 millimeter	About 1 centimeter
• thickness of a thumbtack point • thickness of a dime	• width of a fingertip • thickness of a crayon

About 1 meter	About 1 kilometer
• one big step (for an adult) • width of a front door	• length of 10 football fields (including the end zones) • 1,000 big steps (for an adult)

Units of Mass

About 1 gram	About 1 kilogram
• dollar bill • large paper clip	• wooden baseball bat • 200 U.S. nickels

Units of Capacity and Liquid Volume

About 1 milliliter	About 1 liter
• 20 drops of water • the amount of water that could fit in a centimeter cube*	• 3 regular-size canned drinks • the amount of water that could fit in a thousand cube*

*Note that 1 mL is defined as 1 cm^3, so 1 cm^3 of water is exactly 1 mL, and 1,000 cm^3 of water is exactly 1,000 mL, or 1 L.

U.S. Customary System

Units of Length

About 1 inch	About 1 foot
• width (diameter) of a quarter • width of a man's thumb	• distance from elbow to wrist (for an adult) • length of a piece of paper

About 1 yard	About 1 mile
• one big step (for an adult) • width of a front door	• length of 15 football fields (including the end zones) • 2,000 average-size steps (for an adult)

Units of Weight

About 1 ounce	About 1 pound	About 1 ton
• slice of bread • 11 pennies	• 4 sticks of butter • one adult shoe	• giraffe • car

Units of Capacity and Liquid Volume

About 1 cup	About 1 pint	About 1 quart	About 1 gallon
• individual carton of milk	• small carton of half and half	• medium-size carton of half and half	• large container of milk

three hundred seventy-five

Tables and Charts

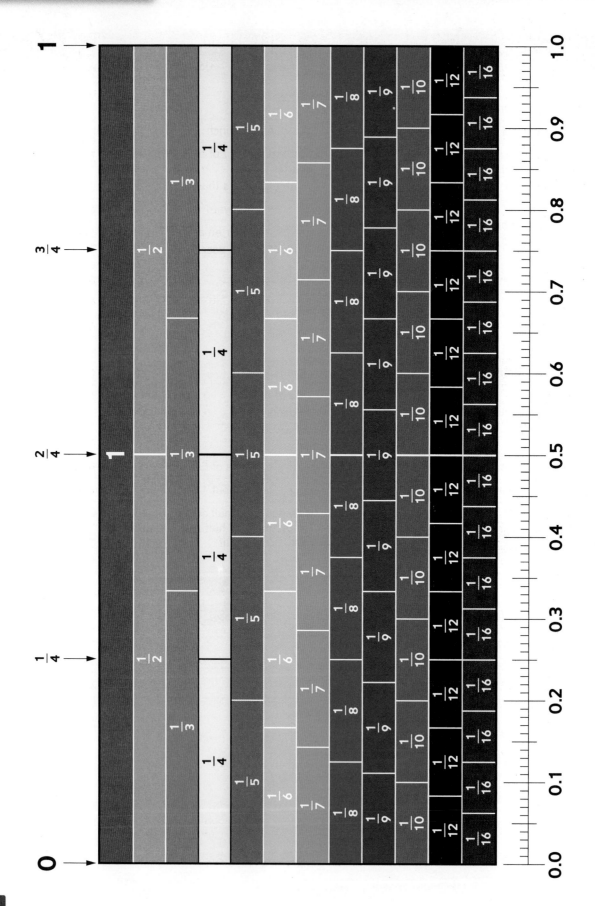

Fraction-Stick and Decimal Number-Line Chart

Equivalent Fractions, Decimals, and Percents

$\frac{1}{2}$	$\frac{2}{4}$	$\frac{3}{6}$	$\frac{4}{8}$	$\frac{5}{10}$	$\frac{6}{12}$	$\frac{7}{14}$	$\frac{8}{16}$	$\frac{9}{18}$	$\frac{10}{20}$	$\frac{11}{22}$	$\frac{12}{24}$	$\frac{13}{26}$	$\frac{14}{28}$	$\frac{15}{30}$	0.5	50%
$\frac{1}{3}$	$\frac{2}{6}$	$\frac{3}{9}$	$\frac{4}{12}$	$\frac{5}{15}$	$\frac{6}{18}$	$\frac{7}{21}$	$\frac{8}{24}$	$\frac{9}{27}$	$\frac{10}{30}$	$\frac{11}{33}$	$\frac{12}{36}$	$\frac{13}{39}$	$\frac{14}{42}$	$\frac{15}{45}$	$0.\overline{3}$	$33\frac{1}{3}$%
$\frac{2}{3}$	$\frac{4}{6}$	$\frac{6}{9}$	$\frac{8}{12}$	$\frac{10}{15}$	$\frac{12}{18}$	$\frac{14}{21}$	$\frac{16}{24}$	$\frac{18}{27}$	$\frac{20}{30}$	$\frac{22}{33}$	$\frac{24}{36}$	$\frac{26}{39}$	$\frac{28}{42}$	$\frac{30}{45}$	$0.\overline{6}$	$66\frac{2}{3}$%
$\frac{1}{4}$	$\frac{2}{8}$	$\frac{3}{12}$	$\frac{4}{16}$	$\frac{5}{20}$	$\frac{6}{24}$	$\frac{7}{28}$	$\frac{8}{32}$	$\frac{9}{36}$	$\frac{10}{40}$	$\frac{11}{44}$	$\frac{12}{48}$	$\frac{13}{52}$	$\frac{14}{56}$	$\frac{15}{60}$	0.25	25%
$\frac{3}{4}$	$\frac{6}{8}$	$\frac{9}{12}$	$\frac{12}{16}$	$\frac{15}{20}$	$\frac{18}{24}$	$\frac{21}{28}$	$\frac{24}{32}$	$\frac{27}{36}$	$\frac{30}{40}$	$\frac{33}{44}$	$\frac{36}{48}$	$\frac{39}{52}$	$\frac{42}{56}$	$\frac{45}{60}$	0.75	75%
$\frac{1}{5}$	$\frac{2}{10}$	$\frac{3}{15}$	$\frac{4}{20}$	$\frac{5}{25}$	$\frac{6}{30}$	$\frac{7}{35}$	$\frac{8}{40}$	$\frac{9}{45}$	$\frac{10}{50}$	$\frac{11}{55}$	$\frac{12}{60}$	$\frac{13}{65}$	$\frac{14}{70}$	$\frac{15}{75}$	0.2	20%
$\frac{2}{5}$	$\frac{4}{10}$	$\frac{6}{15}$	$\frac{8}{20}$	$\frac{10}{25}$	$\frac{12}{30}$	$\frac{14}{35}$	$\frac{16}{40}$	$\frac{18}{45}$	$\frac{20}{50}$	$\frac{22}{55}$	$\frac{24}{60}$	$\frac{26}{65}$	$\frac{28}{70}$	$\frac{30}{75}$	0.4	40%
$\frac{3}{5}$	$\frac{6}{10}$	$\frac{9}{15}$	$\frac{12}{20}$	$\frac{15}{25}$	$\frac{18}{30}$	$\frac{21}{35}$	$\frac{24}{40}$	$\frac{27}{45}$	$\frac{30}{50}$	$\frac{33}{55}$	$\frac{36}{60}$	$\frac{39}{65}$	$\frac{42}{70}$	$\frac{45}{75}$	0.6	60%
$\frac{4}{5}$	$\frac{8}{10}$	$\frac{12}{15}$	$\frac{16}{20}$	$\frac{20}{25}$	$\frac{24}{30}$	$\frac{28}{35}$	$\frac{32}{40}$	$\frac{36}{45}$	$\frac{40}{50}$	$\frac{44}{55}$	$\frac{48}{60}$	$\frac{52}{65}$	$\frac{56}{70}$	$\frac{60}{75}$	0.8	80%
$\frac{1}{6}$	$\frac{2}{12}$	$\frac{3}{18}$	$\frac{4}{24}$	$\frac{5}{30}$	$\frac{6}{36}$	$\frac{7}{42}$	$\frac{8}{48}$	$\frac{9}{54}$	$\frac{10}{60}$	$\frac{11}{66}$	$\frac{12}{72}$	$\frac{13}{78}$	$\frac{14}{84}$	$\frac{15}{90}$	$0.1\overline{6}$	$16\frac{2}{3}$%
$\frac{5}{6}$	$\frac{10}{12}$	$\frac{15}{18}$	$\frac{20}{24}$	$\frac{25}{30}$	$\frac{30}{36}$	$\frac{35}{42}$	$\frac{40}{48}$	$\frac{45}{54}$	$\frac{50}{60}$	$\frac{55}{66}$	$\frac{60}{72}$	$\frac{65}{78}$	$\frac{70}{84}$	$\frac{75}{90}$	$0.8\overline{3}$	$83\frac{1}{3}$%
$\frac{1}{7}$	$\frac{2}{14}$	$\frac{3}{21}$	$\frac{4}{28}$	$\frac{5}{35}$	$\frac{6}{42}$	$\frac{7}{49}$	$\frac{8}{56}$	$\frac{9}{63}$	$\frac{10}{70}$	$\frac{11}{77}$	$\frac{12}{84}$	$\frac{13}{91}$	$\frac{14}{98}$	$\frac{15}{105}$	0.143	14.3%
$\frac{2}{7}$	$\frac{4}{14}$	$\frac{6}{21}$	$\frac{8}{28}$	$\frac{10}{35}$	$\frac{12}{42}$	$\frac{14}{49}$	$\frac{16}{56}$	$\frac{18}{63}$	$\frac{20}{70}$	$\frac{22}{77}$	$\frac{24}{84}$	$\frac{26}{91}$	$\frac{28}{98}$	$\frac{30}{105}$	0.286	28.6%
$\frac{3}{7}$	$\frac{6}{14}$	$\frac{9}{21}$	$\frac{12}{28}$	$\frac{15}{35}$	$\frac{18}{42}$	$\frac{21}{49}$	$\frac{24}{56}$	$\frac{27}{63}$	$\frac{30}{70}$	$\frac{33}{77}$	$\frac{36}{84}$	$\frac{39}{91}$	$\frac{42}{98}$	$\frac{45}{105}$	0.429	42.9%
$\frac{4}{7}$	$\frac{8}{14}$	$\frac{12}{21}$	$\frac{16}{28}$	$\frac{20}{35}$	$\frac{24}{42}$	$\frac{28}{49}$	$\frac{32}{56}$	$\frac{36}{63}$	$\frac{40}{70}$	$\frac{44}{77}$	$\frac{48}{84}$	$\frac{52}{91}$	$\frac{56}{98}$	$\frac{60}{105}$	0.571	57.1%
$\frac{5}{7}$	$\frac{10}{14}$	$\frac{15}{21}$	$\frac{20}{28}$	$\frac{25}{35}$	$\frac{30}{42}$	$\frac{35}{49}$	$\frac{40}{56}$	$\frac{45}{63}$	$\frac{50}{70}$	$\frac{55}{77}$	$\frac{60}{84}$	$\frac{65}{91}$	$\frac{70}{98}$	$\frac{75}{105}$	0.714	71.4%
$\frac{6}{7}$	$\frac{12}{14}$	$\frac{18}{21}$	$\frac{24}{28}$	$\frac{30}{35}$	$\frac{36}{42}$	$\frac{42}{49}$	$\frac{48}{56}$	$\frac{54}{63}$	$\frac{60}{70}$	$\frac{66}{77}$	$\frac{72}{84}$	$\frac{78}{91}$	$\frac{84}{98}$	$\frac{90}{105}$	0.857	85.7%
$\frac{1}{8}$	$\frac{2}{16}$	$\frac{3}{24}$	$\frac{4}{32}$	$\frac{5}{40}$	$\frac{6}{48}$	$\frac{7}{56}$	$\frac{8}{64}$	$\frac{9}{72}$	$\frac{10}{80}$	$\frac{11}{88}$	$\frac{12}{96}$	$\frac{13}{104}$	$\frac{14}{112}$	$\frac{15}{120}$	0.125	$12\frac{1}{2}$%
$\frac{3}{8}$	$\frac{6}{16}$	$\frac{9}{24}$	$\frac{12}{32}$	$\frac{15}{40}$	$\frac{18}{48}$	$\frac{21}{56}$	$\frac{24}{64}$	$\frac{27}{72}$	$\frac{30}{80}$	$\frac{33}{88}$	$\frac{36}{96}$	$\frac{39}{104}$	$\frac{42}{112}$	$\frac{45}{120}$	0.375	$37\frac{1}{2}$%
$\frac{5}{8}$	$\frac{10}{16}$	$\frac{15}{24}$	$\frac{20}{32}$	$\frac{25}{40}$	$\frac{30}{48}$	$\frac{35}{56}$	$\frac{40}{64}$	$\frac{45}{72}$	$\frac{50}{80}$	$\frac{55}{88}$	$\frac{60}{96}$	$\frac{65}{104}$	$\frac{70}{112}$	$\frac{75}{120}$	0.625	$62\frac{1}{2}$%
$\frac{7}{8}$	$\frac{14}{16}$	$\frac{21}{24}$	$\frac{28}{32}$	$\frac{35}{40}$	$\frac{42}{48}$	$\frac{49}{56}$	$\frac{56}{64}$	$\frac{63}{72}$	$\frac{70}{80}$	$\frac{77}{88}$	$\frac{84}{96}$	$\frac{91}{104}$	$\frac{98}{112}$	$\frac{105}{120}$	0.875	$87\frac{1}{2}$%
$\frac{1}{9}$	$\frac{2}{18}$	$\frac{3}{27}$	$\frac{4}{36}$	$\frac{5}{45}$	$\frac{6}{54}$	$\frac{7}{63}$	$\frac{8}{72}$	$\frac{9}{81}$	$\frac{10}{90}$	$\frac{11}{99}$	$\frac{12}{108}$	$\frac{13}{117}$	$\frac{14}{126}$	$\frac{15}{135}$	$0.\overline{1}$	$11\frac{1}{9}$%
$\frac{2}{9}$	$\frac{4}{18}$	$\frac{6}{27}$	$\frac{8}{36}$	$\frac{10}{45}$	$\frac{12}{54}$	$\frac{14}{63}$	$\frac{16}{72}$	$\frac{18}{81}$	$\frac{20}{90}$	$\frac{22}{99}$	$\frac{24}{108}$	$\frac{26}{117}$	$\frac{28}{126}$	$\frac{30}{135}$	$0.\overline{2}$	$22\frac{2}{9}$%
$\frac{4}{9}$	$\frac{8}{18}$	$\frac{12}{27}$	$\frac{16}{36}$	$\frac{20}{45}$	$\frac{24}{54}$	$\frac{28}{63}$	$\frac{32}{72}$	$\frac{36}{81}$	$\frac{40}{90}$	$\frac{44}{99}$	$\frac{48}{108}$	$\frac{52}{117}$	$\frac{56}{126}$	$\frac{60}{135}$	$0.\overline{4}$	$44\frac{4}{9}$%
$\frac{5}{9}$	$\frac{10}{18}$	$\frac{15}{27}$	$\frac{20}{36}$	$\frac{25}{45}$	$\frac{30}{54}$	$\frac{35}{63}$	$\frac{40}{72}$	$\frac{45}{81}$	$\frac{50}{90}$	$\frac{55}{99}$	$\frac{60}{108}$	$\frac{65}{117}$	$\frac{70}{126}$	$\frac{75}{135}$	$0.\overline{5}$	$55\frac{5}{9}$%
$\frac{7}{9}$	$\frac{14}{18}$	$\frac{21}{27}$	$\frac{28}{36}$	$\frac{35}{45}$	$\frac{42}{54}$	$\frac{49}{63}$	$\frac{56}{72}$	$\frac{63}{81}$	$\frac{70}{90}$	$\frac{77}{99}$	$\frac{84}{108}$	$\frac{91}{117}$	$\frac{98}{126}$	$\frac{105}{135}$	$0.\overline{7}$	$77\frac{7}{9}$%
$\frac{8}{9}$	$\frac{16}{18}$	$\frac{24}{27}$	$\frac{32}{36}$	$\frac{40}{45}$	$\frac{48}{54}$	$\frac{56}{63}$	$\frac{64}{72}$	$\frac{72}{81}$	$\frac{80}{90}$	$\frac{88}{99}$	$\frac{96}{108}$	$\frac{104}{117}$	$\frac{112}{126}$	$\frac{120}{135}$	$0.\overline{8}$	$88\frac{8}{9}$%

Note: The decimals for sevenths have been rounded to the nearest thousandth.

Tables and Charts

Formulas	Meaning of Variables
Rectangles	
• Perimeter: $p = (2 * l) + (2 * w)$ • Area: $A = l * w$	p = perimeter; l = length; w = width A = area
Squares	
• Perimeter: $p = 4 * s$ • Area: $A = s^2$	p = perimeter; s = length of side A = area
Parallelograms	
• Area: $A = b * h$	A = area; b = length of base; h = height
Triangles	
• Area: $A = \frac{1}{2} * b * h$	A = area; b = length of base; h = height
Regular Polygons	
• Perimeter: $p = n * s$	p = perimeter; n = number of sides; s = length of side
Mobiles (1 rod, with 2 objects on opposite sides of the fulcrum)	
• When the fulcrum *is* at the center of the rod, the mobile will balance if $W * D = w * d$.	W = weight of one object D = distance of this object from the fulcrum w = weight of the second object d = distance of the second object from the fulcrum
• When the fulcrum is *not* at the center of the rod, the mobile will balance if $(W * D) + (R * L) = w * d$.	R = weight of rod L = distance from center of rod to fulcrum W = weight of object on same side of fulcrum as center D = distance of this object from the fulcrum w = weight of the second object d = distance of second object from the fulcrum
Rectangular Prisms	
• Volume: $V = B * h$, or $V = l * w * h$ • Surface area: $S = 2 * ((l * w) + (l * h) + (w * h))$ The surface area formula is true only when all of the faces of the prism are rectangles.	V = volume; B = area of base; l = length; w = width; h = height S = surface area
Cubes	
• Volume: $V = e^3$ • Surface area: $S = 6 * e^2$	V = volume; e = length of edge S = surface area
Pyramids	
• Volume: $V = \frac{1}{3} * B * h$	V = volume; B = area of base; h = height;
Distances	
• $d = r * t$	d = distance traveled; r = rate of speed; t = time of travel

Appendix

Using Calculators

You can use calculators to help you count and work with whole numbers, fractions, decimals, exponents, and percents.

As with any mathematical tool or strategy, you need to think about when and how to use a calculator. It can help you compute quickly and accurately when you have many problems to do in a short time. Calculators can help you compute with very large and very small numbers that may be hard to do in your head or with pencil and paper. Whenever you use a calculator, estimation should be part of your work. Always ask yourself whether the answer in the display makes sense.

There are many different kinds of calculators. *Four-function calculators* do little more than add, subtract, multiply, and divide whole numbers and decimals. More advanced *scientific calculators* let you find powers and reciprocals and perform some operations with fractions. In higher grades, you may use *graphic calculators* that draw graphs, find data landmarks, and do even more complicated mathematics.

There are many calculators that work well with *Everyday Mathematics*. If the instructions in this book don't work for your calculator or the keys on your calculator are not explained here, refer to the directions that came with your calculator, look them up online, or ask your teacher for help.

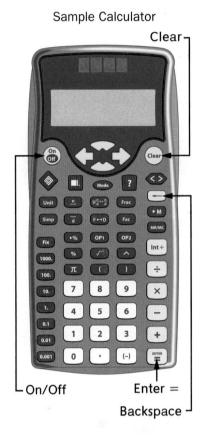

Sample Calculator

Note Many handheld calculators use light cells for power. If you press the ON key and see nothing on the display, hold the front of the calculator toward a light or a sunny window for a moment and then press ON again.

three hundred seventy-nine

Appendix

Basic Operations on a Calculator

Entering and Clearing

Pressing a key on a calculator is called *keying in*, or *entering*. In this book, calculator keys, except numbers and decimal points, are shown in rectangular boxes: [+], [=], [×] and so on. A set of instructions for performing a calculation is called a *key sequence*.

The simplest key sequences turn the calculator on and enter or clear numbers or other characters. These keys are labeled on the calculator on the previous page and summarized below.

Sample Calculator	
Key	Purpose
[On/Off]	Turn the display on.
[Clear] and [On/Off] at the same time	Clear the display and the short-term memory.
[Clear]	Clear only the display.
[←]	Clear the last digit.

Note Calculators have two kinds of memory. *Short-term memory* is for the last number entered. The keys with an "M" on them are for *long-term memory* and are explained on pages 399–405.

Always clear both the display and the memory each time you turn on your calculator.

Many calculators have a backspace key that will clear the last digit or digits you entered without needing to re-enter the whole number.

Example

Enter 123.444. Revise it to 123.456.

Sample Calculator		
	Key Sequence	Display
	123.444	123.444
	[←] [←]	123.4
	56	123.456

Locate and use the backspace key on your calculator.

SRB
380 three hundred eighty

Appendix

Order of Operations and Parentheses on a Calculator

When you use a calculator for basic arithmetic operations, you enter the numbers and operations and press ⊟ or ⌜Enter⌝ to see the answer.

Note Examples for arithmetic calculations using paper and pencil are given in earlier sections of this book.

Try your calculator to see whether it follows the rules for the **order of operations**. Key in 5 ⊕ 6 ⊗ 2 ⊟.

- If your calculator follows the order of operations, it will display 17.

Note See page 203 for more information on the order of operations.

- If it does not follow the order of operations, it will probably do the operations in the order they were entered, adding and then multiplying, and display 22.

Scientific calculators follow the order of operations. If you want the calculator to do operations in a different order, use the **parentheses** keys ⌜(⌝ and ⌜)⌝.

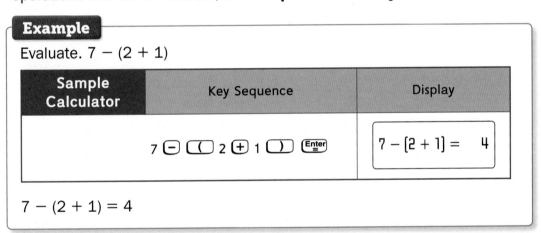

Parentheses can replace the multiplication symbol in some expresions. *Remember to press the multiplication key even when it is not shown.*

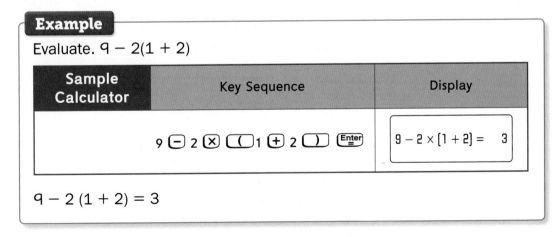

three hundred eighty-one SRB 381

Appendix

Some situations require more than one set of parentheses. *Nested parentheses* are those that are inside each other. Do the operations in the innermost parentheses first. Then do the operations in the next set of parentheses and continue working outward. Different symbols such as brackets [ ] or braces { } can be written to replace the outer parentheses, but they mean the same thing in a mathematical expression.

When using a calculator, the parentheses keys and are used regardless of the kind of grouping symbols written in a number sentence.

Example

Solve. 8 * [16 − (4 + 5)] = ?

Sample Calculator	Key Sequence	Display
	8 ⊗ ⦇ 16 ⊖ ⦇ 4 ⊕ 5 ⦈ ⦈ Enter	8 × [16 − [4 + 5]]= 56

8 * [16 − (4 + 5)] = 56

Note While the symbols * and × both mean multiplication, calculators only have a ⊗ key.

Check Your Understanding

Use your calculator to evaluate each expression.

1. 79 − (4 + 8)
2. 92 − 7(4 + 8)
3. 3 × [(7.6 × 2) − 6] − 20
4. {(46 − 22) * 2}/(6 + 19)

Check your answers in the Answer Key.

Appendix

Negative Numbers on a Calculator

How you enter a negative number depends on your calculator. Use the change sign key, either (−) or +/− depending on your calculator. Both keys change the sign of the number.

Example

Enter −45.

Sample Calculator	Key Sequence	Display
	(−) 45 Enter	−45 = −45

Note For some calculators, if the number on the display is positive, it becomes negative after you press +/−. If the number on the display is negative, it becomes positive after you press +/−. Keys like this are called *toggles*.

Example

What happens if you try to subtract with (−)?
Try it for 38 − 9 = ?

Sample Calculator	Key Sequence	Display
	38 (−) 9 Enter	SYN ERROR

The (−) key is not meant to be used for subtraction.
38 − 9 = 29

Note "SYN" is short for "syntax," which means the ordering and meaning of keys in a sequence.

Note If you try to subtract using the +/− key, it just changes the sign of the first number and adds the digits of the second number to it. This key is not meant to be used for subtraction.

three hundred eighty-three

Appendix

Division with Remainders on a Calculator

The answer to a division problem with whole numbers does not always result in whole number answers. When this happens, most calculators display the answer as a decimal.

Be sure to make an estimate when you divide to see whether your answer makes sense.

Example

Use the division key to solve $39 \div 5 = ?$

Estimate: Since $40 \div 5 = 8$, the quotient will be a little less than 8.

Sample Calculator	Key Sequence	Display
	39 ÷ 5 Enter	39 ÷ 5 = 7.8

$39 \div 5 = $ **7.8** The quotient of 7.8 matches the estimate of a little less than 8.

Some calculators have a second division key that displays the whole-number quotient with a whole-number **remainder.**

Example

Use the division with remainder key to solve $39 \div 5 = ?$

Sample Calculator	Key Sequence	Display
	39 Int÷ 5 Enter	39 ÷ 5 = 7r 4

$39 \div 5 \rightarrow $ **7 R4** This answer makes sense because 7 with 4 left over is close to the estimate of a little less than 8.

"Int" stands for "integer." This kind of division is sometimes called *integer division*.

To write a remainder as a fraction, make the remainder the **numerator** of the fraction and the divisor the **denominator.** For example, the quotient of $39 \div 5$ can be written as $7\frac{4}{5}$.

Try using the division with remainder key on your calculator to see how it works.

Note On some calculators, ÷R means "divide with remainder." You can also divide positive fractions and decimals with ÷R.

Fractions and Percents on a Calculator

Some calculators let you enter, rewrite, and do operations with fractions. Once you know how to enter a fraction, you can add, subtract, multiply, or divide them just like whole numbers and decimals.

Entering Fractions and Mixed Numbers

Most calculators that let you enter fractions use similar key sequences. For fractions, always start by entering the numerator. Then press the appropriate key to tell the calculator to begin writing a fraction.

Example

Enter $\frac{5}{8}$ as a fraction in your calculator.

Sample Calculator	Key Sequence	Display
	5 [n] 8 [d] [Enter]	$\frac{5}{8} = \frac{5}{8}$

Note Pressing [d] after you enter the denominator is optional.

To enter a mixed number, enter the whole number part first. Then press the fraction key or keys to enter the fraction part of the number.

Example

Enter $73\frac{2}{5}$ as a mixed number in your calculator.

Sample Calculator	Key Sequence	Display
	73 [Unit] 2 [n] 5 [d] [Enter]	$73\frac{2}{5} = 73\frac{2}{5}$

Some calculators have different keys for entering fractions and whole numbers. Try entering a mixed number on your calculator.

Appendix

The keys to convert between mixed numbers and fractions greater than one are similar on most fraction calculators.

Example

Convert $\frac{45}{7}$ to a mixed number with your calculator. Then change it back.

Sample Calculator	Key Sequence	Display
	45 [n] 7 [d] [Enter]	$\frac{45}{7} = 6\frac{3}{7}$
	[U$\frac{n}{d}$↔$\frac{n}{d}$]	↑ $\frac{45}{7}$
	[U$\frac{n}{d}$↔$\frac{n}{d}$]	↑ $6\frac{3}{7}$

The key [U$\frac{n}{d}$↔$\frac{n}{d}$] toggles between mixed numbers and fractions greater than one.

Simplifying Fractions

Ordinarily, calculators do not simplify fractions on their own. The steps for simplifying fractions are similar for many calculators, but the order of the steps varies. Two approaches for simplifying fractions on the sample calculator are shown on the next two pages.

Note For the sample calculator, pressing [Enter] is *not* optional in this key sequence. For other calculators, pressing [=] might not be necessary.

Note A tiny up or down arrow on the calculator display indicates that you can scroll up or down to see other lines on the screen.

Appendix

Simplifying Fractions on the Sample Calculator

This calculator lets you simplify a fraction in two ways. Each way divides the numerator and denominator by a common factor. The first approach uses **Simp** to automatically divide by the smallest common factors, and **Fac** to display the factor. If you want to display the fraction in simplest form, you may need to repeat the steps.

Simp simplifies a fraction by a common factor.

Fac displays the common factor used to simplify a fraction.

$\frac{N}{D} \to \frac{n}{d}$ means that the fraction shown is not in simplest form.

Example

Convert $\frac{18}{24}$ to simplest form using smallest common factors.

Sample Calculator	Key Sequence	Display
	18 **n** 24 **d** **Simp** **Enter**	$\frac{18}{24}$ ▶ S $\frac{9}{12}$
	Fac	2
	Fac **Simp** **Enter**	$\frac{9}{12}$ ▶ S $\frac{3}{4}$
	Fac	3
	Fac **Simp** **Enter**	$\frac{3}{4}$ ▶ S $\frac{3}{4}$

$\frac{18}{24} = \frac{9}{12} = \frac{3}{4}$

The smallest common factors are 2 and 3.

When the two fractions on the left and right of the screen are the same, the fraction is in simplest form. In this case $\frac{18}{24}$ in simplest form is $\frac{3}{4}$.

three hundred eighty-seven

Appendix

In the second approach, you can simplify the fraction in one step by telling the calculator to divide by the greatest common factor of the numerator and denominator.

Example

Convert $\frac{18}{24}$ to simplest form in one step by telling the calculator to divide by 6, the greatest common factor of the numerator and denominator.

Sample Calculator	Key Sequence	Display
	18 [n] 24 [d] (Simp) 6 (Enter)	$\frac{18}{24}$ ▸ S6 $\frac{3}{4}$
	(Fac)	↑ 6

$\frac{18}{24} = \frac{3}{4}$

Note Pressing (Fac) toggles between the display of the factor and the display of the fraction.

Try simplifying the fractions in the previous examples to see how your calculator works.

Note If you enter a number that is not a common factor of the numerator and the denominator, the original fraction will show up unchanged in the display with $\frac{N}{D} \rightarrow \frac{n}{d}$ above it.

Appendix

Percents

The calculator shown here, and many others, have a % key, but it is likely that it works differently on different calculators. The best way to learn what your calculator does with percents is to read its manual or look for the information online.

Most calculators include % to solve "percent of" problems.

Sample Calculator

% divides a number by 100.

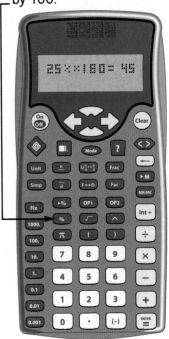

Example

Calculate 25% of 180.

Using the sample calculator, you can multiply 180 and 25% in either order, so both ways are shown.

Sample Calculator	Key Sequence	Display
	180 × 25 % Enter	180 × 25% = 45
	25 % × 180 Enter	25% × 180 = 45

25% of 180 is 45.

You can change percents to decimals with % .

Example

Display 85%, 250%, and 1% as decimals.

Sample Calculator	Key Sequence	Display
	85 % Enter	85% = 0.85
	250 % Enter	250% = 2.5
	1 % Enter	1% = 0.01

85% = 0.85; 250% = 2.5; 1% = 0.01

three hundred eighty-nine

Appendix

You can also use the % to convert percents to fractions.

On many calculators, first change the percent to a decimal as in the previous examples, then use F↔D to change to a fraction.

Example

Display 85%, 250%, and 1% as fractions in simplest form.

Sample Calculator	Key Sequence	Display	
	85 % Enter F↔D Simp Enter	$\frac{85}{100}$ ▶S	$\frac{17}{20}$
	250 % Enter F↔D Simp Enter	$2\frac{5}{10}$ ▶S	$2\frac{1}{2}$
	1 % Enter F↔D		$\frac{1}{100}$

$85\% = \frac{17}{20}$; $250\% = 2\frac{1}{2}$; $1\% = \frac{1}{100}$

Note You many need to use Simp to simplify if the fraction is not already in simplest form. Watch for the $\frac{N}{D} \rightarrow \frac{n}{d}$ symbol in the display.

Try displaying some percents as fractions and decimals on your calculator.

Appendix

Fraction/Decimal/Percent Conversions

You can convert fractions to decimals and percents on any calculator. For example, to rename $\frac{3}{5}$ as a decimal, simply enter 3 ÷ 5 =. The display will show 0.6. To rename the decimal as a percent, just multiply by 100. Then the display will show 60.

You can only convert decimals and percents to fractions on calculators that have special keys for fractions. Such calculators also have keys to change a fraction to its decimal equivalent or a decimal to an equivalent fraction.

Example

Convert $\frac{3}{8}$ to a decimal and back to a fraction.

Sample Calculator	Key Sequence	Display
	3 [n] 8 [d] [Enter]	$\frac{3}{8} = \frac{3}{8}$
	[F↔D]	0.375
	[F↔D]	$\frac{375}{1000}$
	[Simp] [Enter]	$\frac{375}{1000}$ ▶S $\frac{75}{200}$
	[Simp] [Enter]	$\frac{75}{200}$ ▶S $\frac{15}{40}$
	[Simp] [Enter]	$\frac{15}{40}$ ▶S $\frac{3}{8}$

$\frac{3}{8} = 0.375$

$\frac{3}{8} = 0.375 = \frac{375}{1,000} = \frac{75}{200} = \frac{15}{40}$

Note [F↔D] toggles between fraction and decimal notation.

See how your calculator changes fractions to decimals.

Appendix

The tables below show examples of various conversions. Although only one key sequence is shown for each conversion, there are often other key sequences that work as well.

Conversion	Starting Number	Sample Calculator Key Sequence	Display
Fraction to decimal	$\frac{3}{5}$	3 [n] 5 [d] [Enter] [F↔D]	0.6
Decimal to fraction	0.125	.125 [Enter] [F↔D]	$\frac{125}{1000}$
Decimal to percent	0.75	.75 [▶%] [Enter]	.75▶% 75%
Percent to decimal	125%	125 [%] [Enter]	125% = 1.25
Fraction to percent	$\frac{5}{8}$	5 [n] 8 [d] [▶%] [Enter]	$\frac{5}{8}$▶% 62.5%
Percent to fraction	35%	35 [%] [Enter] [F↔D]	$\frac{35}{100}$

Check Your Understanding

Use your calculator to convert between fractions, decimals, and percents.

1. $\frac{3}{16}$ to a decimal
2. 0.185 to a fraction
3. 0.003 to a percent
4. 723% to a decimal
5. $\frac{9}{32}$ to a percent
6. 68% to a fraction

Check your answers in the Answer Key.

Appendix

Other Operations on a Calculator

Your calculator can do more than simple arithmetic with whole numbers, fractions, and decimals. Each calculator model has a variety of functions and particular key sequences for those functions. See the owner's manual or ask your teacher to help you explore these operations. The following pages explain some operations on many calculators.

Rounding

All calculators can round decimals. Decimals must be rounded to fit on the display. For example, if you key in 2 ÷ 3 =,

- The sample calculator shows 11 digits and rounds to the nearest value: 0.6666666667.
- Some other calculators show 8 digits and round down to 0.6666666.

Try 2 ÷ 3 = on your calculator to see how many digits are in the display and how it rounds.

Scientific calculators have a (Fix) key to set, or fix, the place value of decimals on the display. Fixing always rounds to the nearest value.

Example

Clear your calculator and fix it to round to tenths. Round each number to the nearest tenth: 1.34, 812.79, and 0.06.

Sample Calculator	Key Sequence	Display
	(Clear) 8 (Fix) (0.1) (Enter)	Fix 8 = 8.0
	1.34 (Enter)	↑ Fix 1.34 = 1.3
	812.79 (Enter)	↑ Fix 812.79 = 812.8
	.06 (Enter)	↑ Fix .06 = 0.1

1.34 rounds to 1.3.
812.79 rounds to 812.8.
0.06 rounds to 0.1.

Note To turn off fixed rounding on a calculator, press (Fix) then (.).

Note You can fix the sample calculator to round without clearing the display first. It will round the number on the display.

Appendix

The display to round to hundredths is helpful for solving problems about dollars and cents.

Example

Kevin spent $11.23 on music downloads and $14.67 for two e-book downloads. Set your calculator to round to the nearest cent and calculate the total cost of his downloads.

Sample Calculator	Key Sequence	Display
	Fix 0.01	Fix ◀
	11.23 + 14.67 Enter	Fix 11.23 + 14.67 = 25.90

The display shows that his downloads cost $25.90.

If you find the total in the example above with the "Fix" turned off, the display reads 25.9 on most calculators. To show the answers in dollars and cents, fix the display to round to hundredths and you will see 25.90.

Check Your Understanding

Use your calculator to round to the indicated place.

1. 0.67 to tenths
2. 427.88 to ones
3. 4384.4879 to thousandths
4. 0.7979 to hundredths

Check your answers in the Answer Key.

Appendix

Powers and Reciprocals on a Calculator

Powers of numbers can be calculated on all calculators that preserve the order of operations. Look at your calculator to see which key it has for calculating with exponents and finding powers of numbers.

- The key may look like $\boxed{x^y}$ and is read as "x to the y."
- The key may look like $\boxed{\wedge}$ and is called a *caret*.

Example

Find the values of 3^4 and 5^2.

Sample Calculator	Key Sequence	Display
	3 $\boxed{\wedge}$ 4 $\boxed{\text{Enter}}$	3^4 = 81
	5 $\boxed{\wedge}$ 2 $\boxed{\text{Enter}}$	5^2 = 25

$3^4 = 81$

$5^2 = 25$

Note If your calculator has a key similar to $\boxed{x^y}$, follow this sequence for 3^4:
3 $\boxed{x^y}$ 4 $\boxed{=}$.

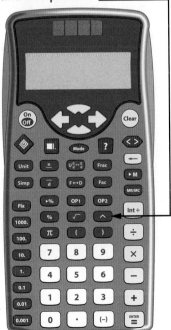

Sample Calculator

$\boxed{\wedge}$ finds powers and reciprocals.

three hundred ninety-five 395

Appendix

On all calculators with keys for entering exponents you can find a reciprocal of a number by raising it to the −1 power. To enter −1, be sure to use the change sign key (−) or +/−, not the subtraction key −.

Example

Find the reciprocals of 25 and $\frac{2}{3}$.

Sample Calculator	Key Sequence	Display
	25 ∧ (−) 1 Enter	25^ − 1 = 0.04
	2 n 3 d ∧ (−) 1 Enter	$\frac{2}{3}$^ − 1 = 1.5
	To rewrite 1.5 as a fraction: F↔D Simp Enter U$\frac{n}{d}$↔$\frac{n}{d}$	$\frac{3}{2}$

The reciprocal of 25 is 0.04.

The reciprocal of $\frac{2}{3}$ is 1.5, or $\frac{3}{2}$.

Note If you press (−) after instead of before the second number, you will get an error message.

Note Many scientific calculators have a reciprocal key 1/x that gives the same result as raising a number to the −1 power.

Appendix

Scientific Notation and Multiplying by Powers of 10 on a Calculator

Scientific notation is a way of writing very large or very small numbers. A number in scientific notation is shown as a product of a number between 1 and 10 and a power of 10. In scientific notation, the 9,000,000,000 bytes of memory on a 9-gigabyte hard drive is written $9 * 10^9$. On scientific calculators, numbers with too many digits to fit on the display are automatically shown in scientific notation.

Different calculators use different symbols for powers of 10. Your calculator may display raised exponents of 10, although some do not. Since the base of the power is always 10, some calculators leave out the 10 and simply put a space between the number and the exponent.

This calculator uses a caret ^ to display scientific notation.

Example

Convert $7 * 10^4$ and $4.35 * 10^5$ to **standard form**.

This is the same as multiplying by a power of 10.

Sample Calculator	Key Sequence	Display
	7 ⊗ 10 ⌃ 4 Enter	7 × 10^4 = 70000
	4.35 ⊗ 10 ⌃ 5 Enter	4.35 × 10^5 = 435000

$7 * 10^4 = 70,000$

$4.35 * 10^5 = 435,000$

This is how some calculators show $9 * 10^9$.

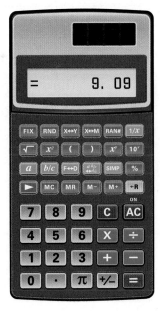

Note Another name for **standard form** is standard notation. The sample calculator and many others do not display large numbers in standard form with a comma like you do with pencil and paper. Some calculators use an apostrophe.

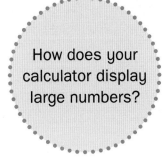

How does your calculator display large numbers?

three hundred ninety-seven

Appendix

Dividing by Powers of 10 on a Calculator

You can divide whole numbers and decimals by powers of 10 on a calculator.

Example

Use your calculator to show $6 \div 10^7 = 0.0000006$.

Sample Calculator	Key Sequence	Display
	6 ÷ 10 ^ 7 Enter	6 ÷ 10^7 = 0.0000006

Example

Use your calculator to show $12.8 \div 10^2$ in standard form.

Sample Calculator	Key Sequence	Display
	12.8 ÷ 10 ^ 2 Enter	12.8 ÷ 10^2 = 0.128

$12.8 \div 10^2$ in standard form is 0.128.

Did You Know?

Numbers divided by powers of 10 sometimes have products that are in the ten-millionths or smaller. These numbers have too many digits for the calculator display. Scientific calculators show the product as a number multiplied by a negative power of 10 in **exponential notation**. You will learn more about negative powers of 10 in future grades.

What do you notice about the location of the decimal point compared to the exponent?

Check Your Understanding

Use your calculator to write the following in standard form.

1. $7 * 10^4$
2. $8.3 * 10^7$
3. $3.726 \div 10^6$
4. $3.4 \div 10^5$

Check your answers in the Answer Key.

Appendix

Using Calculator Memory

Many calculators let you save a number in long-term memory using keys with "M" on them. When you need the number later on, you can recall it from memory. Most calculators display an "M" or similar symbol when there is a number other than 0 in the memory.

Memory Basics

There are two main ways to enter numbers into long-term memory. Some calculators, including most 4-function calculators, have keys such as [M+] and [M-]. Other calculators have keys like those described on this page.

Memory on the Sample Calculator

One way that you can put numbers in the calculator's memory is by using a key to store a value. On this calculator the store key is [▶M] and only works on numbers that have been entered into the display with [Enter].

Sample Calculator	Key Sequence	Purpose
	[MR/MC] [MR/MC]	Clear the long-term memory. This should always be the first step to any key sequence using the memory. Afterward, there will be no "M" in the display. This tells you there is no number in memory.
	[▶M] [Enter]	Store the number entered in the display in memory.
	[MR/MC]	Recall the number stored in memory and show it in the display.

Note If you press [MR/MC] more than twice, you will recall and display the 0 that is now in memory. Press [Clear] to clear the display.

[▶M] stores the displayed number in memory.

[MR/MC] recalls and displays the number in memory. Press it twice to clear memory.

three hundred ninety-nine

Appendix

The following example first shows what happens if you don't enter a number before trying to store it.

Example

Store 25 in memory and recall it to show that it was saved.

Sample Calculator Key Sequence	Display
[MR/MC] [MR/MC] [Clear]	(blank)
25 [▶M] [Enter]	MEM ERROR

Oops. Start again. First, press [Clear] twice.

Sample Calculator Key Sequence	Display
[MR/MC] [MR/MC] [Clear]	(blank)
25 [Enter] [▶M] [Enter]	25 = 25 ᴹ
[Clear]	(blank) ᴹ
[MR/MC]	25 ᴹ

If your calculator is like this one, try the Check Your Understanding problem on the next page.

Appendix

Memory on Some Other Calculators

Some calculators put a 0 into memory when [MC] is pressed. To store a single number in a cleared memory, simply enter the number and press [M+].

Key Sequence	Purpose
[MC]	Clear the long-term memory. This should always be the first step to any key sequence using the memory. Afterward, there will be no "M" in the display. This tells you there is no number in memory.
[MR]	Recall the number stored in memory and show it in the display.
[M+]	Add the number on the display to the number in memory.
[M−]	Subtract the number on the display from the number in memory.

Note Generally, when a calculator turns off, the display clears but a value in memory is *not* erased.

Check Your Understanding

Store 75 in the long-term memory. Clear the display. Use the number stored in memory to compute the area A of a rectangle that has a length of 75 feet and a width of

a. 50 feet   b. 35 feet   c. 63.5 feet

Check your answers in the Answer Key.

four hundred one   SRB 401

Appendix

Using Calculator Memory in Problem Solving

You can use calculator memory to solve problems that take two or more steps to solve.

Note Always be sure to clear the memory after solving one problem and before beginning another.

Example

Compute a 15% tip on a $25 bill. Store the tip in the memory, then find the total bill.

Sample Calculator	Key Sequence	Display
	(MR/MC) (MR/MC) (Clear)	
	15 (%) (×) 25 (Enter)	15% × 25 = 3.75
	(▶M) (Enter)	m 15% × 25 = 3.75
	25 (+) (MR/MC) (Enter)	↑ m 25 + 3.75 = 28.75

The tip is $3.75. The total bill is $28.75.

Check Your Understanding

Compute an 18% tip on an $85 bill. Then find the total bill.

Check your answers in the Answer Key.

Appendix

Example

At the food court, Marguerite ordered 2 pita wraps at $1.49 each and 3 sides at $0.89 each. How much change will she receive from a $10 bill?

Sample Calculator Key Sequence	Display
MR/MC MR/MC Clear	
2 × 1.49 Enter ►M Enter	2 × 1.49 = 2.98 (M)
3 × .89 Enter ►M +	3 × .89 = 2.67 (↑ M)
10 − MR/MC Enter	10 − 5.65 = 4.35 (↑ M)

Marguerite will receive $4.35 in change.

Note The key sequence ►M + is a shortcut to add the displayed number to memory. Similarly, ►M − subtracts a number from memory.

Appendix

Example

At the movie theater, Mrs. Hiller bought 2 adult tickets at $8.25 each and 3 child tickets at $4.75 each. She redeemed a $5 coupon. How much did she pay for tickets?

Sample Calculator Key Sequence	Display
(MR/MC) (MR/MC) (Clear)	
2 (×) 8.25 (Enter) (▶M) (Enter)	2 × 8.25 = 16.5
3 (×) 4.75 (Enter) (▶M) (+)	3 × 4.75 = 14.25
(MR/MC) (−) 5 (Enter)	30.75 − 5 = 25.75

Mrs. Hiller paid $25.75 for the tickets.

Note If you fix the rounding to hundredths, all the values will be displayed as dollars and cents.

Appendix

Note If you fix the rounding to hundredths, all the values will be displayed as dollars and cents.

Example

Juan bought the following tickets to a baseball game for himself and 6 friends: 2 bleacher seats at $15.25 each and 5 mezzanine seats at $27.50 each. What is the mean (average) cost of each ticket?

Sample Calculator Key Sequence	Display
(MR/MC) (MR/MC) (Clear)	
2 (×) 15.25 (Enter) (▶M) (Enter)	M 2 × 15.25 = 30.5
5 (×) 27.50 (Enter) (▶M) (+)	↑ M 5 × 27.50 = 137.5
(MR/MC) (÷) 7 (Enter)	↑ M 168 ÷ 7 = 24

The mean (average) cost of a ticket is $24.00.

Check Your Understanding

1. How much would 2 shirts and 2 hats cost if shirts cost $18.50 each and hats cost $13.25 each?
2. How much would it cost to take a family of 2 adults and 4 children to a matinee if tickets cost $6.25 for adults and $4.25 for children?

Check your answers in the Answer Key.

four hundred five

Appendix

Skip Counting on a Calculator

You may have used a four-function calculator or a scientific calculator to skip count in earlier grades.

Recall that you need to tell the calculator:

1. What number to count by
2. Whether to count up or down
3. What the starting number is
4. When to count

Here's how to skip count on the sample calculator.

Sample Calculator

Op1 and **Op2** allow you to program and repeat operations.

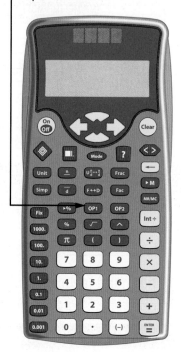

Example

Starting at 3, count by 7s.

Sample Calculator		
Purpose	Key Sequence	Display
Tell the calculator to count up by 7. **Op1** is programmed to do any operation with any number that you enter between presses of **Op1**.	**Op1** ⊕ 7 **Op1**	Op1 +7
Tell the calculator to start at 3 and do the first count.	3 **Op1**	Op1 3+7 1 10
Tell the calculator to count again.	**Op1**	Op1 10+7 2 17
Keep counting by pressing **Op1**.	**Op1**	Op1 17+7 3 24

To count back by 7, begin with **Op1** ⊖ 7 **Op1**.

Note You can use **Op2** to define a second constant operation. **Op2** works in exactly the same way as **Op1**.

Note The number in the lower left corner of the display shows how many counts you have made.

SRB
406   four hundred six

You can also skip count by fractions on a calculator.

> **Example**
>
> Starting at $\frac{1}{2}$, count by $\frac{1}{4}$s.
>
> | Sample Calculator | | |
> |---|---|---|
> | Purpose | Key Sequence | Display |
> | Tell the calculator to count up by $\frac{1}{4}$. **Op1** is programmed to do any operation with any number that you enter between presses of **Op1**. | **Op1** **+** 1 **n** 4 **d** **Op1** | Op1 $+\frac{1}{4}$ |
> | Tell the calculator to start at $\frac{1}{2}$ and do the first count. | 1 **n** 2 **d** **Op1** | Op1   1 $\frac{3}{4}$ |
> | Tell the calculator to count again. | **Op1** | ↑ Op1   2   1 |
> | Keep counting by pressing **Op1**. | **Op1** | ↑ Op1   3   1$\frac{1}{4}$ |
>
> To count back by $\frac{1}{4}$s, begin with **Op1** **−** 1 **n** 4 **d** **Op1**.

> **Check Your Understanding**
>
> Use your calculator to do the following counts. Write five counts each.
> 1. Starting at $\frac{1}{3}$ count on by $\frac{2}{3}$s.
> 2. Starting at 10, count back by $\frac{3}{4}$s.
>
> Check your answers in the Answer Key.

four hundred seven

Appendix

Predicting Body Sizes

Anthropometry is the study of human body sizes and proportions. An *anthropometrist* gathers data on the size of the body and its components. Body-size data are useful to engineers, architects, industrial designers, interior designers, clothing manufacturers, and artists.

- Automotive engineers use body-size data to design vehicles and to set standards for infant and child safety seats.
- Architects take body-size data into account when designing stairs, planning safe kitchens and bathrooms, and providing access space for people who use wheelchairs.
- Clothing manufacturers use body-size data to create sewing patterns.

Not even identical twins are exactly alike. Body sizes and proportions differ depending on age, sex, and ethnic or racial attributes.

- There are no perfect rules that can be used to exactly predict one body measurement given another body measurement. For example, no rule can exactly predict a person's weight given the person's height, or height given arm length.
- There are imperfect rules and rules of thumb that can be useful in relating one body measurement to another.

Appendix

One imperfect rule is sometimes used to predict the height of an adult when the length of the adult's tibia is known. The *tibia* is the shinbone. When the measurements are in inches, the rule is

Height = (2.6 * Length of Tibia) + 25.5.

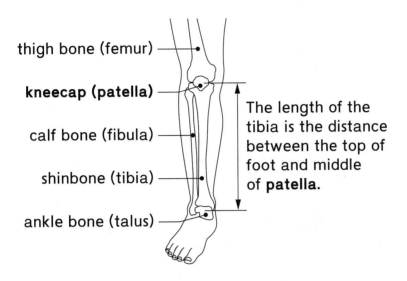

The length of the tibia is the distance between the top of foot and middle of **patella**.

Another rule relates the circumference of a person's neck to the circumference of the person's wrist:

Circumference of Neck = 2 * Circumference of Wrist.

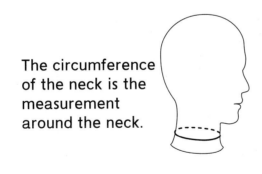

The circumference of the neck is the measurement around the neck.

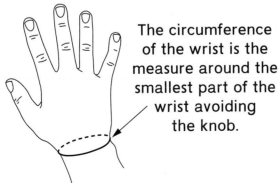

The circumference of the wrist is the measure around the smallest part of the wrist avoiding the knob.

four hundred nine

Glossary

1-dimensional (1-D) Having length, but not area or volume. A 1-dimensional figure can bend or be straight.

2-dimensional (2-D) Having *area* but not volume. A 2-dimensional *surface* can be flat like a piece of paper or curved like a dome.

3-dimensional (3-D) Having length, width, and thickness. Solid objects take up *volume* and are 3-dimensional. A figure whose points are not all in a single plane is 3-dimensional.

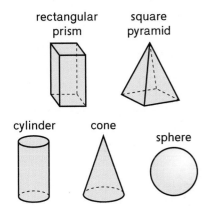

absolute value The distance between a number and 0 on the *number line*. The absolute value of a *positive number* is the number itself. The absolute value of a *negative number* is the *opposite* of the number. The absolute value of 0 is 0. For example, the absolute value of 3 is 3, and the absolute value of −6 is 6. The notation for the absolute value of a number n is $|n|$.

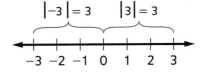

accurate As correct as possible for the situation.

acute angle An angle with a measure less than 90°.

Acute angles

acute triangle A triangle with three acute angles.

Acute triangle

addend Any one of a set of numbers that are added. For example, in $5 + 3 + 1 = 9$, the addends are 5, 3, and 1.

Additive Identity The number zero (0). The additive identity is the number that when added to any other number, gives that other number.

address box A place where the address of a spreadsheet *cell* is shown when the cell is selected.

B3		
A	B	C

The address bar displays B3.

algebra The branch of mathematics that uses letters and other symbols to stand for quantities that are *unknown* or vary. Algebra is used to describe patterns, express numerical relationships, and *model* real-world situations.

algebraic expression An *expression* that contains a *variable*. For example, letting M represent Maria's height in inches, if Maria is 2 inches taller than Joe, then the algebraic expression $M - 2$ represents Joe's height.

algorithm A set of step-by-step instructions for doing something, such as carrying out a computation or solving a problem.

four hundred eleven SRB 411

Glossary

angle A figure that is formed by two rays or line segments with a common endpoint. The rays or segments are called the sides of the angle. The common endpoint is called the *vertex* of the angle. Angles are measured in *degrees* (°).

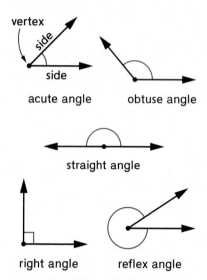

apex In a *pyramid* or cone the point at the tip opposite the *base*. In a pyramid, all the nonbase faces meet at the apex.

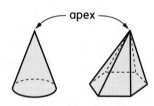

approximate Close to exact. In many situations it is not possible to get an exact answer, but it is important to be close to the exact answer.

area The amount of *surface* inside 2-dimesional shape. The measure of the area is how many units, such as square inches or square centimeters, cover the surface.

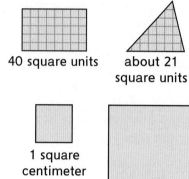

area model A *model* for multiplication problems in which the length and width of a *rectangle* represent the *factors*, and the *area* of the rectangle represents the *product*.

Area model for 3 * 5 = 15

array An arrangement of objects in a regular pattern, usually in rows and columns. In *Everyday Mathematics*, an array is a rectangular array unless specified otherwise.

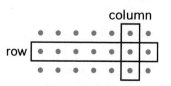

A rectangular array

Associative Property of Addition A *property* of addition (but not subtraction) that says that when you add three numbers, you can change the grouping without changing the *sum*. For example:
(4 + 3) + 7 = 4 + (3 + 7)

Associative Property of Multiplication A *property* of multiplication (but not division) that says that when you multiply three numbers, you can change the grouping without changing the *product*. For example:
(5 * 8) * 9 = 5 * (8 * 9)

average A typical value for a set of numbers. The word *average* usually refers to the *mean* of a set of numbers.

axis of coordinate grid Either of the two number lines that intersect to form a *coordinate grid*.

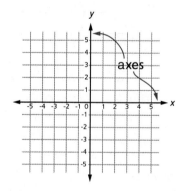

balance point In *Everyday Mathematics,* for data, the balance point is the *mean*.

SRB
412 four hundred twelve

Glossary

bar graph A graph that uses horizontal or vertical bars to represent data.

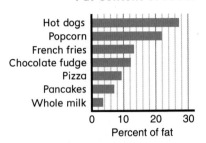

bar model A diagram that represents an *equation* with two equal length bars.

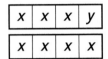

Bar model of $3x + y = 4$

base The *side* of a *polygon* or *face* of a polyhedron from which the *height* is measured.

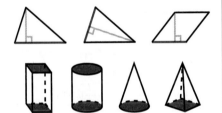

Bases are shown in red.

base (in exponential notation) The number that is raised to a power. For example, in 5^3, the base is 5. See *exponential notation* and *power of a number* n.

base ten Our system for writing numbers that uses 10 symbols called *digits*. The digits are 0, 1, 2, 3, 4, 5, 6, 7, 8, and 9. You can write any number using only these 10 digits. Each digit has a value that depends on its place in the number. In this system, moving a digit one place to the left makes that digit worth 10 times as much. And moving a digit one place to the right makes that digit worth one-tenth as much. See *place value*.

basic facts The addition facts (whole-number addends of 10 or less) and their related subtraction facts, and the multiplication facts (whole number factors of 10 or less) and their related division facts. Facts are organized into fact families.

benchmark A well-known number or measure that can be used to check whether other numbers, measures, or estimates make sense. For example, a benchmark for length is that the width of a man's thumb is about one inch. The numbers 0, $\frac{1}{2}$, 1, $1\frac{1}{2}$ may be useful benchmarks for fractions.

bimodal distribution A set of values clustered in two locations; *distributions* that have two modes. Traffic on weekdays is usually bimodal because it clusters at morning and evening rush hour.

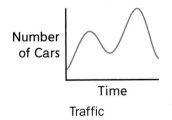

Traffic

bin An interval for collecting, aggregating, organizing, or graphing *data*.

box plot A plot displaying the spread, or distribution, of a *data* set using 5 *landmarks*: usually the minimum, lower quartile, *median*, upper quartile, and maximum. Also called a box-and-whiskers plot.

Landmark	Tomato Plant Heights
Minimum	14
Lower quartile	16
Median	20
Upper quartile	25
Maximum	32

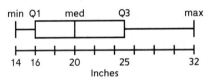

##

capacity (1) The amount a container can hold. Capacity is usually measured in units such as cups, fluid ounces, and liters. (2) The amount something can hold. For example, a computer harddrive may have a capacity of 64 TB, or a scale may have a capacity of 400 lbs.

category A group whose members are defined by a shared attribute. For example, triangles are polygons that share the attribute of having three sides.

Glossary

cell A location at the intersection of a column and a row in a *spreadsheet* or table that can contain data. Cells are often named by their *addresses*. The address of a cell in a spreadsheet program is the column letter followed by the row number. For example, cell B3 in column B, row 3, is highlighted below.

	A	B	C	D
1				
2				
3				
4				

circle In 2-dimensions, the set of all points that are the same distance from a fixed point. The fixed point is the center of the circle, and the distance is the radius. The center and interior of a circle are not part of the circle. A circle together with its interior is called a disk or a circular region. See *diameter*.

Circle

close-but-easier numbers Numbers that are close to the original numbers in the problem, but easier for solving problems. For example, to estimate 494 + 78, you might use the close-but-easier numbers 490 and 80.

coefficient The number, or *constant factor*, in a *variable term*. For example, in 3c + 8, 3 is a coefficient of c.

column addition A method for adding numbers in which the *addends'* digits are first added in each *place-value* column separately, and then 10-for-1 trades are made until each column has only one digit. Lines are drawn to separate the place-value columns.

```
       100s  10s  1s
         2    4   8
    +    1    8   7
       ─────────────
         3   12   15
         3   13    5
       ─────────────
         4    3    5
       248 + 187 = 435
```

common denominator For two or more fractions, a number that is a *multiple* of both or all *denominators*. For example, the fractions $\frac{1}{2}$ and $\frac{2}{3}$ have the common denominators 6, 12, 18, and so on. If the fractions have the same denominator, that denominator is called a common denominator. See *quick common denominator*.

common factor A *counting number* is a common *factor* of two or more counting numbers if it is a factor of each of those numbers. For example, 4 is a common factor of 8 and 12.

common multiple A number that is a *multiple* of two or more given numbers. For example, common multiples of 6 and 8 include 24, 48, and 72. See *least common multiple (LCM)*.

common numerator Same as *like numerator*.

Commutative Property of Addition A *property* of addition (but not of subtraction) that says that changing the order of the numbers being added does not change the *sum*. This property is often called the *turn-around rule* in *Everyday Mathematics*. For example: 5 + 10 = 10 + 5

Commutative Property of Multiplication A *property* of multiplication (but not of division) that says that changing the order of the numbers being multiplied does not change the *product*. This property is often called the *turn-around rule* in *Everyday Mathematics*. For example: 3 * 8 = 8 * 3.

Glossary

compose To make a number or shape by putting together smaller numbers or shapes. For example, you can compose a 10 by putting together ten 1s: 1 + 1 + 1 + 1 + 1 + 1 + 1 + 1 + 1 + 1 = 10. You can compose a pentagon by putting together an equilateral triangle and a square.

A composed pentagon

composite number A *counting number* that has more than two different *factors*. For example, 4 is a composite number because it has three factors: 1, 2, and 4.

congruent figures Figures that have the same shape and the same size. 2-dimensional figures will make an exact fit if one is placed on top of the other and lined up.

Congruent pentagons

conjecture A statement that is thought to be true based on information or mathematical thinking.

consecutive Following one after another in an uninterrupted order. For example, A, B, C, and D are four consecutive letters of the alphabet; 6, 7, 8, 9, and 10 are five consecutive whole numbers.

constant A *quantity* that does not change.

continuous Without gaps or breaks. For example, *real numbers* are continuous because there are always more real numbers between any two real numbers. Heights of a tree are continuous because the tree goes through every height possible between its shortest and tallest measurements. *Whole numbers* and counts are not continuous because there are spaces between each number or count.

coordinate (1) One of the two numbers in an ordered p*air*. The number pair is used to locate a point on a coordinate grid. (2) A number used to locate a point on a number line.

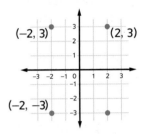

Coordinates

coordinate grid (rectangular coordinate grid) A grid formed by two number lines that intersect at their *zero points* and form right angles. Each number lines is referred to as an *axis*. You can locate points on the grid with *ordered pairs* of numbers called coordinates.

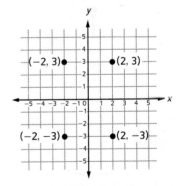

Coordinate grid

corresponding sides Sides in the same relative position in similar or congruent figures. Corresponding sides "match up."

Glossary

corresponding terms Terms that are in the same position within two lists. For example, in the table below, the third term in the "in" column corresponds to the third term in the "out" column. They are corresponding terms.

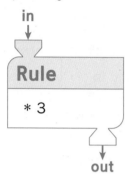

in	out
1	3
2	6
3	9
4	12

counting numbers The numbers used to count things. The set of counting numbers is {1, 2, 3, 4, . . .}. Compare *whole numbers*.

counting-up subtraction A subtraction strategy in which you count up from the smaller to the larger number to find the *difference*. For example, to solve 16 − 9, count up from 9 to 16.

cube (1) A polyhedron with 6 square *faces*. A cube has 8 *vertices* and 12 *edges*.

(2) The smallest base-10 block called a centimeter cube.

Cubes

cubic unit A unit used in measuring volume, such as a cubic centimeter or a cubic foot.

data Information that is gathered by counting, measuring, questioning, or observing.

decimal A number written in standard, *base ten* notation that contains a *decimal point*, such as 2.54. A whole number is a decimal, but is usually written without a decimal point.

decimal point A dot used to separate the ones and tenths places in *decimals*.

decompose To separate a number or shape into smaller numbers or shapes. For example, 14 can be decomposed into 1 ten and 4 ones. A square can be decomposed into two isosceles right triangles. Any even number can be decomposed into two equal parts: $2n = n + n$.

degree (°) (1) A unit of measure for *angles* based on dividing a *circle* into 360 equal parts. Latitude and longitude are measured in degrees, and these degrees are based on angle measures. (2) A unit of measure for *temperature*. A small raised circle (°) can be used to show degrees, as in a 70° angle or 70°F for room temperature.

denominator The number below the line in a *fraction*. A fraction may be used to name part of a *whole*. If the whole is divided into equal parts, the denominator represents the number of equal parts into which the whole is divided. The denominator determines the size of each part. In the fraction $\frac{a}{b}$, b is the denominator.

dependent (responding) variable A *variable* whose value is dependent on the value of at least one other variable. Compare *independent (manipulated) variable*.

diameter A line segment that goes through the center of a circle and has endpoints on the circle. Sometimes diameter refers to the length of this line segment. The diameter of a sphere is defined in the same way. The diameter of a circle or sphere is twice the length of its radius.

Glossary

difference The result of subtracting one number from another. See *minuend* and *subtrahend*.

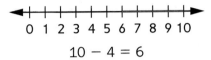

$10 - 4 = 6$

digit One of the number symbols 0, 1, 2, 3, 4, 5, 6, 7, 8, and 9 in the standard, *base ten* system.

displacement A method for measuring the *volume* of an object that involves submerging, or placing it under water and measuring how much water it moves.

display bar A place where the formula for a *spreadsheet cell* is shown when the cell is selected.

distribution The arrangement and frequency of *data* values in a data set. A distribution may be summarized with *measures of center* and *variability*. For example, distribution of adult men's heights is clumped in the middle and tapers to both sides. This distribution can be summarized by *average* height of about 69 inches with a *range* of about 85 inches.

Distributive Property of Multiplication over Addition and Subtraction A *property* that relates multiplication and addition or subtraction. This property gets its name because it "distributes" a *factor* over terms inside parentheses.

$a * (b + c) = (a * b) + (a * c)$,

so

$2 * (5 + 3) =$
$(2 * 5) + (2 * 3) =$
$10 + 6 =$
16

and

$a * (b - c) = (a * b) - (a * c)$,

so

$2 * (5 - 3) =$
$(2 * 5) - (2 * 3) =$
$10 - 6 =$
4

dividend The number in division that is being divided. For example, in $35 \div 5 = 7$, the dividend is 35.

divisible by If one *counting number* can be divided by a second counting number with a *remainder* of 0, then the first number is divisible by the second number. For example, 28 is divisible by 7 because 28 divided by 7 is 4, with a remainder of 0.

Division of Fractions Property A fact that makes division with *fractions* easier: division by a fraction is the same as multiplication by that fraction's *reciprocal*. For example, because the reciprocal of $\frac{1}{2}$ is 2, the division problem $4 \div \frac{1}{2}$ is equivalent to the multiplication problem $4 * 2$. See *multiplicative inverses*.

divisor In division, the number that divides another number. For example, in $35 \div 5 = 7$, the divisor is 5.

dot plot See *line plot*.

edge Any *side* of a *polyhedron's faces*.

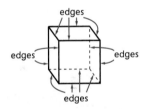

efficient strategy A method that can be applied easily and quickly.

empty set The *set* containing no members. For example, the set of solutions to $3 < x < -3$, or values of x that are both greater than 3 and less than -3, is the empty set. Same as null set.

four hundred seventeen

Glossary

enlarge To increase the size of an object or a figure without changing its shape. See *scale factor*.

equal See *equivalent*.

equal parts *Equivalent* parts of a *whole*. For example, dividing a pizza into 4 equal parts means each part is $\frac{1}{4}$ of the pizza and is equal in size to each of the other 3 parts.

4 equal parts, each $\frac{1}{4}$ of a pizza

equation A *number sentence* that contains an equal sign. For example, $15 = 10 + 5$ is an equation.

equilateral polygon A *polygon* in which all sides are the same length.

Equilateral polygons

equilateral triangle A *triangle* with all three sides equal in length. In an equilateral triangle, all three angles have the same measure.

An equilateral triangle

equivalent *Equal* in value but possibly in a different form. For example, $\frac{1}{2}$, 0.5, and 50% are all equivalent.

equivalent equations *Equations* that have the same *solution* set. For example, $2 + x = 4$ and $6 + x = 8$ are equivalent equations because the solution set for each is $x = 2$.

equivalent expressions *Expressions* that name the same number or set of possible numbers. For example, 17 and $10 + 7$ are *equivalent* expressions. The expressions $4(a + b)$ and $4a + 4b$ are also equivalent.

equivalent fractions *Fractions* that name the same number. For example, $\frac{1}{2}$ and $\frac{4}{8}$ are equivalent fractions.

equivalent fractions rule A rule stating that if the *numerator* and *denominator* of a *fraction* are each multiplied or divided by the same nonzero number, the result is a fraction *equivalent* to the original fraction.

equivalent problem A division problem solved by writing an *equivalent expression*. For example, to solve 35.6/0.5, you may solve the equivalent problem 356/5.

equivalent ratios *Ratios* that make the same comparison. Two or more ratios are equivalent if they can be named as equivalent fractions. For example, the ratios 12 to 20, 6 to 10, and 3 to 5 are equivalent because $\frac{12}{20} = \frac{6}{10} = \frac{3}{5}$.

estimate An answer close to an exact answer. To estimate means to give an answer that should be close to an exact answer.

evaluate a numerical expression To carry out the *operations* in a numerical *expression* to find a single value for the expression.

even number A *counting number* that can be divided by 2 with no remainder. The even numbers are 2, 4, 6, 8, and so on. 0, −2, −4, −6, and so on are also usually considered even.

expanded form A way of writing a number as the *sum* of the values of each *digit*. For example, in expanded form, 356 is written $300 + 50 + 6$. Compare *standard form*.

exponent A number used in *exponential notation* to tell how many times the *base* is used as a *factor*. For example, in 5^3, the base is 5, the exponent is 3, and $5^3 = 5 * 5 * 5 = 125$. See *power of a number* n.

exponential notation
A way to show repeated multiplication by the same *factor*. For example, 2^3 is exponential notation for $2 * 2 * 2$. The small raised 3 is the *exponent*. It tells how many times the number 2, called the *base*, is used as a factor.

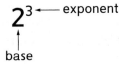

expression A group of mathematical symbols that represents a number—or can represent a number if values are assigned to any *variables* in the expression. An expression may include numbers, variables, *operation symbols,* and *grouping symbols* — but *not relation symbols* (=, >, <, and so on). Any expression that contains one or more variables is called an algebraic expression.

extended facts Variations of basic facts involving *multiples* of 10, 100, and so on. For example, $30 + 70 = 100$, $40 * 5 = 200$, and $560 / 7 = 80$ are extended facts.

face A flat *surface* on a *3-dimensional* shape.

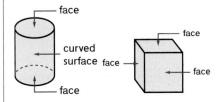

factor Whenever two or more numbers are multiplied to give a *product,* each of the numbers that is multiplied is called a factor. For example, in $4 * 1.5 = 6$, 6 is the product and 4 and 1.5 are called factors. Compare *factor of a counting number* n.

factor of a counting number n A *counting number* whose product with another counting number equals *n*. For example, 2 and 3 are *factors* of 6 because $2 * 3 = 6$. But 4 is not a factor of 6 because $4 * 1.5 = 6$ and 1.5 is not a counting number.

factor pair Two *factors* of a *counting number* whose *product* is the number. A number may have more than one factor pair. For example, the factor pairs for 18 are 1 and 18, 2 and 9, and 3 and 6.

factor tree A way to get the *prime factorization* of a *counting number*. Write the original number as a product of counting-number factors. Then write each of these factors as a product of factors, and so on, until the factors are all prime numbers. A factor tree looks like an upside-down tree, with the root (the original number) at the top and the leaves (the factors) beneath it.

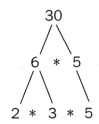

factorial (!) A *product* of a whole number and all the smaller whole numbers except 0. An exclamation point (!) is used to write factorials. For example, "five factorial" is written as 5! and is equal to $5 * 4 * 3 * 2 * 1 = 120$. 0! is defined to be equal to 1.

false number sentence A *number sentence* that is not true. For example, $8 = 5 + 5$ is a false number sentence.

first quartile (Q1) (1) The middle value of the numbers below the *median* in a *data set,* also called the *lower quartile.* (2) Informally, the *interval* between this middle point of the lower half of the data and the *minimum* of the data set.

five-number summary A list containing the *minimum, first quartile, median, third quartile,* and *maximum* of a *data* set.

Glossary

formula A general rule for finding the value of something. A formula is often written using letters, called *variables*, which stand for the quantities involved. For example, the formula for the area of a rectangle may be written as $A = l * w$, where A represents the area of the rectangle, l represents its length, and w represents its width.

fourth quartile (Q4) (1) Informally, the *interval* between the *third quartile* and the *maximum*. (2) Sometimes the maximum.

fraction (1) A number in the form $\frac{a}{b}$ where a and b are *integers* and b is not 0. A fraction may be used to name part of a whole, or to compare two quantities. A fraction may also be used to represent division. For example, $\frac{2}{3}$ can be thought of as 2 divided by 3. See *numerator* and *denominator*. (2) A fraction that satisfies the definition above, but includes a unit in both the numerator and the denominator. This definition of fraction includes any rate that is written as a fraction. For example, $\frac{50 \text{ miles}}{1 \text{ gallon}}$ and $\frac{40 \text{ pages}}{10 \text{ minutes}}$. (3) Any number written using a fraction bar, where the fraction bar is used to indicate division. For example, $\frac{2.3}{6.5}, \frac{(1\frac{4}{5})}{12}$, and $\frac{\frac{3}{4}}{\frac{5}{8}}$.

fraction stick A diagram used to represent simple fractions.

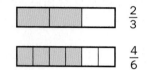

frequency graph A graph showing how often each value occurs in a *data* set. Dot plots are usually frequency graphs.

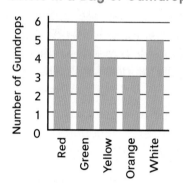

frequency table A table in which *data* are tallied and organized, often as a first step toward making a *frequency graph*.

Color	Number of Gumdrops
red	|||||
green	|||||	
yellow	||||
orange	|||
white	|||||

front-end estimation An estimation method that keeps only the left-most digit in the numbers and puts 0s in for all others. For example, the front-end *estimate* for $45{,}600 + 53{,}450$ is $40{,}000 + 50{,}000 = 90{,}000$.

function machine An imaginary machine that uses a rule to pair input numbers put in (inputs) with numbers put out (outputs). Each input is paired with exactly one output. Function machines are used in "What's My Rule" problems.

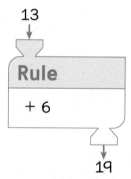

Function machine for the rule +6.

geometric solid A *3-dimensional* shape, such as a *prism*, *pyramid*, *cylinder*, *cone*, or *sphere*. Despite its name, a geometric solid is hollow; it does not contain the points in its *interior*.

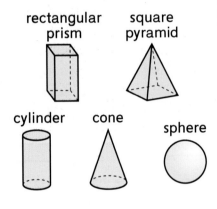

Geometry Template An *Everyday Mathematics* tool that includes a millimeter ruler, a ruler with sixteenth inch intervals, half-circle and full-circle *protractors*, a percent circle, pattern-block shapes, and other geometric figures. The template can also be used as a compass.

great span The distance from the tip of the thumb to the tip of the little finger (pinkie) when the hand is stretched as far as possible.

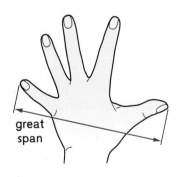

greatest common factor (GCF) The largest *factor* that two or more *counting numbers* have in common. For example, the common factors of 24 and 36 are 1, 2, 3, 4, 6, and 12. The greatest common factor of 24 and 36 is 12.

grouping symbols Symbols such as parentheses ( ), brackets [ ], and braces { } that tell the order in which operations in an *expression* are to be done. For example, in the expression (3 + 4) ∗ 5, the operation in the parentheses should be done first. The expression then becomes 7 ∗ 5 = 35.

height (1) The length of the shortest line segment connecting a *vertex* of a shape to the line containing the base opposite it. (2) The length of the shortest line segment connecting a vertex or apex of a solid to the plane containing the base opposite it. (3) The line segment itself.

Heights/altitudes of 2-D figures are shown in blue.

Heights/altitudes of 2-D figures are shown in blue.

hierarchy of shapes A way to organize shapes into categories and subcategories. All attributes for a category are attributes of each of its subcategories. Subcategories have more attributes. A hierarchy is often shown in a diagram with the most general category at the top and arrows or lines connecting categories to their subcategories.

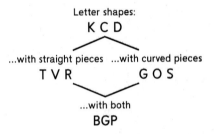

histogram A *bar graph* of numerical *data* that are grouped into *intervals*, called *bins*, along a number line. The number of the values within an interval determines the height of the bar. Many histograms have fixed intervals, or equal-width bins.

Glossary

image The reflection of an object that you see when you look in a mirror. Also, a figure that is produced by a transformation (a reflection, translation, or rotation, for example) of another figure. Compare *preimage*.

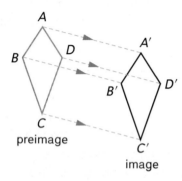

independent (manipulated) variable A *variable* whose value does not rely on the values of other variables. Compare *dependent (responding) variable*.

inequality A *number sentence* with >, <, ≥, ≤, or ≠. For example, the sentence 8 < 15 is an inequality.

integer A number in the set {. . . , −4, −3, −2, −1, 0, 1, 2, 3, 4, . . .}; a *whole number* or the opposite of a whole number, where 0 is its own opposite.

interior of a figure The inside of a closed figure. The interior is usually not considered to be part of the figure.

interpolate To *estimate* an unknown value between known values. Graphs are often useful tools for interpolation.

interquartile range (IQR) (1) The distance between the *lower* and *upper quartiles* in a *data* set. It is sometimes illustrated by a box in a box-and-whiskers plot. (2) The *interval* between the lower and upper quartiles.

interval The set of all numbers between two numbers, *a* and *b*, which may include *a* or *b* or both.

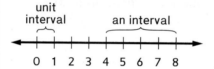

inverse operations Two *operations* that undo the effects of each other. Addition and subtraction are inverse operations, as are multiplication and division.

inverse-operations strategy An *equation-solving strategy* that involves isolating the *unknown* by applying *inverse operations* to both sides of the equation. For example, $2x + 3 = 5$ can be solved by subtracting 3 then dividing by 2 on both sides of the equation:

$2x + 3 = 5$
$2x + 3 - 3 = 5 - 3$
$2x = 2$
$2x / 2 = 2 / 2$
$x = 1$

irrational numbers Numbers that cannot be written as *fractions* where both the numerator and the denominator are *integers* and the denominator is not zero. For example, 1.10100100010000. . . is an irrational number.

isosceles trapezoid A *trapezoid* with a pair of base angles that have the same measure.

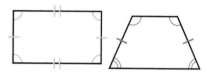

Isosceles trapezoids

isosceles triangle A *triangle* with at least two *sides* equal in length. In an isosceles triangle, at least two *angles* have the same measure. A triangle with all three sides the same length is an isosceles triangle, but is usually called an *equilateral triangle*.

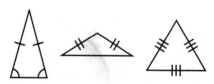

Isosceles triangles

Glossary

kite A *quadrilateral* with two pairs of adjacent equal length *sides*. The four sides can all have the same length, so a rhombus is a kite.

Kites

landmark A notable feature of a *data* set. Some landmarks are the *mean*, *median*, *mode*, *maximum*, *minimum*, and *range*.

lattice multiplication An old way to multiply multidigit numbers in a diagram that looks like a lattice.

least common denominator (LCD) The *least common multiple* of the *denominators* in each of two or more *fractions*. For example, the least common denominator of $\frac{1}{2}$, $\frac{4}{5}$, and $\frac{3}{8}$ is 40.

least common multiple (LCM) The smallest number that is a *multiple* of two or more numbers. For example, while some common multiples of 6 and 8 are 24, 48, and 72, the least common multiple of 6 and 8 is 24.

length The distance between two points along a path.

like Equal or the same.

like denominators Same as *common denominator*.

like numerator A number that is the numerator of two or more fractions. For example, the fractions $\frac{3}{11}$ and $\frac{3}{7}$ have *common numerator* of 3.

like terms In an *algebraic expression,* either the *constant* terms or any terms that contain the same variable(s) raised to the same power(s). For example, $4y$ and $7y$ are like terms in the expression $4y + 7y - z$.

line A straight path that goes on forever in both directions.

Line *RP* or *PR*

line graph A graph that uses line segments to connect *data* points. Line graphs are often used to show how something changed over a period of time.

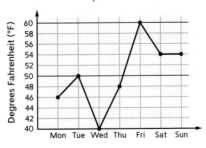

line of reflection (mirror line) A line halfway between a figure (preimage) and its reflected image. In a reflection, a figure is "flipped over" the line of reflection.

line plot A sketch of *data* in which check marks, Xs, or other marks above a labeled line show the frequency of each value.

Test Scores

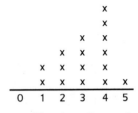

Number Correct

line segment A straight path joining two points. The two points are called endpoints of the segment.

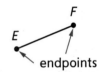

Line segment *EF* or *FE*

line symmetry A figure has line symmetry if a line can be drawn through it so that it is divided into two parts that are mirror images of each other. The two parts look alike but face in opposite directions. See *line of symmetry*.

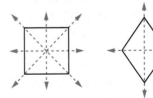

These figures have line symmetry.

lower quartile Same as *first quartile (Q1)* definition (1).

four hundred twenty-three

Glossary

magnitude of a number The size of a number; the number's distance from 0. The absolute value of a number is its magnitude. For example, the magnitude of −3 is 3 because it is 3 away from 0.

map scale A tool that helps you estimate real distances between places shown on a map. It relates distances on the map to distances in the real world. For example, a map scale may show that 1 inch on a map represents 100 miles in the real world. See *scale*.

mass A measure of how much matter is in object. Mass is often measured in grams or kilograms.

mathematical argument An explanation of why a claim is true or false using words, pictures, symbols, or other representations. For example, if you claim that $\frac{1}{2} + \frac{3}{5} = \frac{4}{7}$ is not true, you say that $\frac{3}{5}$ is more than $\frac{1}{2}$, so the answer to $\frac{1}{2} + \frac{3}{5}$ is greater than 1. Since $\frac{4}{7}$ is less than 1, $\frac{1}{2} + \frac{3}{5} = \frac{4}{7}$ must not be true.

mathematical practices Ways of working with mathematics. Mathematical practices are habits or actions that help people use mathematics to solve problems.

mathematical structure A relationship among mathematical objects, operations, or relations; a mathematical pattern, category, or *property*. For example, the Distributive Property of Multiplication over Addition is a structure of arithmetic. The number grid illustrates some patterns and structures that exist in our number system.

maximum The largest amount; the greatest number in a set of *data*.

mean A typical value for a set of numerical data, often called the average. The *mean* is found by dividing the sum of the numbers by the number of numbers in the set. The mean is often referred to simply as the average.

mean absolute deviation (m.a.d.) In a *data* set, the *average* of the distances individual data points are from the average of the whole data set. It is a measure of how spread out a distribution is.

measure of center A value representing what is typical or central to a *data* set. *Mean* and *median* are both measures of center.

measure of variability A measure of how spread out a set of data is. Some measures of variability include *range*, *interquartile range*, and *mean absolute deviation*.

median The middle value in a set of numerical *data* when the numbers are listed in order from smallest to largest, or from largest to smallest. If there are an even number of data points, the median is the *mean* of the two middle values. The median is a *measure of center*.

metric system A measurement system based on decimals and multiples of 10. The metric system is used by scientists and people in most countries around the world except the United States.

midpoint The point halfway between two other points. The midpoint of a line segment is the point halfway between the endpoints.

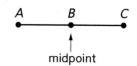

length of $\overline{AB}$ = length of $\overline{BC}$

minimum The smallest amount; the smallest number in a set of *data*.

minuend In subtraction, the number from which another number is subtracted. For example, in $19 - 5 = 14$, the minuend is 19. See *difference* and *subtrahend*.

mixed number A number that is written using both a *whole number* and a *fraction*. For example, $2\frac{1}{4}$ is a mixed number equal to $2 + \frac{1}{4}$.

mode The value or values that occur most often in a *data* set.

model A representation of a real-world object or situation. Number sentences, diagrams, and pictures can be models.

multiple of a number n (1) A *product* of n and a *counting number*. For example, the multiples of 7 are 7, 14, 21, 28, ... (2) a product of n and an *integer*. The multiples of 7 are . . . , −21, −14, −7, 0, 7, 14, 21, ...

multiplicative identity The number 1. The multiplicative identity is the number that when multiplied by any other number is that other number.

multiplicative comparison Statements about quantities that use multiplication as the comparison. For example, "Bev has three times as much money as Biff" is a multiplicative comparison.

negative numbers A number that is less than zero; a number to the left of zero on a horizontal *number line* or below zero on a vertical number line. The symbol − may be used to write a negative number. For example, "negative 5" is usually written as −5.

net A 2-dimensional figure created to represent a *3-dimensional* figure by cutting and unfolding or separating its faces and sides. A 2-dimensional figure that can be folded to form all the faces of a closed 3-dimensional figure is called a net. For example, if a cereal box is cut along some of its edges and laid out flat, it will form a net for the box.

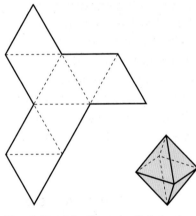

Net for octahedron   Octahedron

normal distribution A set of values that is clustered around the center of the data and trails off to both sides. This *distribution* is also called a bell curve. The data shown below are normally distributed.

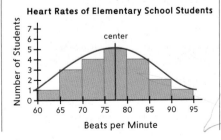

null set Same as *empty set*.

number line A *line* with numbers marked in order on it.

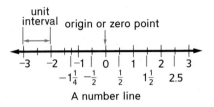

A number line

number model A *number sentence* or *expression* that models or fits a *number story* or situation. For example, the story *Sally had $5, and then she earned $8*, can be modeled as the number sentence $5 + 8 = 13$, or as the expression $5 + 8$.

number sentence At least two numbers or *expressions* separated by a *relation* symbol ($=, >, <, \geq, \leq, \neq$). Most number sentences contain at least one operation symbol ($+, -, \times, *, \div,$ or $/$). Number sentences may also have grouping symbols, such as parentheses and brackets.

number story A story with a problem that can be solved using arithmetic.

numerator The number above the line in a *fraction*. A fraction may be used to name part of a *whole*. If the whole is divided into equal parts, the numerator represents the number of equal parts being considered. In the fraction $\frac{a}{b}$, a is the numerator.

Glossary

obtuse angle An angle with measure greater than 90° and less than 180°.

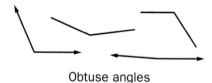

Obtuse angles

obtuse triangle A triangle with an obtuse angle.

An obtuse triangle

open sentence A *number sentence* with a *variable* that is either true or false depending on what value is substituted for the variable. For example, $5 + x = 13$ is an open sentence. The sentence is true if 8 is substituted for *x*. The sentence is false if 4 is substituted for *x*.

operation An action performed on numbers, *expressions,* or variables to produce other numbers, expressions, *variables.* Addition, subtraction, multiplication, and division are the four basic arithmetic operations.

operation symbol A symbol used to stand for a mathematical operation. Common operation symbols are +, −, *, ÷, and /.

opposite of a number *n* A number that is the same distance from 0 on the number line as a given number, but on the opposite site of 0. For example, the opposite of +3 is −3, and the opposite of −5 is +5.

order of operations Rules that tell in what order to perform operations in arithmetic and algebra.

ordered pair Two numbers that are used to locate a point on a rectangular coordinate grid. The first number gives the position along the horizontal axis, and the second number gives the position along the vertical axis. The numbers in an ordered pair are called coordinates. Ordered pairs are usually written inside parentheses: (5, 3).

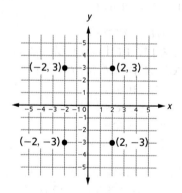

Ordered pairs

origin (1) The point (0, 0) where the two axes of a coordinate grid meet. (2) The 0 point on a *number line.*

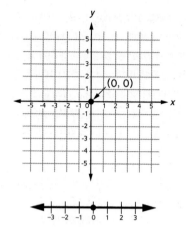

The points at 0 and (0, 0) are origins.

outlier A value far from most of the others in a *data* set. Outliers are commonly much larger or much smaller than other values.

pan balance A tool used to weigh objects or compare *weights.* The pan balance is also used as a model in balancing and solving equations.

pan-balance problem In *Everyday Mathematics,* a problem in which *equations* are represented with pan balances and weights. The pans are kept balanced by performing the same *operations* on both sides.

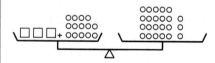

A pan-balance problem

SRB
426 four hundred twenty-six

Glossary

parallel Lines, line segments, or rays in the same *plane* are parallel if they never cross or meet, no matter how far they are extended. Two planes are parallel if they never cross or meet. A line and a plane are parallel if they never cross or meet. The symbol ∥ means *is parallel to*.

parallel bases

parallelogram A *quadrilateral* with two pairs of *parallel* sides. Opposite sides of a parallelogram are congruent. Opposite angles in a parallelogram have the same measure. All parallelograms are *trapezoids*.

Parallelogram

parentheses Grouping *symbols*, ( ), used to tell which parts of an expression should be calculated first.

partial-products diagram A rectangular grid that is used to organize the *partial products* in a multiplication problem. Each factor is decomposed and written along one side the rectangle. Partial products are recorded in each section of the grid then added to find the total product.

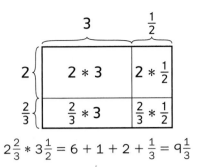

$$2\tfrac{2}{3} * 3\tfrac{1}{2} = 6 + 1 + 2 + \tfrac{1}{3} = 9\tfrac{1}{3}$$

partial-products multiplication (1) A way to multiply in which the value of each digit in one factor is multiplied by the value of each digit in the other factor. The final *product* is the sum of these *partial products*. (2) A similar method for multiplying mixed numbers.

partial-quotients division A way to divide in which the dividend is divided in a series of steps. The *quotients* for each step (called partial quotients) are added to give the final answer.

partial-sums addition A way to add in which *sums* are computed for each place (ones, tens, hundreds, and so on) separately. The partial-sums are then added to give the final answer.

part-to-part ratio A *ratio* that compares a part of a whole to another part of the same whole. For example, the statement "There are 8 boys for every 12 girls" expresses a part-to-part ratio. Compare *part-to-whole ratio*.

part-to-whole ratio A *ratio* that compares a part of a whole to the whole. For example, the statements "8 out of 20 students are boys" and "12 out of 20 students are girls," both express part-to-whole ratios. Compare *part-to-part ratio*.

percent (%) Per hundred or out of a hundred. For example, "48% of the students in the school are boys" means that 48 out of every 100 students in the school are boys; $48\% = \tfrac{48}{100} = 0.48$.

perimeter The distance around the boundary of a shape. The perimeter of a circle is called its circumference.

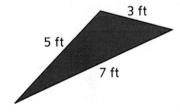

Perimeter = 5 ft + 3 ft + 7 ft = 15 ft

Glossary

perpendicular Being part of two lines that cross or meet at right angles. Planes that cross or meet at right angles are perpendicular. The symbol ⊥ means "is perpendicular to."

Perpendicular lines

Perpendicular planes

Perpendicular rays

per-unit rate A *rate* with 1 in the denominator. Per-unit rates tell how many of one thing there are for one of another thing. For example, "2 dollars per gallon" is a per-unit rate. "12 miles per hour" and "4 words per minute" are also examples of per-unit rates.

place value A system that gives a *digit* a value according to its position in a number. In our *base-ten* system for writing numbers, moving a digit one place to the left makes that digit worth 10 times as much. And moving a digit one place to the right makes that digit worth one tenth as much. For example, in the number 456, the 4 in the hundreds place is worth 400; but in the number 45.6, the 4 in the tens place is worth 40.

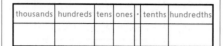

A place-value chart

plane A flat *surface* that extends forever.

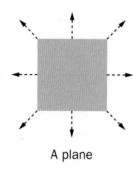

A plane

plane figure A set of points that is entirely contained in a single plane.

plateau distribution A set of values that is evenly distributed within some range. The data shown below are uniformly distributed.

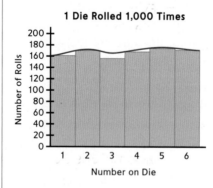

plot To draw on a number line, coordinate grid, or graph. The points plotted may come from *data*.

point An exact location in space. The center of a circle is a point. Lines have an unlimited number of points on them.

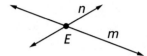

Lines *m* and *n* intersect at point *E*.

polygon A 2-dimensional figure that is made up of *line segments* joined end to end to make one closed path. The line segments of a polygon may not cross.

Polygons

polyhedron A *3-dimensional* figure whose surfaces (*faces*) are all flat and formed by *polygons*. Each face consists of a polygon and the *interior* of that polygon. The faces may meet but not cross. A polyhedron does not have any curved surface.

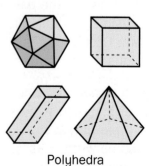

Polyhedra

SRB 428 four hundred twenty-eight

Glossary

population In *data* collection, the collection of people or objects that is the focus of study.

population density The number of people living in a geographic area, usually given as a *unit rate*, such as 876 people per square mile.

positive numbers A number that is greater than zero; a number to the right of zero on a horizontal *number line*, or above zero on a vertical number line. A positive number may be written using the + symbol, but is usually written without it. For example, +10 = 10. Compare *negative numbers*.

power Same as *exponent*.

power of 10 (1) A whole number that can be written as a product of 10s. For example, 100 is equal to 10 * 10, or 10^2. 100 is called "the second power of 10," "10 to the second power," or "10 squared." (2) A number that can be written as a product of $\frac{1}{10}$s is also a power of 10. For example, $10^{-2} = \frac{1}{10^2} = \frac{1}{10} * 10 = \frac{1}{10} * \frac{1}{10}$ is a power of 10.

power of a number *n* The *product* of *factors* that are all the same. For example, 5 * 5 * 5 (or 125) is called "5 to the third power" or "the third power of 5" because 5 is a factor three times. 5 * 5 * 5 can also be written as 5^3. See *exponent*.

precise Exact. The smaller the *unit* used in measuring, the more precise the measurement is. For example, a measurement to the nearest inch is more precise than a measurement to the nearest foot. A ruler with $\frac{1}{8}$ inch markings is more precise than a ruler with $\frac{1}{4}$ inch markings.

preimage A geometric figure that is changed (by a reflection, or rotation, for example) to produce another figure. Compare *image*.

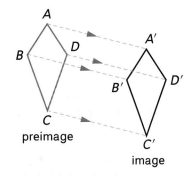

preimage   image

prime factorization A *counting number* written as a product of prime *factors*. Every counting number greater than 1 has a unique prime factorization. For example, the prime factorization of 24 is 2 * 2 * 2 * 3. The factorization of a prime number is that number. For example, the prime factorization of 13 is 13.

prime number A *counting number* that has exactly two different *factors*: itself and 1. For example, 5 is a prime number because its only factors are 5 and 1. The number 1 is not a prime number because 1 has only a single factor, the number 1 itself.

prism A polyhedron with two *parallel faces*, called *bases* that are the same size and shape. The other faces connect the bases and are shaped like *parallelograms*. The edges that connect the bases are parallel. Prisms get their names from the shape of their bases.

Triangular prism

Rectangular prism

Hexagonal prism

product The result of multiplying two or more numbers, called *factors*. For example, in 4 * 3 = 12, the product is 12.

property (1) A general statement about a mathematical relationship, such as the turn-around rule or the Distributive Property of Multiplication over Addition. (2) Same as attribute.

Glossary

protractor A tool on the *Geometry Template* that is used to measure and draw *angles*. The half-circle protractor can be used to measure and draw angles up to 180°; the full-circle protractor, to measure angles up to 360°.

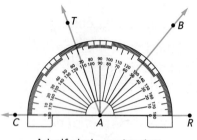

A half-circle protractor

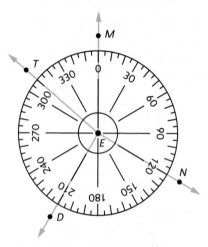

A full-circle protractor

pyramid A *polyhedron* in which one *face*, the *base*, may have any polygon shape. All of the other faces are triangular and come together at a point called the *apex*. A pyramid takes its name from the shape of its base.

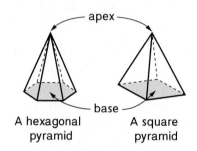

A hexagonal pyramid   A square pyramid

Q

quadrangle A polygon that has four angles. Same as *quadrilateral*.

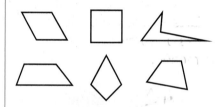

quadrant One of the four sections of a *rectangular coordinate grid*. The quadrants are typically numbered I, II, III, and IV counterclockwise beginning at the upper right.

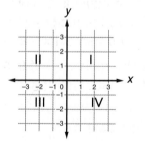

quadrilateral A *polygon* that has four sides. Same as *quadrangle*.

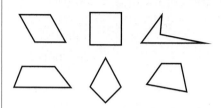

quantity A number with a unit, usually a measurement or count.

quartile See *first*, *second*, *third*, *fourth*, *lower*, and *upper quartile*, *interquartile range*, and *median*.

quick common denominator (QCD) The *product* of the *denominators* of two or more *fractions*. For example, the quick common denominator of $\frac{1}{4}$ and $\frac{3}{6}$ is $4 * 6$, or 24. As the name suggests, this is a quick way to get a *common denominator* for a collection of fractions, but it does not necessarily give the *least common denominator*.

quotient The result of dividing one number by another number. For example, in $35 \div 5 = 7$, the quotient is 7.

$$\text{dividend} \searrow \overset{\text{divisor}}{\downarrow} \swarrow \text{quotient}$$
$$35/5 = 7$$

$$\text{dividend} \searrow \overset{\text{divisor}}{\downarrow} \swarrow \text{quotient}$$
$$40 \div 8 = 5$$

quotient → 3
divisor → 12)36 ← dividend

SRB
430   four hundred thirty

Glossary

radius A line segment from the center of a circle (or sphere) to any point on the circle (or sphere). Also the length of this line segment.

random sample A *sample* that gives all members of the *population* the same chance of being selected.

range The *difference* between the *maximum* and the *minimum* in a *data* set. The range is a measure of how spread out a distribution is.

rate A comparison by division of two quantities with unlike *units*. For example, a speed such as 55 miles per hour is a rate that compares distance with time. See *ratio*.

ratio A comparison of two quantities using division. Ratios can be expressed with fractions, decimals, percents, or words. Sometimes they are written with a colon between the two numbers that are being compared. For example, if a team wins 3 games out of 5 games played, the ratio of wins to total games can be written as $\frac{3}{5}$, 0.6, 60%, 3 to 5, or 3:5. See *rate*.

ratio comparison Same as *multiplicative comparison*.

ratio/rate table A way of displaying ratio or *rate* information. In a ratio/rate table, the fractions formed by the two numbers in each column are *equivalent fractions*.

miles	35	70	105	140	175	210
gallons	1	2	3	4	5	6

rational numbers Any number that can be written as an *integer* divided by a nonzero integer. Most of the numbers you have used are rational numbers. For example, $\frac{2}{3}$, $-\frac{2}{3}$, 60% = $\frac{60}{100}$, and $-1.25 = -5/4$ are all rational numbers.

ray A straight path that starts at one endpoint and goes on forever in one direction.

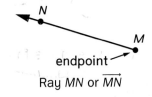

Ray MN or $\overrightarrow{MN}$

real numbers Any *rational* or *irrational numbers*.

reciprocals Two numbers whose product is 1. For example, the reciprocal of 5 is $\frac{1}{5}$, and the reciprocal of $\frac{1}{5}$ is 5; the reciprocal of 0.4 ($\frac{4}{10}$) is $\frac{10}{4}$, or 2.5, and the reciprocal of 2.5 is 0.4. See *multiplicative inverse*.

rectangle A *parallelogram* whose corners are all right *angles*.

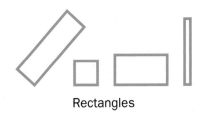

Rectangles

rectangle method A method for finding area in which rectangles are drawn around a figure or parts of a figure. The rectangles form regions with boundaries that are rectangles or triangular halves of rectangles. The area of the original figure can be found by adding or subtracting the areas of these regions.

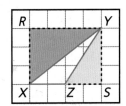

Area of △XYZ = area of rectangle RYSX − area of △XRY − area of △YSZ

rectangular prism A *prism* with rectangular *bases*. The four *faces* that are not bases are either *rectangles* or other *parallelograms*. A rectangular prism may model a shoebox.

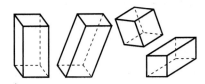

Rectangular prisms

four hundred thirty-one | SRB 431

Glossary

rectangular pyramid A *pyramid* with a rectangular *base*.

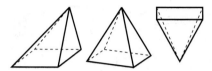

Rectangular pyramids

reflection The "flipping" of a figure over a line (the line of reflection) so that its image is the mirror image of the original figure (preimage). A reflection of a solid figure is a mirror-image "flip" over a plane.

A reflection

regular polygon A *polygon* whose sides are all the same length and whose interior *angles* are all the same measure. For example, a square is a regular polygon.

regular polyhedron A *polyhedron* whose *faces* are formed by congruent *regular polygons*.

A tetrahedron (4 equilateral triangles)   A cube (6 squares)   An octahedron (8 equilateral triangles)

A dodecahedron (12 regular pentagons)   An icosahedron (20 equilateral triangles)

relation symbol A symbol used to express a relationship between two quantities. Some relationship symbols are <, >, and =.

relatively prime Having no *factors* in common other than 1. For example, 8 and 21 are relatively prime because the only number that divides them both without remainder is 1.

remainder An amount left over when one number is divided by another number. For example, if 38 books are divided into 5 equal piles, there will be 7 books per pile, with 3 books left over. We may write $38 \div 5 \rightarrow 7$ R3, where R3 stands for the remainder.

repeating decimal A *decimal* in which one digit or a group of digits is repeated without end. For example, 0.3333... and $23.\overline{147} = 23.147147...$ are repeating decimals. Compare *terminating decimal*.

represent To show, symbolize, or stand for something. For example, numbers can be represented using base-10 blocks, spoken words, or written numerals.

rhombus A *quadrilateral* whose sides are all the same length. All rhombuses are *parallelograms* and *kites*. Every square is a rhombus, but not all rhombuses are squares.

Rhombuses

right angle An *angle* with a measure of 90°.

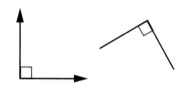

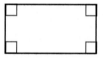

Right angles

right triangle A triangle that has a right angle (90°).

Right triangle

rotation A movement of a figure around a fixed point; a "turn."

A rotation

round To change a number slightly to make it easier to work with or to make it better reflect the level of precision of the data. Often numbers are rounded to the nearest multiple of 10, 100, 1,000, and so on. For example, 12,964 rounded to the nearest thousand is 13,000.

rubric A tool used to rate work based on its quality.

sample A part of a group chosen to *represent* the whole group. See *population* and *random sample*.

scale (1) The ratio of a distance on a map, globe, or drawing to an actual distance. (2) A system of ordered marks at fixed intervals used in measurement; or any instrument that has such marks. For example, a ruler with scales in inches and centimeters, and a thermometer with scales in °F and °C. See *map scale* and *scale drawing*. (3) A tool for measuring *weight* or *mass*.

scale drawing A drawing of an object or a region in which all parts are drawn to the same *scale*. Architects and builders use scale drawings.

Scale drawing of an 8-inch woodpecker.

scale factor The *ratio* of the size of a drawing or model of an object to the actual size of the object. See *scale model* and *scale drawing*.

scale model A model of an object in which all parts are in the same proportions as in the actual object. For example, many model trains and airplanes are *scale* models of actual vehicles.

scale of a number line The spacing of the marks on a *number line*. The scale of the number line below is halves.

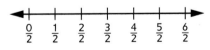

scalene triangle A *triangle* with *sides* of three different lengths.

Scalene triangle

scientific notation A system for writing numbers in which a number is written as the product of a *power of 10* and a number that is at least 1 and less than 10. Scientific notation allows you to write big and small numbers with only a few symbols. For example, $4 * 10^{12}$ is scientific notation for 4,000,000,000,000.

second quartile (Q2) (1) Same as *median*. (2) Informally, the *interval* between the median and the *first quartile*.

sequence A list of numbers, often created by a rule that can be used to extend the list.

set A collection or group of objects, numbers, or other items.

side (1) One of the *line segments* of a *polygon*. (2) One of the rays or segments that make up an *angle*. (3) One of the *faces* of a 3-dimensional figure.

Glossary

side-by-side bar graph A *bar graph* that uses pairs of bars to compare two related *data* sets. The graph below compares road miles and air miles from Los Angeles to different cities.

A side-by-side bar graph

significant digits The *digits* in a number that convey useful and reliable information. A number with more significant digits is more *precise* than a number with fewer significant digits.

similar figures Figures that have the same shape, but not necessarily the same size.

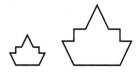

Similar figures

simplify an expression To rewrite an *expression* by combining *like terms* and *constants*. For example, $7y + 4 + 5 + 3y$ simplifies to $10y + 9$ and $3(2k + 5) - k$ simplifies to $5k + 15$. Equations with simplified expressions are often easier to solve.

skewed distribution A set of values clustered in one area then trailing off in one direction. The data shown below are right skewed because the trailing off data is to the right of the cluster.

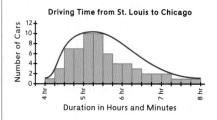

solid See geometric solid.

solution of an open sentence A value that makes an *open sentence* true when it is substituted for the *variable*. For example, 7 is a solution of $5 + n = 12$.

solution set The set of all solutions of an equation or inequality. For example, the solution set of $x^2 = 25$ is $\{5, -5\}$ because substitution of either 5 or -5 for x makes the sentence true.

spreadsheet program A computer application in which numerical information is arranged in *cells* in a grid. The computer can use the information in the grid to perform mathematical operations and evaluate *formulas*. When a value in a cell changes, the values in all other cells that depend on it are automatically changed.

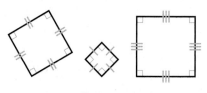

A spreadsheet

square A *rectangle* whose sides are all the same length. A rectangle that is also a *rhombus*.

Squares

square number A number that is the *product* of a counting number with itself. For example, 25 is a square number because $25 = 5 * 5$. The square numbers are 1, 4, 9, 16, 25, and so on. A square number can be represented by a square array.

square of a number n The product of a number with itself. For example, 81 is the square of 9 because $81 = 9 * 9$. And 0.64 is the square of 0.8 because $0.64 = 0.8 * 0.8$.

square pyramid A *pyramid* with a *square* base.

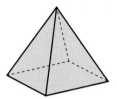

Square pyramid

square root of a number *n* The square root of a number *n* is a number that, when multiplied by itself, gives *n*. For example, 4 is the square root of 16 because 4 * 4 = 16.

square unit A *unit* used in measuring *area*, such as a square centimeter or a square foot.

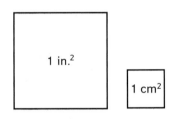

Square units

standard form The most familiar way of representing numbers. In standard form, numbers are written using the *base-ten place-value* system. For example, standard form for three hundred fifty-six is 356. Compare *expanded form*.

standard notation Same as *standard form*.

standard unit Measurement *units* that are the same size no matter who uses them and when or where they are used.

statistical question A question that would be answered by collecting or analyzing *data*.

statistics (1) The study of numerical data: collecting, organizing, and analyzing *data* to interpret it and answer questions. See *statistical question*. (2) The numerical data themselves.

subcategory A more specific category within a given category. Subcategories are usually defined by an attribute shared by some, but not all, of the members of the larger category. For example, triangles are a subcategory of the larger category of polygons because some, but not all, polygons have three sides.

substitute To replace one thing with another. In a *formula*, to replace *variables* with numerical values.

subtrahend In subtraction, the number being subtracted. For example, in 19 − 5 = 14, the subtrahend is 5. See *difference* and *minuend*.

sum The result of adding two or more numbers. For example, in 5 + 3 = 8, the sum is 8.

surface (1) The boundary of a *3-dimensional* object. Common surfaces include the top of a body of water, the outermost part of a ball, and the topmost layer of ground that covers Earth.

(2) Any *2-dimensional* layer, such as a *plane* or the *faces* of a polyhedron.

surface area The total area of all the *surfaces* of a *3-dimensional* object. The surface area of a rectangular prism is the sum of the areas of its six faces.

tally chart A chart that uses marks, called tallies, to show how many times each value appears in a set of data.

Number of Pull-Ups	Number of Children
0	||||| /
1						
2	////
3	//
4	
5	///
6	/

tape diagram A diagram used to show *ratios*. For example, the number story "Sally plants 20 flowers with a ratio of 2 irises for every 3 tulips" can be represented with this tape diagram.

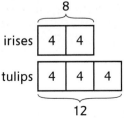

four hundred thirty-five 435

Glossary

temperature A measure of how hot or cold something is.

term In an expression, a number or a product of a number and one or more *variables*. In the equation $5y + 3k = 8$, the terms are $5y$, $3k$, and 8. The 8 is a *constant* term, or simply a constant, because it has no variable part. (2) An element in a *sequence*. In the sequence of multiples of 10, the terms are 10, 20, 30, 40, and so on.

terminating decimal A *decimal* that ends. For example, 0.5 and 2.125 are terminating decimals. Compare *repeating decimal*.

third quartile (Q3) (1) The middle value of the numbers above the *median* in a *data* set, also called the *upper quartile*. (2) Informally, the *interval* between this middle point of the upper half of the data and the median of the data set.

tool Anything that can be used for performing a task. Calculators, rulers, fraction circle pieces, and tape diagrams are examples of mathematical tools.

trade-first subtraction A subtraction method in which all trade are done before any subtractions are carried out.

trapezoid A *quadrilateral* that at least one pair of *parallel* sides.

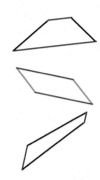

Trapezoids with parallel bases marked in the same color

trial-and-error method A method for finding the solution of an *equation* by trying several test numbers.

triangle A *polygon* that has 3 sides and 3 angles.

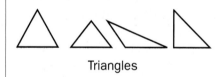

Triangles

triangular prism A *prism* whose *bases* are triangles.

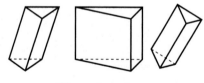

Triangular prisms

true number sentence A *number sentence* in which the relation symbol accurately connects the two sides. For example, $15 = 5 + 10$ and $25 > 20 + 3$ are both true number sentences.

turn-around rule A rule for solving addition and multiplication problems based on the *Commutative Property*. For example, if you know that $6 * 8 = 48$, then, by the turn-around rule, you also know that $8 * 6 = 48$.

U.S. customary system The measuring system most frequently used in the United States. See the Tables of Measures on page 374.

U.S. traditional addition algorithm An addition method that involves adding digits by place-value columns starting at the right.

```
  1   1
  3   4   8
+ 2   6   3
─────────────
  6   1   1
```

U.S. traditional long division algorithm A division method that relies on estimating products of the dividend with the digits in the divisor, starting from the left.

```
       163
    ┌──────
  6 ) 978
     -6
     ───
      37
     -36
     ───
      18
     -18
     ───
       0
```

Glossary

U.S. traditional multiplication algorithm A multiplication method that produces partial sums based on multiplying the values of each digit starting from the right.

```
      1 1
      1 1
      1 2 2
  *       7 5
  ─────────────
        6 1 0
  + 8 5 4 0
  ─────────────
    9 1 5 0
```

U.S. traditional subtraction algorithm A subtraction method that involves subtracting digits by place-value columns starting from the right, making 10-for-1 trades as needed.

```
      3 14
    4̷ 4̷ 7
  −   1 6 5
  ─────────
      2 8 2
```

uniform distribution See plateau distribution.

unit A label used to put a number in context. In measuring length, for example, the inch and the centimeter are units. In a problem about 5 apples, *apple* is the unit.

unit conversion A change from one measurement *unit* to another using a fixed relationship such as 1 yard = 3 feet or 1 inch = 2.54 centimeters.

unit cube A *cube* with edge lengths of 1.

unit fraction A *fraction* whose *numerator* is 1. For example, $\frac{1}{2}$, $\frac{1}{3}$, $\frac{1}{8}$, and $\frac{1}{20}$ are unit fractions.

unit rate A rate in which one of the quantities being compared is 1. For example, 70 miles per hour is a unit rate because it is the number of miles traveled in 1 hour. Similarly, 1 chaperone for every 10 students in a unit rate. See per-unit rate.

unit ratio A rate in which one of the quantities being compared is 1. For example, 70 miles per hour is a unit rate because it is the number of miles traveled in 1 hour. Similarly, 1 chaperone for every 10 students is a unit rate.

unit square A *square* with side lengths of 1.

unknown A quantity whose value is not known. An unknown is sometimes represented by a _____, a ?, or a letter.

unlike Unequal or not the same.

upper quartile Same as *third quartile (Q3)* definition (1).

variability of a distribution How spread out the values in a set of data are.

variable A letter or other symbol that can be replaced by any number. In the *number sentence* $5 + n = 9$, any number may be substituted for *n*, but only 4 makes the sentence true. In the inequality $x + 2 < 10$, any number may be substituted for *x*, but only numbers less than 8 make the sentence true. In the equation $a + 3 = 3 + a$, any number may be substituted for a, and every number makes the sentence true. See *unknown*.

variable term A *term* that contains at least one *variable*.

vertex The point where the *sides* of an *angle*, the sides of a polygon, or the *edges* of a *polyhedron* meet. Plural is vertexes or vertices. Informally called a corner.

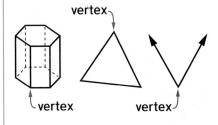

vertical Upright; perpendicular to the horizon.

SRB
four hundred thirty-seven **437**

Glossary

volume A measure of how much space a solid object takes up. Volume is often measured in liquid units, such as liters, or cubic units, such as cubic centimeters or cubic inches. The volume or capacity of a container is a measure of how much the container will hold.

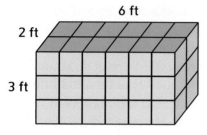

36 cubic feet

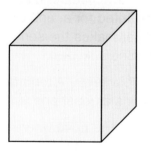

1 cubic inch

1 cubic centimeter

If the cubic centimeter were hollow, it would hold exactly 1 milliliter ($\frac{1}{1000}$ liter).

weight A measure of how heavy something is.

"What's My Rule?" A type of problem with "in" numbers, "out" numbers, and a rule that changes the in numbers to the out numbers. Sometimes you have to find the rule. Other times, you use the rule to figure out the in or out numbers. Sometimes, you can graph in and out numbers as ordered pairs.

whole An entire object, collection of objects, or quantity being considered; 100%.

whole numbers The *counting numbers*, together with 0. The set of whole numbers is {0, 1, 2, 3, ...}

x-axis In a *coordinate grid*, the horizontal number line.

x-coordinate The first number in a pair of numbers used to locate a point on a *coordinate grid*. The x-coordinate gives the position of the point along the horizontal axis. For example, 2 in the point (2, 3) is the x-coordinate.

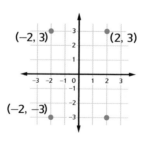

Coordinates

yard A U.S. customary unit of length equal to 3 feet, or 36 inches.

y-axis In a *coordinate grid*, the vertical number line.

y-coordinate The second number in a pair of numbers used to locate a point on a *coordinate grid*. The y-coordinate gives the position of the point along the vertical axis. For example, 3 in the point (2, 3) is the y-coordinate.

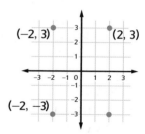

Coordinates

Answer Key

Page 12

The sum is $5\frac{3}{4}$. Sample answers: With a number line:

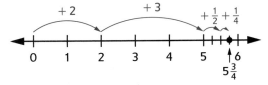

With number sentences: $2\frac{1}{2} + 3\frac{1}{4}$ is the same as $2 + \frac{1}{2} + 3 + \frac{1}{4}$. I can use the Commutative and Associative Properties to simplify the addition. $2 + \frac{1}{2} + 3 + \frac{1}{4} = 2 + 3 + \frac{1}{2} + \frac{1}{4} = 5 + (\frac{1}{2} + \frac{1}{4})$. I can think about fraction circles in my head and see that the sum of the fraction parts is $\frac{3}{4}$. So, $5 + (\frac{1}{2} + \frac{1}{4}) = 5 + \frac{3}{4} = 5\frac{3}{4}$.

The representations are similar because they both showed adding the 2 and 3 before adding the fractions. The representations are different because when I added the fractions on the number line, I thought about going one half and then another fourth of a unit distance. In the number sentence representation, I thought about putting together $\frac{1}{2}$ and $\frac{1}{4}$ fraction circle pieces and could see $\frac{3}{4}$ in my head.

Page 16

1. Area of the rectangle: $7 * 3 = 21$ cm^2; Area of large triangle: $(7 * 3) / 2 = 10.5$ cm^2; Area of smaller triangle: $(4 * 3) / 2 = 6$ cm^2; Area of $\triangle KLM = 21 - 10.5 - 6 = 4\frac{1}{2}$ cm^2

2. Yes, the areas are both $4\frac{1}{2}$ cm^2. Yes, their conjecture works for $\triangle KLM$.

Page 28

Sample answer: Lockers 2, 3, 5, 7, 11, 13, 17, 19, 23, 29, 31, 37, 41, 43, 47, 53, 59, 61, 67, 71, 73, 79, 83, 89, and 97. If a locker has exactly two stickers, the number of the locker has exactly two factors, and those factors are 1 and the number itself. If the only factors that a number has are 1 and the number itself, that means there are no other factors, and the number is only a multiple of itself and 1. These types of numbers are called prime numbers.

Answer Key

Page 40
1. Part-to-part ratio; Sample answer: 3 cups of flour are needed for every 2 cups of sugar.
2. Part-to-whole ratio; Sample answer: 2 out of every 5 cars of a set are red.
3. Rate; Sample answer: 32 miles are traveled on 1 gallon of gas.

Page 53
1. Sample answers: 1 inch by $1\frac{1}{2}$ inches; 6 inches by 9 inches; 10 inches by 15 inches
2. a and c

Answer Key

Page 58
1. a. Sample answers: $\frac{1}{4}$, or $\frac{25}{100}$; 0.25
   b. Sample answers: $\frac{2}{5}$, $\frac{40}{100}$, or $\frac{4}{10}$; 0.4
   c. Sample answers: $\frac{7}{10}$, or $\frac{70}{100}$; 0.70
   d. Sample answers: $\frac{500}{100}$, or $\frac{5}{1}$; 5.00, 5.0, or 5
   e. Sample answers: $\frac{300}{100}$, or $\frac{30}{10}$; 3.00, 3.0, or 3
2. a. 65%   b. 90%   c. 300%

Page 59
1. 40 miles; Sample explanation: 1% of 200 is 2, so 20% of 200 is $20 * 2 = 40$.
2. 66 students; Sample explanation: 1% of 300 is 3, so 22% of 300 is $22 * 3 = 66$.

Page 60
1. $72   2. $12 discount; $36

Page 62
80% of the telephone pole is seen above ground.

Page 63
1. a. 75%   b. 34%
   c. 40%   d. 250%
2. a. 60%; No.   b. 80%; Yes.
   c. 75%; No.

Page 64
16 attempted free-throw shots

Page 71
1. 108 inches   2. 4 pints
3. 500 millimeters

Page 93
1. opposite: 4.5; absolute value: 4.5
2. opposite: $-7\frac{2}{3}$; absolute value: $7\frac{2}{3}$
3. opposite: −74; absolute value: 74
4. opposite: 1,452; absolute value: 1,452
5. 12   6. 15

Page 97
1. twenty-six thousand, four hundred eighty-two; 6,000
2. forty-five million, six hundred seventy-eight thousand, nine hundred ten; 600,000
3. two hundred seven thousand, four hundred sixty-four; 60
4. eight million, seven hundred sixty-five thousand, four hundred thirty-two; 60,000

Page 98
1. 36   2. 64   3. 100,000
4. 9   5. 50,625   6. 161,051

Page 100
1. 83,405   2. 2,673,952
3. $(8 * 10^3) + (7 * 10^2) + (4 * 10^1) + (4 * 10^0)$
4. $(1 * 10^6) + (4 * 10^5) + (5 * 10^4) + (6 * 10^3) + (9 * 10^2)$

Page 101
1. False   2. True   3. False   4. False

Page 102
1. 1, 3, 17, 51
2. 1, 3, 9, 27, 81
3. 1, 2, 4, 11, 22, 44
4. 1, 2, 3, 4, 6, 8, 9, 12, 18, 24, 36, 72
5. 1, 2, 4, 8, 16, 32, 64, 128
6. 1, 2, 5, 10, 25, 50

four hundred forty-one

Answer Key

Page 103
Divisible by 2: 4,470; 526; 13,860
Divisible by 3: 105; 4,470; 621; 13,680
Divisible by 5: 105; 4,470; 13,680
Divisible by 6: 4,470; 13,680
Divisible by 9: 621; 13,680
Divisible by 10: 4,470; 13,680

Page 104
The prime factors can be given in any order.

1. $3 * 2 * 2$
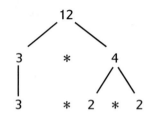

2. $2 * 2 * 2 * 2 * 2$
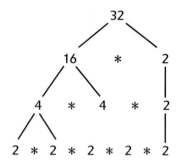

3. $7 * 2 * 3$
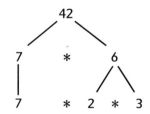

4. $2 * 2 * 2 * 3$
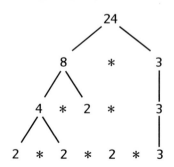

5. $2 * 5 * 5$
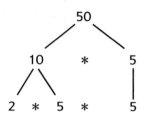

6. $5 * 5 * 2 * 2$
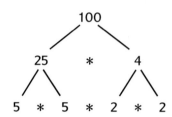

Page 108
1. 4 tenths; zero point four
2. 1 and 65 hundredths; one point six five
3. 872 thousandths; zero point eight seven two
4. 16 and 4 hundredths; sixteen point zero four
5. 3 thousandths; zero point zero zero three
6. 59 and 61 thousandths; fifty-nine point zero six one

Page 114
1. Sample estimate: $16 - 9 = 7$
2. Sample estimate: $10 * 3 = 30$
3. Sample estimate: $36 / 12 = 3$
4. Sample estimate: $150 + 390 + 210 = 750$
5. Sample estimate: $700 * 40 = 28,000$
6. Sample estimate: $750 / 30 = 25$

442 four hundred forty-two

Answer Key

Page 124
1. 46; Sample estimate: $80 - 40 = 40$
2. 291; Sample estimate: $650 - 350 = 300$
3. 242; Sample estimate: $540 - 300 = 240$
4. 72; Sample estimate: $825 - 750 = 75$
5. 14.15; Sample estimate: $21.5 - 7.5 = 14$
6. 8.21; Sample estimate: $10 - 2 = 8$
7. 4.318; Sample estimate: $7.4 - 3 = 4.4$
8. 0.74; Sample estimate: $1.2 - 0.5 = 0.7$

Page 127
1. 12.7; Sample estimate: $22 - 9 = 13$
2. 8.21; Sample estimate: $10 - 1.8 = 8.2$
3. 0.037; Sample estimate: $7.8 - 7.8 = 0$
4. 0.24; Sample estimate: $0.75 - 0.50 = 0.25$
5. 593; Sample estimate: $650 - 50 = 600$
6. 404; Sample estimate: $750 - 350 = 400$
7. 174; Sample estimate: $450 - 275 = 175$
8. 3,798; Sample estimate: $5,200 - 1,400 = 3,800$

Page 128
1. 1 m

Page 132
1. 2,400
2. 180,000
3. 3,200,000
4. 2,100,000

Page 134
1. 716; Sample estimate: $200 * 4 = 800$
2. 2,368; Sample estimate: $40 * 60 = 2,400$
3. 12.88; Sample estimate: $3 * 4 = 12$
4. 19.825; Sample estimate: $(3 * 6) + (3 * \frac{1}{2}) = 19.5$
5. 1.5756; Sample estimate: $\frac{1}{2} * 3 = 1.5$

Page 135
1. 21,024; Sample estimate: $50 * 400 = 20,000$
2. 1.7496; Sample estimate: $5 * 0.4 = 2.0$
3. 0.15111; Sample estimate: $0.07 * 2 = 0.14$

Answer Key

Page 138
1. 6,622; Sample estimate: 80 * 80 = 6,400
2. 7,448; Sample estimate: 75 * 100 = 7,500
3. 4,536; Sample estimate: 7 * 600 = 4,200
4. 74.25; Sample estimate: 16 * 5 = 80
5. 68.51; Sample estimate: 4 * 17 = 68
6. 283.03; Sample estimate: 8 * 34 = 272

Page 139
1. 84
2. 7,000
3. 50
4. 50

Page 140
1. 7.98
2. 0.0078
3. $0.360
4. 0.008

Page 146
1. 8.04; Sample estimate: 4 / 0.5 = 40 / 5 = 8
2. 460; Sample estimate: 7 / 0.02 = 700 / 2 = 350
3. 840; Sample estimate: 200 / 0.25 = 20,000 / 25 = 800

Page 147
1. 120; Sample estimate: 880 / 8 = 110
2. 164; Sample estimate: $6 \overline{)900}^{150}$
3. 134 R3; Sample estimate: $4 \overline{)560}^{140}$
4. 880; Sample estimate: 5,400 / 6 = 900

Page 150
1. 26; Sample estimate: 750 / 25 = 30
2. 194 R3; Sample estimate: 3,500 / 20 = 175
3. 447 R8; Sample estimate: $15 \overline{)6,000}^{400}$
4. 110 R6; Sample estimate: $50 \overline{)5,000}^{100}$

Page 151
1. $1.21; Sample estimate: 7.20 / 6 = 72 / 6 (÷10) = 12 (÷10) = 1.2
2. $1.23; Sample estimate: $7 \overline{)8.40}^{1.2}$
3. $0.80 R2¢; Sample estimate: $7 \overline{)5.60}^{0.8}$
4. $2.68; Sample estimate: 8.10 / 3 = 2.7

Page 152
1. 2.07; Sample estimate: 8.4 / 4 = 84 / 40 = 2.1
2. 2.41; Sample estimate: 10 / 4 = 5 / 2 = 2.5
3. 1.445; Sample estimate: $9 / 6 = 1\frac{1}{2}$
4. 7.73; Sample estimate: 40 / 5 = 8

Page 154
1. 1,120; Sample estimate: 770 / 0.7 = 7,700 / 7 = 1,100
2. 24.6; Sample estimate: 40 / 2 = 20
3. 156; Sample estimate: 4.8 / 0.04 = 480 / 4 = 120
4. 610; Sample estimate: 3 / 0.005 = 3,000 / 5 = 600

Answer Key

Page 163
1. $12\frac{3}{4}$
2. $8\frac{2}{3}$
3. $6\frac{4}{5}$
4. $3\frac{12}{16}$, or $3\frac{3}{4}$
5. $\frac{19}{4}$
6. $\frac{11}{3}$
7. $\frac{29}{6}$
8. $\frac{7}{3}$
9. $5\frac{5}{3}, 4\frac{8}{3}, 3\frac{11}{3}, 2\frac{14}{3}, 1\frac{17}{3}, \frac{20}{3}$

Page 164
1. 0.75
2. 0.4
3. 1.5
4. 0.55
5. 0.12

Page 165
1. 0.7
2. about 0.62 or 0.63
3. about 3.33
4. 0.75
5. about 1.44
6. about 0.43

Page 167
1. 0.875
2. $0.\overline{3}$
3. $0.41\overline{6}$
4. 0.3125
5. $0.\overline{2}$
6. $1.1\overline{6}$

Page 168
1. >
2. >
3. <
4. >
5. >

Page 169
1. Sample answer: $\frac{7}{8}$
2. Sample answer: $\frac{3}{6}$

Page 172
1. $\frac{1}{9}$
2. $\frac{7}{9}$
3. $\frac{3}{7}$
4. $\frac{13}{16}$
5. $\frac{11}{12}$

Page 173
Sample answers:
1. $\frac{2}{6}$ and $\frac{5}{6}$
2. $\frac{15}{20}$ and $\frac{12}{20}$
3. $\frac{7}{10}$ and $\frac{15}{10}$
4. $\frac{5}{20}$ and $\frac{6}{20}$
5. $\frac{32}{48}$ and $\frac{42}{48}$

Page 175
1. 32 posters
2. 20 counters
3. $4,000 per month

Page 183
1. Sample estimate: $2 + 8 = 10$; $9\frac{8}{8}$, or 10
2. Sample estimate: $4 + 2\frac{1}{2} = 6\frac{1}{2}$; $5\frac{13}{10}$, or $6\frac{3}{10}$
3. Sample estimate: $7 + 4 = 11$; $9\frac{17}{12}$, or $10\frac{5}{12}$
4. Sample estimate: $9 + 7 = 16$; $14\frac{17}{10}$, or $15\frac{7}{10}$
5. Sample estimate: $5\frac{1}{2} - 1 = 4\frac{1}{2}$; $4\frac{1}{15}$
6. Sample estimate: $6 - 3 = 3$; $3\frac{1}{8}$
7. Sample estimate: $4\frac{1}{2} - 1 = 3\frac{1}{2}$; $3\frac{7}{9}$
8. Sample estimate: $8\frac{1}{2} - 4 = 4\frac{1}{2}$; $4\frac{5}{6}$

Page 187
1. >
2. <
3. >
4. =

Page 188
1. Prediction: <5; $3\frac{3}{4}$
2. Prediction: <6; 4
3. Prediction: <4; $3\frac{1}{5}$
4. Prediction: <6; $2\frac{2}{5}$
5. Prediction: <6; $4\frac{1}{2}$

Answer Key

Page 192
1. Sample estimate: $\frac{1}{4} * 1 = \frac{1}{4}$; $\frac{3}{8}$
2. Sample estimate: $3 * 5 = 15$; $\frac{40}{3}$, or $13\frac{1}{3}$
3. Sample estimate: $3 * 3 = 9$; $\frac{85}{10}$, or $8\frac{1}{2}$

Page 195
1. 18 people; $9 \div \frac{1}{2} = \frac{18}{2} \div \frac{1}{2} = 18$
2. 30 bracelets; $10 \div \frac{1}{3} = \frac{30}{3} \div \frac{1}{3} = 30$
3. 28; Rewrite $\frac{1}{4} * \square = 7$ as $7 \div \frac{1}{4} = \square$. Then $7 \div \frac{1}{4} = \frac{28}{4} \div \frac{1}{4} = 28$.

Page 196
1. $\frac{12}{5}$, or $2\frac{2}{5}$
2. $\frac{15}{5}$, or 3
3. $\frac{7}{21}$, or $\frac{1}{3}$
4. $\frac{8}{12}$, or $\frac{2}{3}$
5. $\frac{28}{10}$, or $2\frac{4}{5}$

Page 199
1. 125 miles
2. $y = 5$

Page 202
1. $m - 3$ inches
2. $2H$ minutes
3. $\$3 + \$0.50 * R$
4. $2(4 - m)$
5. $7q + 46$
6. $\frac{1}{4} * (15 + x)$
7. $(8 + n) / 12$

Page 203
1. 31
2. 13.5
3. 0
4. 6
5. 240
6. 40

Page 205
1. $(6 * 100) + (6 * 40) = 840$
2. $(35 * 6) - (15 * 6) = 120$
3. $(4 * 80) - (4 * 7) = 292$
4. $1.23 * (456 + 789) = 1.23 * (1,245) = 1,531.35$; $(1.23 * 456) + (1.23 * 789) = 560.88 + 970.47 = 1,531.35$

Page 206
b and d

Page 207
1. True
2. True
3. False
4. True
5. False
6. False

Page 209
1. $c = 14$
2. $z = 7$
3. $f = 10.5$
4. $\$20 - \$12.49 = c$, or $\$20 = \$12.49 + c$; $c = \$7.51$
5. $\$10$ per week $* w = \$90$, or $\$90 / \$10$ per week $= w$; $w = 9$ weeks

Page 210
1. Any number less than 8
2. Any number greater than 0
3. [number line from −2 to 11 with point at 8]

Page 213
1. 576 in.², or 4 ft²
2. 1,200 feet
3. 72 m²
4. $i = \$1,000 * r * t$

Page 218
1. Sample answer: $4B + 2M = 1M + 8B$
2. $6C + 2P = 8P + 3C$
3. $n = 1$
4. $w = 5$
5. $y = 4$

Page 220
1. 12 is the solution because when 12 is substituted for b both sides of the equation are equal to 44.
$5(12 + 3) - 3(12) + 5 = 75 - 36 + 5 = 44$
$4(12 - 1) = 4 * 11 = 44$
2. $x = 4$
3. $s = -2$
4. $b = 3$

Answer Key

Page 221

1.
in	out
n	$\frac{n}{3}$
9	3
36	12

2.
in	out
k	k − 4
11	7
28	24

3.
in	out
x	x * 30
4	120
10	300

Page 222

1. 500 words
2. $31.50

Page 224

1. Sample answer: The input or *in* number x is independent. The output or *out* variable y is dependent. $y = \frac{1}{2} * x$
2. The volume (V) of the cube depends on the edge length (s), so V is dependent and s is independent. $V = s^3$

Page 226

1. Balanced
2. Unbalanced

Page 227

1. Unbalanced
2. Balanced
3. x = 2; weight on left = $\frac{1}{2}x$, or 1 unit

Page 230

1. Labels (of food items)
2. 4.50; A number representing the unit price of 1 quart of macaroni salad
3. B1
4. 5; A number representing the number of packages of hamburger buns
5. A4, A5, A6, A7, A8, C4, C5, C6, C7, C8
6. Column D

Answer Key

Page 239
1. a. Hexagon b. Quadrilateral
   c. Octogon
2. Sample answers:
   a.
   b.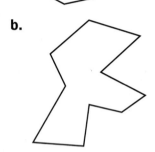

Page 240
1.

   △KLJ, △LJK, △LKJ, △KJL, △JLK
2. Sample answer: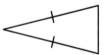
3. Sample answer: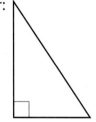

Page 242
1. Sample answer:
   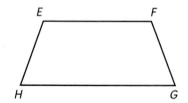
2. No.

Page 245
Sample answers:
1. a. Both have a curved surface and at least one circular face.
   b. A cylinder has 3 surfaces, while a cone has 2 surfaces. A cylinder has 2 circular bases, 2 edges, and no vertices. A cone has exactly 1 circular base, 1 edge, and 1 vertex.
2. a. Both have a flat base and one vertex
   b. A cone has a curved surface, while all of the surfaces of a pyramid are flat faces. The base of a pyramid is a polygon, while the base of a cylinder is a circle. A cone has only one vertex, while a pyramid has at least four vertices.

Page 246
1. a. 5 faces b. 1 face
2. a. 5 faces b. 2 faces

Page 247
1. a. 4 faces b. 6 edges
   c. 4 vertices
2. Pentagonal pyramid
3. Sample answers:
   a. The faces of each are polygons. The base of each is used to name them.
   b. Aside from the base, the faces of a pyramid are always triangles. Aside from the 2 bases, the faces of a prism are always parallelograms.

Page 248
1. a. 7 faces b. 12 edges
   c. 10 vertices

Page 249
1. a. 12 edges b. 8 vertices
2. Sample answers:
   a. Their faces are congruent equilateral triangles.
   b. Tetrahedrons have 4 faces, 6 edges, and 4 vertices. Octahedrons have 8 faces, 12 edges, and 6 vertices.

Answer Key

Page 251
1. 24 feet
2. 90 yards

Page 253
1. 40 in.^{2}
2. 26 yd^2

Page 257
20 cm^2

Page 263
1. C
2. B
3. A

Page 266
1–5.

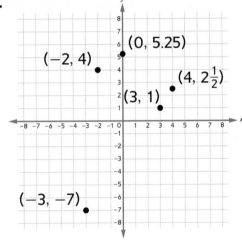

Page 280
1. Statistical question
2. Not a statistical question
3. Statistical question
4. Statistical question

Page 283
1. Number of Hits Made by Players in a Baseball Game

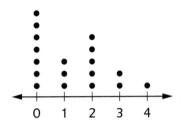

2. 0 hits
3. 5 players

Page 286
1. 78 degrees
2. 81 degrees

Page 288
1. Mean score: 85
2. Sample answer: I added the scores and divided the sum by the number of test scores: [75 + 4(80) + 2(85) + 2(90) + 2(95)] / 11 = 85.

Answer Key

Page 290
1. Mean: 19 pennies; median: 19 pennies
2. Mean: 16 pennies; median: 18 pennies
3. At first Maribel's mean and median were the same (19 pennies). With a new result of 4 pennies, the mean was affected the most, dropping to 16 pennies. The median changed by one penny to 18 pennies.
4. The median best represents a typical number, since it seems to be in the middle of most of her data. The mean was affected so much by the outlier of 4 pennies that it is not a typical value.

Page 296
1. About 18 cars
2. Sedans are most likely to weigh around 3,000 pounds, because that is where the bars on this histogram are the highest. No cars weigh 1,000 pounds, and only a few cars weigh close to 2,000 pounds or close to 4,000 pounds, since those bars on the histogram are the lowest.

Page 299
Sample answer:

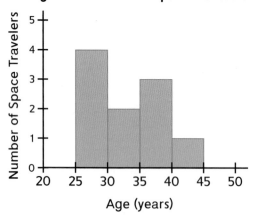

Ages of the First 10 Space Travelers

Page 307
1. Left-skewed
2. There isn't a tail on the right side because the highest possible score is 10 points.
3. The mean is 7.16, the median is 8, and the mode is 9. No, they are not all in the same location on the graph.

Page 358
1. India
2. Greenland

Page 361
1. 7 men's finishing times
2. 15 minutes and 55 seconds

Page 369
1. 113.11 meters
2. Approximately 15
3. Approximately 1 day and $8\frac{1}{4}$ hours

Page 382
1. 67
2. 8
3. 7.6
4. 1.92

Page 392
1. 0.1875
2. $\frac{185}{1,000}$, or $\frac{37}{200}$
3. 0.3%
4. 7.23
5. 28.125%
6. $\frac{68}{100}$, or $\frac{17}{25}$

Page 394
1. 0.7
2. 428
3. 4384.488
4. 0.80

Page 398
1. 70,000
2. 83,000,000
3. 0.000003726
4. 0.000034

Answer Key

Page 401
- a. 3,750 square feet
- b. 2,625 square feet
- c. 4,762.5 square feet

Page 402
tip $15.30; total $100.30

Page 405
1. $63.50
2. $29.50

Page 407
1. Sample calculator: 1, $1\frac{2}{3}$, $2\frac{1}{3}$, 3, $3\frac{2}{3}$;
2. Sample calculator: $9\frac{1}{4}$, $8\frac{2}{4}$, $7\frac{3}{4}$, 7, $6\frac{1}{4}$;

Index

A

Absolute value
  defined, 92
  distance as, 92–93, 96
  for ordered pairs, 272
  practicing with *Absolute Value Sprint,*
      318–319
Accuracy, 23–25
Acre, 374
Acute angle, 236
Acute triangle, 240
Addends, 120
Addition
  Addition Property of Zero, 231
  Associative Property of Addition, 10, 11,
      206, 231
  with base-10 blocks, 117
  column addition, 119
  Commutative Property of Addition, 10, 11,
      206, 231
  with decimals
    with base-10 blocks, 117
    methods for, 118–120
    practicing, 334, 350
    precision in, 128
    problem solving with, 4–8
  Distributive Property (of Multiplication over
      Addition), 11, 204, 206, 231
  with fractions
    Addition of Fractions Property, 232
    estimating with benchmarks, 176
    methods for, 45, 46
    practicing, 328, 329, 337
    process of, 179
  front-end estimation in, 113 (See also under
      Estimation)
  Identity Property of Addition, 231
  with mixed numbers, 177, 180, 337
  order of operations and, 203, 373
  partial-sums addition, 118
  tying to subtraction, 219
  U.S. traditional addition, 120
  with whole numbers, 118–120, 350
Addition of Fractions Property, 232
Addition Property of Zero, 231
Addition Top-It, 350
Address (for cells in spreadsheets), 229
Address box, 229
Airplanes, scale models of, 84
Algebra, foundations for, 198–199, 225
Algebra Election, 320–321
Algebraic expressions, 200–201, 206, 220.
        See also Expressions
Analytic geometry, 234, 235. See also
        Coordinate grids

Angles
  defined, 236, 237
  measuring, 236
  notation (markings) for, 236, 239
  in quadrilaterals, 242
  in triangles, 240, 269
  types of, 236
Apex, 245
Approximating a solution, 215
Architecture, scale models in, 83
Area. See also Surface area
  composing/decomposing shapes, 257
  defined, 252
  of parallelograms, 212, 254, 378
  percentage of, 63
  real-world applications of, 252
  of rectangles (See also Area models)
    counting squares, 253
    formulas for, 253, 260, 261, 378
    spreadsheet for, 199
    using distributive property for, 204–205
  of squares, 212, 223
  of triangles, 13–16, 255–256, 260, 261, 378
  unit conversions in, 374
  units for, 213, 252
  world land area (real-world data), 358
Area models
  for multiplying fractions, 188, 189
  for multiplying mixed numbers, 9–10
Arguments, 13–16
Arrays
  defined, 45
  rectangular arrays, 102
  solving ratio problems with, 45, 46
  square arrays, 98
The arts
  models in, 86
  paper folding, 273–278
Associative Property of Addition, 10, 11,
      206, 231
Associative Property of Multiplication, 132,
      206, 231
Athletes, international (real-world data), 364–365
Auber, Daniel, 86
Average, 21, 284. See also Mean
Axes, 95, 265
Ayoob, Joe, 278

B

Balance point, 285–286
Bar graphs
  defined, 295
  real-world data, 357, 364, 365
  scale for, 308

four hundred fifty-three

Index

Bar model, for solving equations, 216
Base
   in area formulas, 212, 253, 254, 255–256, 260, 261, 378
   exponential, 98
   geometric
      of cones, 244
      of cylinders, 244
      of prisms, 212, 248, 259
      of pyramids, 247, 259
      of triangles, 15, 16, 255
   in volume formulas, 224, 260, 261, 378
Base-10 blocks, 111, 117
Base-10 system, 109, 129–130
Basic facts, 132, 133. See also Extended facts
Benchmark fractions, 168, 176–177
Bimodal distribution, 306
Bins, 296, 303–304
Biodiversity (real-world data), 366–367
Birds
   counts of, 281
   worldwide diversity (real-world data), 366
Boston Marathon records (real-world data), 361
Box-and-whiskers plots, 300
Box plots
   analyzing, 303
   comparing, 302
   constructing, 300, 301–302
   data distribution in, 305, 307
   data landmarks in, 300
   defined, 300
   skewed distributions with, 307
Brackets, 207, 208
Buildings, scale models of, 83, 85
Build-It, 322

C

Calculators
   checking answers with, 320, 327, 328
   computations with, 288, 323, 332
   exponential notation on, 98
   in problem solving, 35, 36
   renaming fractions as decimals with, 165, 167, 168
   repeating decimals on, 167
   square roots with, 90
   variables in programming for, 199
Capacity. See Volume
Carrying, in addition, 120
Categories, in hierarchies, 241, 243
Cell, in spreadsheets, 229
Census, 309–314
Centimeter (cm), 66, 250, 374, 375
Century, 374
Charts and tables. See Tables

Circle graphs, 370
Close-but-easier numbers, 114, 143
Closed path, 238
Codes, as numbers, 88
Coefficient, 201
Collins, John, 278
Column addition, 119. See also Addition
Combine like terms, 206, 220
Commas, for numbers, 97
Common denominators
   adding/subtracting fractions, 179
   adding/subtracting mixed numbers, 180, 181, 182
   comparing fractions, 168
   dividing fractions, 195
   finding, 173
Commutative Property of Addition, 10, 11, 206, 231
Commutative Property of Multiplication, 132, 206, 231
Comparing
   box plots, 302
   data, 302
   decimals, 111–112
   fractions, 41, 168, 172, 322, 330
   measurements, 88, 101
   numbers, 94, 101
   problem-solving strategies, 4, 8
   products, 187
   ratios, 42, 344
Composite numbers, 104
Computers
   in the census, 311–312, 313
   in modeling, 85
   spreadsheets for, 229–231
   variables in programs for, 199
Concave polygon, 239. See also Polygons
Cones, 244, 245
Congruent figures, 268, 269
Conjectures, 13–16
Constant, 206, 219
Constraints, 211
Continuous data, 296. See also Data
Convex polygon, 239. See also Polygons
Convex polyhedron, 249. See also Polyhedrons
Coordinate geometry, 234. See also Coordinate grids
Coordinate grids
   calculating distance on, 96, 266, 333
   congruent figures on, 268
   defined, 95, 265
   graphing rules on, 199
   introduction to, 265
   plotting points on
      connecting algebra and geometry by, 235
      finding locations on, 234

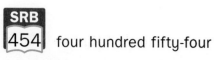
four hundred fifty-four

Index

polygons on, 267
practice with, 333
problem solving by, 222
process of, 95–96, 265–266
ratio/rate graphs with, 51–53
reflections of, 271–272
similar figures on, 269
Coordinates, 95
Coordinate system, 235
Corresponding sides, 268, 269
Counting, with numbers, 88
Counting numbers
defined, 89
divisibility tests for, 103
factors of, 102
GCF of, 105
GCF × LCM, 106
prime versus composite, 104
as rational numbers, 90
square numbers as, 98
Counting-up subtraction, 123–124
Country land area (real-world data), 358
Cubes
edges of, 245
faces of, 244
net for, 264
origami for, 274
as polyhedrons, 249
surface area of, 264, 378
volume of, 258, 262, 378
Cubic centimeters (cm^3), 258, 374, 375
Cubic decimeters (dm^3), 374
Cubic feet (ft^3), 258, 374
Cubic inches (in.^{3}), 258, 374
Cubic meters (m^3), 258, 374
Cubic yards (yd^3), 258, 374
Cubit, 65
Cup (c), 67, 374, 375
Cylinders, 224, 244, 245

D

Daring Division, 323
Data. *See also* Real-world data
analyzing, 21, 280, 300
benefits of, 314
in census taking, 310–311, 313
collecting, 20, 84, 280, 281–282, 304
comparing, 302
continuous, 296
defined, 280
distribution of, 305–307
estimation of, 113
graphing
bar graphs, 295
box plots
analyzing, 303
comparing, 302
constructing, 300, 301–302
data distribution in, 305, 307
data landmarks in, 300
defined, 300
skewed distributions with, 307
dot plots
data organization with, 283
finding measures of center with, 284, 285, 286, 288
finding measures of spread with, 291
problem solving with, 21–22
histograms
bimodal distributions with, 306
data analysis with, 299, 303–304
defined, 296
making, 297–298
normal distributions with, 305
purpose of, 296
skewed distributions with, 307
persuasion with, 308
reading and analyzing graphs, 303–304
landmarks for (*See also specific landmarks*)
finding, 335–336
measures of center, 284–290, 303
measures of spread and variability, 291–294, 300
organizing, 283
processing of, 312
typical values, 289–290
Day (d), 71, 374
Decade, 374
Decagon, 238
Decennial census, 309, 311, 312
Decimal point
defined, 108
placement of
in dividing decimals, 140, 143, 151, 323
in multiplying decimals, 130–131, 135
practice with, 350
Decimals
adding, 4–8, 117–119, 128, 334, 350
comparing, 111–112
on coordinate grids, 266
defined, 107
dividing
decimals by decimals, 145–146, 154
decimals by whole numbers, 140, 144, 151–152, 323, 350
whole numbers by decimals, 153
equivalencies among, 56
estimating with (*See under* Estimation)
evaluating inequalities with, 4
expanded form for, 109–110
grid for, 377
identifying number of decimal places, 332

four hundred fifty-five

Index

Decimals (*continued*)
  multiplying
    methods of, 134–136, 138
    by powers of 10, 130–131
    practicing, 325, 348, 350
  use close-but-easier numbers, 114 on number grids, 56
  on number lines, 112, 376
  percent as, 58
  place value of, 109–110, 334, 373
  powers of 10 for, 110, 135
  purpose of, 107
  reading, 108
  remainders as, 142
  renaming fractions as, 107, 159, 164–167
  subtracting
    with base-10 blocks, 117
    methods of, 122, 124, 126
    practicing, 334, 350
    precision in, 128
  terminating versus repeating, 164
  uses of, 89, 107
  writing, 108, 109
Decimal Top-It, 350
Decimeter (dm), 250, 374
Deepest lakes of the world (real-world data), 368
Deforestation, 252
Degree, 236
Degrees Celsius, 66
Degrees Fahrenheit, 67
Degree symbol, 236
Denominators. See *also* Fractions
  common
    adding/subtracting fractions, 179
    adding/subtracting mixed numbers, 180, 181, 182
    comparing fractions, 168
    dividing fractions, 195
    finding, 173
  comparing fractions, 168
  defined, 156
  dividing, 160, 195
  Equivalent Fractions Property, 232
  least common multiple method for, 173
  multiplying, 145, 160, 173, 189
  quick common, 173
  remainders as fractions, 142
  renaming fractions, 161, 173
  renaming mixed numbers, 162–163
  unit fraction problem solving, 174–175
Dependent variable, 223–224. See *also* Variables
Descartes, Rene, 235
Design, scale models in, 83
Deviation, mean absolute, 293–294
Digits, 97, 207. See *also* Numbers

Dippel, George, 278
Display bar, 229
Distance
  as absolute value, 92–93, 96
  on coordinate grids, 96, 266, 333
  formula for, 199, 213, 378
  map scales for, 79, 80, 159
  natural measures for, 65
  in number stories, 20–22
  variables that describe, 223
Distribution of data, 305–307
Distributive Property (of Multiplication over Addition), 11, 204, 206, 231
Distributive Property (of Multiplication over Subtraction), 205, 206, 231
Dividend
  decimals as, 151–152, 154
  defined, 140, 141
  equal sharing and, 194
Divisibility, 103
Divisibility Dash, 324
Division
  counting numbers and, 89
  with decimals
    decimals by decimals, 145–146, 154
    decimals by whole numbers, 140, 144, 151–152, 323, 350
    whole numbers by decimals, 153
  divisibility tests, 324
  finding equivalent fractions with, 160
  finding equivalent ratios with, 44
  finding factors, 102
  finding unit ratios with, 50
  with fractions
    Division of Fractions Property, 196, 232
    equal groups, 193
    equal sharing, 194
    methods for, 193, 194
    missing factors and, 193
    practicing, 330
    using common denominators, 195
  fractions as, 145, 155, 166, 201
  interpreting remainders in, 141–142 (See *also* Remainders)
  with mixed numbers, 196
  multiplication and division table, 373
  notation for, 141, 145, 158, 201
  order of operations, 203, 373
  partial-quotients method, 143–144, 166
  by powers of 10, 139
  ratios and, 38
  renaming decimals as fractions with, 166
  renaming fractions as decimals with, 166, 167
  short division, 148
  tying to multiplication, 140, 193, 194, 219
  U.S. traditional long division, 147–154

four hundred fifty-six

Index

with whole numbers
  extended division facts, 139
  fractions as, 89
  methods of, 143–144, 147–150
  practicing, 323, 350
Division of Fractions Property, 196, 232
Division Top-It, 350
Divisor
  decimals as, 153, 154
  defined, 141
  multidigit, 149–150
  multiples of, 324
  single-digit, 147–148
Dodecagon, 238
Doggone Decimal, 325
Dot plots
  data organization with, 283
  finding measures of center with, 284, 285, 286, 288
  finding measures of spread with, 291
  problem solving with, 21–22

E

Earth, model of, 81
Edges, 245, 248, 258, 262
Education, modeling in, 86
Electoral votes, 320–321
Electricity consumption (real-world data), 370
Electronic computer, 311
Elements (Euclid), 235
Endpoints, 236, 238
Engineering, scale models in, 84–85
Enlargement, 72, 73, 269
Enumeration, 309. *See also* U.S. Census (photo essay)
Enumerators, 310, 312, 313
Equal groups, for dividing fractions, 193
Equal-sharing division, 194
Equal sign (=), 101, 207
Equations
  defined, 207
  set notation for, 208–209
  simplifying, 220
  solving
    approximating a solution, 215
    bar model for, 216
    inverse-operations strategy for, 219
    pan-balance method for, 217–218
    practice with, 320–321, 327
    trial and error method for, 214–215
Equilateral triangles
  defined, 240
  hexaflexagons and, 277
  hierarchy for, 241
  as regular polygons, 239
  in regular polyhedrons, 249

Equivalent division problem, 145, 153, 154, 193
Equivalent equations, 218, 219
Equivalent expressions, 206, 216
Equivalent fractions
  comparing fractions, 168
  comparing ratios, 42
  converting ratios to unit ratios, 50
  defined, 160
  Equivalent Fractions Property, 232
  finding (identifying), 56, 160, 172, 329
  grid for, 377
  percent as, 60
  ratios as, 41
  renaming fractions as, 173
  renaming fractions as decimals, 164
Equivalent Fractions Property, 232
Equivalent ratios
  defined, 41
  percents as, 55
  practice with, 345
  problem solving with, 43–44, 48, 51–53, 63
  proportions as, 41
  ratio/rate graphs for, 51–53
  ratio/rate tables for, 43–44
  tape diagrams for, 48
  unit conversions with, 66, 67
Escher, M. C., 275
Estimate, 113
Estimation
  close-but-easier numbers, 114, 143
  with decimals
    adding, 5, 116, 118, 119, 120
    dividing, 144, 146, 151–154, 323
    multiplying, 114, 134, 136, 138, 325
    subtracting, 122, 124, 126
  with fractions
    adding, 328
    benchmark fractions, 176–177
    multiplying, 184, 185, 189, 191
    rounding in, 178
  front-end estimation, 113
  importance of, 113
  of length, 250
  with mixed numbers
    adding, 177, 180, 328
    multiplying, 191
    subtracting, 176–177, 181–182, 183
  in problem solving, 23
  with ratio/rate graphs, 53
  reasonable estimates, 280, 282
  rounding as, 115, 143
  in unit conversions, 71
  with whole numbers
    adding, 114, 118, 119, 120
    dividing, 143, 144, 147–150
    finding mystery numbers, 332

four hundred fifty-seven

Index

Estimation (*continued*)
    multiplying, 133, 136, 137, 325, 338
    subtracting, 121, 123, 125, 127
Euclid, 235
Euler's Theorem, 245
Even numbers, 27, 103
Expanded form, 100, 109
Exponential notation
  defined, 98, 99, 110
  extended multiplication facts and, 132
  multiplying whole numbers with, 129
  notation for, 98, 99, 110
Exponents, 98, 110, 130, 203, 373
Expressions. *See also* Algebraic expressions
  defined, 200
  equivalent expressions, 200, 216
  evaluating, 202–203
  naming numbers with, 340
  in problem solving, 30
  renaming, 216
  simplifying, 206, 220
  variables in, 31, 200
  writing, 9
Extended facts
  defined, 132
  division with, 139
  multiplication with, 132, 133
  practice with, 349

F

Face, 244, 247, 248, 249, 263, 264
Factor Captor, 326
Factor pairs, 28, 102
Factor rainbows, 28, 102
Factors
  of counting numbers, 102
  decomposing, 11
  divisibility and, 103
  with exponents, 98
  finding, 326
  greatest common, 105
  identifying, 27, 28, 102, 105
  missing factors, in dividing fractions, 193
  partial-products multiplication and, 133
  Powers of a Number Property, 232
  prime versus composite numbers, 104
  repeated factors, 129
Factor tree, 104
Fair share, mean as, 287–288
Fathom, 65
Fat yard, 250. *See also* Meter (m)
Finger stretch, 65
First to 100, 327
Five-number summary, 300
Flexagon, 276

Flips. *See* Reflections
Fluid ounce (fl oz), 67, 374
Foot (ft), 65, 67, 68, 73, 250, 374, 375
Foreign-born population (real-world data), 357
Formulas. *See also* specific formulas
  chart of, 378
  defined, 212
  evaluating, 212
  in spreadsheets, 229
  units in, 213
  variables in, 199, 212
Fraction Action, Fraction Friction, 328
Fractional edge lengths, 262
Fraction Capture, 329
Fraction circles
  comparing fractions, 168
  dividing fractions, 195
  identifying mixed numbers, 157
  reading/writing fractions, 156–157, 158
  renaming fractions, 161
  renaming mixed numbers, 162
Fractions
  adding, 176, 179, 232, 328
  Addition of Fractions Property, 232
  benchmark fractions, 168, 176–177
  comparing, 41, 168, 172, 322, 330
  on coordinate grids, 266
  defined, 89
  dividing, 193–196, 330
  division as, 145, 155, 166, 201
  equal to one, 156
  equivalencies among, 56, 57
  equivalent fractions
    comparing fractions, 168
    comparing ratios, 42
    converting ratios to unit ratios, 50
    defined, 160
    Equivalent Fractions Property, 232
    finding (identifying), 56, 160, 172, 329
    grid for, 377
    percent as, 60
    ratios as, 41
    renaming fractions as, 173
    renaming fractions as decimals, 164
  estimating with, 176–178
  exponential notation and, 110
  finding a fraction of a fraction, 185
  finding a fraction of a number, 184
  finding fractions between fractions, 169–170
  greater than one, 157, 161 (*See also* Mixed numbers)
  meanings of, 158
  in measurement, 58, 89, 155, 159
  multiplying
    fractions by fractions, 185, 186, 189, 330, 348

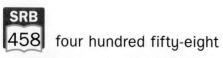

Index

fractions by whole numbers, 184, 188, 348
   mixed numbers by fractions, 191
   whole numbers by fractions, 187, 188, 331, 348
on number grids, 56
on number lines, 89, 158
ordering, 322
origin of, 155
purpose of, 107, 155
rates as, 155, 158
as rational numbers, 91
ratios as, 39, 40, 41, 42, 89, 155, 158
reading, 156–157
reciprocals of, 196
remainders as, 142, 161
renaming
   as decimals, 89, 159, 164–167, 168
   decimals as, 107, 159, 164–167
   as equivalent fractions, 160
   fractions greater than one, 161
   mixed numbers as, 162–163, 181, 182, 183, 192
   as percent, 57, 60, 159
simplifying, 331
subtracting, 176, 177, 179, 232, 337
Subtraction of Fractions Property, 232
unit fractions, 161, 174–175, 184
uses of, 89, 159
whole numbers as, 89, 189, 329
writing, 156–157
Fraction sticks, 165, 171–172, 376
Fraction strips, 169, 170, 185
Fraction Top-It, 330
Fraction/Whole Number Top-It, 331
Front-end estimation, 113. *See also* Estimation
Fulcrum, 226, 227, 378
Full-circle protractor, 236
Function machines, 199, 221, 223

G

Gallon (gal), 67, 374, 375
Games
   Absolute Value Sprint, 318–319
   Addition Top-It, 350
   Algebra Election, 320–321
   Build-It, 322
   Daring Division, 323
   Decimal Top-It, 350
   Divisibility Dash, 324
   Division Top-It, 350
   Doggone Decimal, 325
   Factor Captor, 326
   First to 100, 327
   Fraction Action, Fraction Friction, 328
   Fraction Capture, 329
   Fraction Top-It, 330

Fraction/Whole Number Top-It, 331
Getting to One, 332
Hidden Treasure, 333
High-Number Toss, 334
introduction to, 316–317
Landmark Shark, 335–336
Mixed-Number Spin, 337
Multiplication Bull's Eye, 338
Multiplication Top-It, 349
Multiplication Wrestling, 339
Name That Number, 340
Percent Spin, 341–342
Polygon Capture, 343
Ratio Comparison, 344
Ratio Dominoes, 345
Ratio Memory Match, 346
Solution Search, 346
Spoon Scramble, 348
Subtraction Top-It, 350
Gardner, Martin, 276
Geometric solids, 244–245, 259, 263. *See also* specific solids
Geometry
   angles
      defined, 236, 237
      measuring, 236
      notation (markings) for, 236, 239
      in quadrilaterals, 242
      in triangles, 240, 269
      types of, 236
   introduction to, 234–235
   lines and line segments, 237 (*See also* Line segments)
   paper folding (photo essay), 273–278
   parallelograms
      area of, 212, 254, 378
      as quadrilaterals, 242, 243
      as rectangles, 254
      as rhombuses, 243
      as trapezoids, 242
   polygons
      congruency of, 268
      on coordinate grids, 267
      defined, 238
      flexagons as, 276
      perimeter of, 251, 378
      prefixes for, 238, 373
      properties of, 343
      reflecting, 271
      sides of, 238, 251
      as 2-dimensional figures, 244
      types of, 238, 239
      vertex (vertices) of a, 238
   quadrilaterals
      angle/side markings for, 239
      congruency of, 268

four hundred fifty-nine

Index

Geometry (continued)
    on coordinate grids, 267
    defined, 242
    hierarchy for, 243
    as polygons, 238, 239
    similarity of, 269
  rays, 236, 237
  solids (See Geometric solids)
  triangles
    area of, 13–16, 255–256, 260, 261, 378
    converting to parallelograms, 255
    dimensionality of, 234
    enlarging, 269
    equilateral (See Equilateral triangles)
    hierarchy for, 241
    notation for, 240
    parts of, 240
    perimeter of, 128
    as polygons, 238, 239
    similarity of, 269
    theorem for, 235
    types of, 240
Getting to One, 332
Goals for Mathematical Practice (GMPs), 3, 4, 5, 6, 7, 8, 9, 10, 12, 14, 16, 18, 19, 21, 22, 24, 25, 27, 28, 31
Gossage, Howard, 278
Gram (g), 66, 67, 374, 375
Graphing
  bar graphs, 295
  box plots
    analyzing, 303
    comparing, 302
    constructing, 300, 301–302
    data distribution in, 305, 307
    data landmarks in, 300
    defined, 300
    skewed distributions with, 307
  coordinate graphs, 17–18, 333 (See also Coordinate grids)
  dot plots
    data organization with, 283
    finding measures of center with, 284, 285, 286, 288
    finding measures of spread with, 291
    problem solving with, 21–22
  histograms
    bimodal distributions with, 306
    data analysis with, 299, 303–304
    defined, 296
    making, 297–298
    normal distributions with, 305
    purpose of, 296
    skewed distributions with, 307
  persuasion with, 308
  problem solving with, 222

  reading and analyzing graphs, 303–304
  solution sets for inequalities, 210–211
  of variables, 224
Greater-than-or-equal-to symbol (≥), 207
Greater-than symbol (>), 101, 207
Greatest common factor (GCF), 105, 106, 160
Great International Paper Airplane Book, The (Mander, Dippel, and Gossage), 278
Great Pyramid of Giza, 247, 263
Great span, 65
Grid method, 105, 106
Grids, 63, 64. See also Coordinate grids; Number grids
Grouping symbols, 207. See also Brackets; Parentheses

H

Half-circle protractor, 236
Handheld computers, 313
Handshakes problem, solving, 33–36
Height
  of cylinders, 224
  of parallelograms, 212, 254
  of prisms, 212, 259
  of pyramids, 259
  of triangles, 15, 16, 255
  as width, 253
  world mountains (real-world data), 368
  world trees (real-world data), 369
Heptagon, 238
Hexaflexagon, 277
Hexagon, 238, 239, 276
Hexagonal prism, 246, 248
Hexagonal pyramid, 246, 247
Hexahexaflexagon, 277
Hidden Treasure, 333
High-Number Toss, 334
Hines, Jim, 107
Histograms
  bimodal distributions with, 306
  data analysis with, 299, 303–304
  defined, 296
  making, 297–298
  normal distributions with, 305
  purpose of, 296
  skewed distributions with, 307
HO scale (model railroads), 82
Hour (hr), 71, 374
Hundredths, 57, 58, 108

I

Identification, with numbers, 88
Identity Property of Addition, 231
Identity Property of Multiplication, 231
Image, 72–74, 271, 272

four hundred sixty

Improper fractions, 157. See also Fractions, greater than one
Inch (in.), 67, 73, 250, 374, 375
Independent variable, 223–224. See also Variables
Inequalities
   defined, 207, 210
   evaluating, 4
   infinite solutions for, 208
   set notation for, 208–209, 210
   solving, 210–211, 337, 347
   variables in, 4–8, 210
Infinite, 208
In/out table (rule), 69, 199, 221
Integers, 89, 90
Interior, 238
International athletes (real-world data), 364–365
Interpreting remainders, 141–142. See also Remainders
Interquartile range (IQR), 293, 300
Intersecting lines, 237
Intervals, 296
Inverse-operations strategy, for solving equations, 219
Irrational numbers, 90, 91
Isosceles right triangle, 241
Isosceles triangle, 240, 241

J
Joint (natural measure), 65

K
Kilogram (kg), 66, 374, 375
Kiloliter (kL), 374
Kilometer (km), 250, 374, 375
Kites, 243

L
Labels (in spreadsheets), 229
Lakes of the world (real-world data), 368
Landmark Shark, 335–336
Largest reservoirs of the world (real-world data), 359
Lattice, 136
Lattice multiplication, 136
Laws of perspective, 235
Layering, 259
Leading-digit estimation. See Front-end estimation
Least common denominator, 173. See also Common denominators
Least common multiple (LCM), 106, 173. See also Multiples
Length
   as base, 253
   defined, 250

   of edges, 262
   measuring, 65
   personal references for, 66, 67, 250, 375
   for scale models, 75
   for sides of regular polygons, 239
   unit conversions, 250, 374
   unit ratios, 73, 74
Less-than-or-equal-to symbol ($\leq$), 207
Less-than symbol ($<$), 101, 207
Like terms, 206, 220
Line of reflection, 271, 272
Line of symmetry, 270
Line plots, 283
Lines, 51, 237
Line segments
   in angles, 236
   congruency of, 268
   defined, 237
   in polygons, 238
   in triangles, 234
Liquid volume. See Volume
Liter (L), 66, 374, 375
Long division, 147–154
Lower quartile (Q1), 292, 300, 301–302, 303
Lowest terms, 160

M
Magnitude, 92
Major League Baseball international athletes (real-world data), 365
Mammals (real-world data), 366
Mander, Jerry, 278
Maps
   dimensionality of, 234
   map scales for, 79, 80, 159
   as real-world data, 353, 356
Mass
   personal references for, 66, 375
   unit conversions, 374
Mathematical models, 17–19
Mathematical practices, 1–36
Mathematical reasoning, 12, 42, 71, 177
Mathematical representations, 9–12
Mathematical tools, 20–22
Maximum data value
   with box plots, 300, 301–302
   defined, 291
   with histograms, 297, 298
   reading and analyzing graphs, 303
McCarroll, June, 237
Mean
   balance point as, 285–286
   box plots and, 300
   data analysis with, 289
   defined, 284
   fair shares as, 287–288

Index

Mean (*continued*)
  finding, 21, 335–336
  mean absolute deviation, 293
  normal distributions and, 305
  skewed distributions and, 307
Mean absolute deviation (m.a.d.), 293–294
Measurement
  comparisons in, 88, 101
  decimals in, 89
  fractions in, 89, 155, 159
  natural measures in, 65
  numbers for, 88
  personal references for, 66, 67, 250, 375
  precision in, 128
  ratios in, 65–67
  standard units in, 65
  tools for, 20–22
  unit conversions, 66, 67, 68, 69, 70–71, 128, 374
Measures of center, 284–290, 300, 303. See also specific measures
Measures of spread and variability, 291–294, 300. See also specific measures
Median
  box plots and, 300, 301–302
  data analysis with, 289
  defined, 284
  finding, 335–336
  normal distributions and, 305
  quartiles and, 292
  reading and analyzing, 303
  skewed distributions and, 307
Meter (m), 66, 68, 250, 374, 375
Metric system, 66, 250, 374, 375
Metric ton (t), 374
Middle quartile (Q2), 292
Mile (mi), 67, 250, 374, 375
Milligram (mg), 374
Milliliter (mL), 66, 374, 375
Millimeter (mm), 68, 250, 374, 375
Minimum data value
  with box plots, 300, 301–302
  defined, 291
  with histograms, 297, 298
  reading and analyzing graphs, 303
Minuend, 127. See also Subtraction
Minute (min), 71, 374
Missing factors, in dividing fractions, 193
Mixed numbers
  adding, 177, 180, 337
  decomposing, 190
  defined, 157
  dividing, 196
  estimating with (See under Estimation)
  multiplying, 9–12, 186, 190–192, 339
  purpose of, 107, 157
  renaming as fractions, 162–163, 181, 182, 183, 192
  subtracting, 181–182, 337
Mixed-Number Spin, 337
Mobiles, 226–227, 378
Möbius strips, 275
Mode
  box plots and, 300
  data analysis with, 289
  defined, 284
  finding, 335–336
  skewed distributions and, 307
  spread and, 306
Modeling. See also Area models
  mathematical models, 17–19
  scale models (scale drawings), 75–77, 78, 81–86
Modular origami, 274
Molecules, scale models of, 81, 85
Money
  adding decimals, 7
  decimals as, 89
  dividing decimals, 151
  in number stories, 17–19, 23–25
  percent applications with, 54, 59, 60
Month (mo), 374
Most-visited museums (real-world data), 360
Mountains of the world (real-world data), 368
Multiples
  defined, 106
  extended facts and, 132
  finding, 26–27, 106
  least common, 106
  recognizing, 324
Multiplication
  Associative Property of Multiplication, 132, 206, 231
  Commutative Property of Multiplication, 132, 206, 231
  with decimals
    methods of, 134–136, 138
    by powers of 10, 130–131
    practicing, 325, 348, 350
    use close-but-easier numbers, 114
  Distributive Property (of Multiplication over Addition), 11, 204, 231
  Distributive Property (of Multiplication over Subtraction), 205, 206, 231
  exponential notation as, 100, 129
  finding equivalent fractions with, 160
  finding equivalent ratios with, 44
  finding factors, 102
  finding unit ratios with, 50
  with fractions
    fractions by fractions, 185, 186, 189, 330, 348

Index

fractions by whole numbers, 184, 188, 348
  Multiplication of Fractions Property, 189, 232
  whole numbers by fractions, 187, 188, 331, 348
front-end estimation in, 113 (See also under Estimation)
GCF × LCM, 106
Identity Property of Multiplication, 231
lattice multiplication, 136
with mixed numbers, 9–12, 186, 190–192, 339
multiplication and division table, 373
Multiplication Property of One, 231
notation for, 133, 201, 220
number models for, 102
order of operations, 203, 373
partial-products multiplication
  for decimals, 130–131, 134
  for mixed numbers, 9–10, 12, 190–191, 339
  for whole numbers, 133, 134
powers of 10 for, 100, 129–131, 135
predicting product size, 186–187
process of, 133
ratios and, 38, 41
with reciprocals, 196
square numbers, 98
tying to division, 140, 193, 194, 219
using addition to multiply, 188
U.S. traditional multiplication, 137–138
with whole numbers
  extended multiplication facts, 132
  methods of, 133, 134, 136, 137
  by powers of 10, 129–130
  practicing, 325, 338, 349
Multiplication Bull's Eye, 338
Multiplication of Fractions Property, 189, 232
Multiplication Property of One, 231
Multiplication Top-It, 349
Multiplication Wrestling, 339
Multiplicative comparison (relationships), 41, 65
Multi-step unit conversions, 70–71
Museums of the world (real-world data), 360
Mute Girl of Portici, The (Auber), 86
Mystery numbers, finding, 332

N

Name-collection boxes, 206
Name That Number, 340
National Basketball Association international athletes (real-world data), 364
National parks (real-world data), 353
Native American population (real-world data), 356
Natural measures, 65
Natural yard, 65. See also Yard (yd)
Negative numbers, 90, 92, 94, 231, 265
Nets, 249, 263, 264

Nonagon, 238
Nonconvex polygon, 239. See also Polygons
Non-Euclidean geometry, 235
Non-repeating decimals, 90
Normal distribution, 305
Notation
  for absolute value, 92
  for angles, 236, 239
  for congruency, 268
  for degrees, 236
  for division, 141, 145, 158, 201
  expanded form, 100, 109
  exponential notation, 98, 99, 110
  for GCF, 105
  for LCM, 106
  for lines and line segments, 237
  for multiplication, 133, 201, 220
  for percent, 54
  for positive/negative numbers, 90
  for remainders, 141, 161
  set notation, 208–209, 210
  for sides, 239
  standard notation, 98, 99, 100, 110
  for triangles, 240
Not-equal sign (≠), 101, 207
Number grids, 56, 107
Number lines
  for box plots, 301
  comparing fractions on, 168
  comparisons on, 94, 112
  for coordinate grids, 95, 265
  decimals on, 91, 112, 376
  distance on, 92, 93
  finding a fraction of a fraction, 185
  finding fractions between fractions, 169–170
  fractions on, 89, 158
  for histograms, 296
  for inequality solution sets, 210–211
  multiplying fractions with, 188
  origin on, 94
  positive and negative numbers on, 92, 93, 94
  real number line, 91
  renaming mixed numbers on, 162
  rounding with, 115
  zooming in on, 112, 170
Number models. See also Number sentences
  defined, 209
  for division, 141
  for multiplication, 102
  in problem solving, 18–19, 209
  simplifying, 288
Numbers
  absolute value of, 318–319
  commas in, 97
  comparing, 94, 101
  divisibility tests for, 324

Index

Numbers (*continued*)
  even numbers, 27, 103
  expanded form for, 100
  exponential notation for, 98, 99, 110
  kinds of, 89–90, 91
  magnitude of, 92
  naming with expressions, 340
  odd numbers, 27
  opposites, 90, 92, 94
  positive versus negative, 90
  prime versus composite, 104
  properties of (See Properties of operations)
  reading, 97, 99
  rounding, 115–116
  in spreadsheets, 229
  square numbers, 27, 28
  standard notation for, 98, 99, 100, 110
  uses of, 88
Number sentences. *See also* Equations; Inequalities
  defined, 207
  finding factors, 102
  for inequalities, 210–211
  for multiplying mixed numbers, 9, 11
  open sentences, 208, 210
  solving
    bar models for, 216
    defined, 208
    trial and error method for, 214–215
  variables in, 207
  writing, 337
Number stories. *See also* Problem solving
  for distance, 20–22
  with money, 17–19, 23–25
Numerators. *See also* Fractions
  comparing fractions, 168
  defined, 156
  dividing, 160, 195
  Equivalent Fractions Property, 232
  multiplying, 145, 160, 173, 189
  remainders as fractions, 142
  renaming fractions, 161, 173
Numerical expressions, 200, 202. *See also* Expressions

O

Obtuse angle, 236
Obtuse triangle, 240
Octagon, 238, 239
Odd numbers, 27
Olympic medal counts (real-world data), 363
1-figure width, 65
One whole. *See* The whole
Open sentence, 208, 210. *See also* Number sentences

Operational symbols, 207
Opposite of Opposites Property, 232
Opposites, 90, 92, 94
Opposites Property, 231
Ordered pairs. *See also* Coordinate grids, plotting points on
  defined, 95, 265
  graphing (plotting), 17, 51–53, 95–96, 333
Ordering fractions, 322
Order of operations
  practice with, 202, 203, 340
  rules for, 203, 373
Origami, 273–274
Origin, 94, 95, 265
Ounce (oz), 67, 374, 375
Outliers, 291, 303

P

Pan-balance method, for solving equations, 217–218
Paper airplanes, 278
Paper folding (photo essay), 273–278
Parallel lines, 235, 237
Parallel line segments, 237, 242, 248
Parallelograms
  area of, 212, 254, 378
  as quadrilaterals, 242, 243
  as rectangles, 254
  as rhombuses, 243
  as trapezoids, 242
Parentheses
  multiplication and, 201
  order of operations and, 202, 203, 373
  removing by simplifying, 220
Partial-products diagram, 134, 190, 191
Partial-products multiplication. *See also* Multiplication
  for decimals, 130–131, 134
  for mixed numbers, 9–10, 12, 190–191, 339
  for whole numbers, 133, 134
Partial quotient, 143
Partial-quotients division, 143–144, 166. *See also* Division
Partial-sums addition, 118. *See also* Addition
Partitioning, of rectangles, 9
Part-to-part ratios, 38, 39, 40, 50, 55, 57
Part-to-whole ratios, 38, 39, 40, 55, 57, 61
Patterns
  in algebra, 225
  in data, 305–307
  in exponents, 110
  as mathematical structure, 26–28
  in powers of 10, 130
  in problem solving, 35
Pentagon, 238, 239

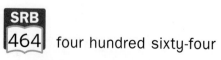

Pentagonal prism, 248
Pentagonal pyramid, 246, 247
Percent
   decimals as, 58
   defined, 54, 159
   equivalencies among, 56, 57
   finding (calculating), 61, 342
   finding the whole, 64
   fractions as, 57, 60, 159
   grid for, 377
   interpreting, 54–55
   multiplying, 348
   notation for, 54
   on number grids, 56
   part-to-whole ratios as, 55, 57
   practice with, 341–342
   problem solving with, 59–64
   unit percent, 59, 60
Percent Spin, 341–342
Perimeter
   defined, 251
   formulas for, 251, 378
   for triangles, 128
Perpendicular lines, 237
Perpendicular line segments, 237, 254, 255, 270
Personal references for measurement, 66, 67, 250, 375
Perspective, 235
Persuasive graphs, 308
Photo essays
   paper folding, 273–278
   scale models, 81–86
   U.S. census, 309–314
Pint (pt), 67, 70, 374, 375
Place, of digits, 97
Place value
   of decimals, 109–110, 334, 373
   expanded form and, 100, 109
   making comparisons with, 111
   powers of 10 and, 129 (See also Powers of 10)
   rounding and, 116
   of whole numbers, 97, 373
Plane geometry, 234
Planes, 235
Plants (real-world data), 367
Plateau distribution, 306
Points. *See also* Vertex (vertices)
   on coordinate grids
     connecting algebra and geometry with, 235
     finding locations on, 234
     polygons on, 267
     practice with, 333
     problem solving by, 222
     process of, 95–96, 265–266
   coordinates of, 95

   defined, 237
   on number lines, 92, 93, 94
   for ratio/rate graphs, 51–53
Polygon Capture, 343
Polygons
   congruency of, 268
   on coordinate grids, 267
   defined, 238
   flexagons as, 276
   perimeter of, 251, 378
   prefixes for, 238, 373
   properties of, 343
   reflecting, 271
   sides of, 238, 251
   as 2-dimensional figures, 244
   types of, 238, 239
   vertex (vertices) of a, 238
Polyhedrons
   defined, 246
   Euler's Theorem for, 245
   nets for, 249
   origami for, 274
   regular polyhedrons, 249
Population
   defined, 282
   real-world data, 356, 357
Positive numbers, 90, 92, 94, 231, 265
Postulates, 235
Pound (lb), 67, 374, 375
Powers of 10
   for decimals, 110
   defined, 99, 110
   dividing by, 139, 140
   expanded form and, 100, 109–110
   extended facts and, 132, 139
   measurement interrelationships with, 66
   multiplying by, 100, 129–131, 135
   renaming fractions with, 164
Powers of a Number Property, 232
Precision, 23–25, 128
Prefixes
   for metric measurements, 66, 373
   for polygons, 238, 373
Preimage, 271, 272
Prime factorization, 104
Prime numbers, 104
Prisms, 212, 248, 259, 263, 264. *See also specific prisms*
Problem solving
   accuracy in, 23–25
   arguments in, 13–16
   conjectures in, 13–16
   diagram for, 32
   making sense of problem, 4–8, 16
   mathematical models in, 17–19
   mathematical practices for, 4–7, 32–36

Index

Problem solving (*continued*)
  persevering in, 6–7, 8
  precision in, 23–25
  process for, 32–36
  representations in, 9–12
  structure in, 26–28
  tools for, 20–22
  using diagrams (drawings) in, 9, 13–16, 20, 27, 29, 33
  using graphs in, 222
  using rules in, 222
  using tables in, 17, 27, 29, 31, 35, 222
Product, 186–187
Projective geometry, 235
Properties of operations
  Addition of Fractions Property, 232
  Addition Property of Zero, 231
  Associative Property of Addition, 10, 11, 206, 231
  Associative Property of Multiplication, 123, 206, 231
  Commutative Property of Addition, 10, 11, 206, 231
  Commutative Property of Multiplication, 132, 206, 231
  Distributive Property (of Multiplication over Addition), 11, 204, 206, 231
  Distributive Property (of Multiplication over Subtraction), 205, 206, 231
  Division of Fractions Property, 196, 232
  Equivalent Fractions Property, 232
  Identity Property of Addition, 231
  Identity Property of Multiplication, 231
  Multiplication of Fractions Property, 189, 232
  Multiplication Property of One, 231
  Opposite of Opposites Property, 232
  Opposites Property, 231
  Powers of a Number Property, 232
  Subtraction of Fractions Property, 232
  variables in, 198, 204–205, 231
Proportions, equivalent ratios as, 41
Protractor, 236
Punch card machine, 310
Pyramids
  edges of, 245
  faces of, 244
  as regular tetrahedrons, 249
  surface area of, 263, 264
  types of, 246, 247
  volume of, 259, 378

Q

Q1. See Lower quartile (Q1)
Q3. See Upper quartile (Q3)
Quadrangle, 242

Quadrants, 95, 96, 265, 267
Quadrilaterals
  angle/side markings for, 239
  congruency of, 268
  on coordinate grids, 267
  defined, 242
  hierarchy for, 243
  as polygons, 238, 239
  similarity of, 269
Quantity, 212
Quart (qt), 67, 70, 374, 375
Quartiles, 292–293
Quick common denominator, 173. See also Common denominators
Quotients
  defined, 141
  estimating, 323
  in fraction division, 330
  non-zero remainders, 103

R

Random sample, 282
Range
  on box plots, 300
  defined, 291
  finding, 335–336
  on histograms, 306
  interquartile range, 293
Rates. See also Unit rate
  defined, 39, 158
  distance formula, 199, 213
  fractions as, 89, 158
  map scales as, 79
  scale models and, 75, 77
  units for, 40
  writing, 40
Ratio Comparison, 344
Ratio Dominoes, 345
Ratio Memory Match, 346
Rational numbers, 90, 91. See also Fractions
Ratio/rate graphs, 51–53
Ratio/rate tables
  making scale drawings/models with, 76–77, 78
  with percents, 60, 61, 64
  problem solving with, 43–44, 51
  size changes with, 73, 74, 76, 77
  for unit conversions, 68, 70
Ratios
  comparing, 42, 344
  defined, 38
  equivalent ratios
    defined, 41
    percents as, 55
    practice with, 345

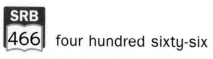

four hundred sixty-six

    problem solving with, 43–44, 48, 51–53, 63
    ratio/rate graphs for, 51–53
    ratio/rate tables for, 43–44
    tape diagrams for, 48
    unit conversions with, 66, 67
  as fractions, 39, 40, 41, 42, 89, 155, 158
  in measurement, 65–67
  notation for, 346
  part-to-part, 38, 39, 40, 50, 55, 57
  part-to-whole, 38, 39, 40, 55, 57, 61
  practice with, 344–346
  problem solving with, 42, 43–53
  ratio/rate graphs, 51–53
  ratio/rate tables
    making scale drawings/models with, 76–77, 78
    with percents, 60, 61, 64
    problem solving with, 43–44, 51
    size changes with, 73, 74, 76, 77
    for unit conversions, 68, 70
  representations of, 39–40
  scale models (photo essay), 81–86
  size changes with, 72–74
  tape diagrams for, 45, 46, 47, 48, 62
  unit ratios
    defined, 49
    equivalent fractions and, 50
    map scales and, 79
    in money conversions, 43
    problem solving with, 50
    size-change factors and, 73–74
Rays, 236, 237
Real number line, 91. See also Number lines
Real numbers, 91
Real-world data
  biodiversity, 366–367
  Boston Marathon records, 361
  data sources, 371–372
  electricity consumption, 370
  foreign-born population, 357
  introduction to, 352
  lakes of the world, 368
  largest reservoirs of the world, 359
  most-visited museums, 360
  mountains of the world, 368
  Native American population, 356
  Olympic medal counts, 363
  science and nature, 366–370
  sporting records, 361–365
  state facts, 354–355
  tallest buildings of the world, 359
  Tour de France records, 362
  trees of the world, 369
  Utah national parks, 353
Reasonable estimates, 280, 282
Reasoning (concept of), 12

Reciprocal, 196
Rectangle method, to find area of triangles, 13, 15–16
Rectangles
  area (See also Area models)
    counting squares, 253
    formulas for, 253, 260, 261, 378
    spreadsheet for, 199
    using distributive property for, 204–205
  bar model for solving equations, 216
  hierarchy for, 243
  parallelograms converted as, 254
  partitioning, 9
  perimeter of, 251, 378
Rectangular arrays, 102. See also Arrays
Rectangular prisms
  as geometric solids, 246, 248
  net for, 263
  surface area for, 378
  volume for, 260, 262, 378
Rectangular pyramid, 246, 261
Reduction, 72, 74, 75, 79, 269
Reference systems, 88
Reflections, 234, 271–272
Reflex angle, 236
Regular dodecahedron, 249
Regular hexagon, 239. See also Hexagon
Regular icosahedron, 249
Regular nonagon, 239. See also Nonagon
Regular octagon, 239. See also Octagon
Regular octahedron, 249
Regular pentagon, 239. See also Pentagon
Regular polygon, 239, 251. See also Polygons
Regular polyhedron, 249. See also Polyhedrons
Regular tetrahedron, 249
Relational symbols, 200, 207. See also Equal sign (=); Greater-than symbol (>); Less-than symbol (<)
Relatively prime numbers, 105, 106
Remainders. See also Division
  decimals as, 142
  defined, 141
  divisibility and, 103
  finding factors and, 102
  fractions as, 142, 161
  interpreting, 141–142
  non-zero, 103
  notation for, 141, 161
  writing, 141
Renewable energy (real-world data), 370
Repeated factors, 129
Repeating decimals, 164, 167
Representations
  creating, 9–10
  defined, 9
  making connections between, 12

Index

Representations (*continued*)
    making sense of, 10–12
    visual, 346
Reptiles (real-world data), 367
Reservoirs of the world (real-world data), 359
Rhombus, 243
Right angle, 236
Right-skewed distribution, 307
Right triangle, 240
Rotations, 234
Rounding
    in division, 143
    in estimating with fractions and mixed numbers, 178
    estimation as, 115
    of numbers, 115–116

S

Sample, 282
Sampling
    in the census, 310, 312
    in data collection, 282
Scale drawing, 78
Scale factor, 75, 76, 77, 269
Scale models, 75–77, 81–86
Scalene right triangle, 241
Scalene triangle, 240, 241
Science
    real-world data, 366–370
    scale models in, 85, 86
Second (sec), 374
Self-enumeration forms, 312
Set notation, 208–209, 210
Shapes, 29–30, 257. *See also specific shapes*
Short division, 148
Sides
    of angles, 236
    corresponding, 268, 269
    of polygons, 238, 251
    of quadrilaterals, 242
    of triangles, 240
Similar figures, 269
Simplest form, 160
Simplifying
    expressions, 206, 220
    fractions, 331
Size-change factor, 72–74, 75, 78, 79
Skewed distributions, 307
Slides. *See* Translations
Solid geometry, 234
Solution, 208. *See also* Equations; Inequalities; Number sentences
Solution Search, 347
Solution set, 210–211, 218. *See also* Equations; Inequalities

Space shuttle, model of, 82
Spheres, 244, 245
Spoon Scramble, 348
Sports records (real-world data), 361–365
Spread, 291, 293, 305
Spreadsheets (spreadsheet program)
    history of, 228
    purpose of, 228
    using, 229–231
    variables in, 199
Square arrays, 98. *See also* Arrays
Square centimeters (cm^2), 252, 374
Square decimeters (dm^2), 374
Square feet (ft^2), 252, 374
Square inches ($in.^2$), 252, 374
Square kilometers (km^2), 252, 374
Square meters (m^2), 252, 374
Square miles (mi^2), 252, 374
Square numbers, 27, 28, 98
Square pyramid, 247
Square root, 90
Squares
    area of, 212, 223, 378
    geometry of, 239, 243
    of numbers, 90, 225 (*See also* Square numbers)
    perimeter of, 251, 378
Square yards (yd^2), 252, 374
Stacked bar graph, 295
Standard notation (form), 98, 99, 100, 110
Standards for Mathematical Practice, 1–36
Standard units, 65, 66, 67
State facts (real-world data), 354–355
Statistical landmarks
    measures of center, 284–290, 303
    measures of spread and variability, 291–294, 300
Statistical questions
    collecting data, 280, 281–282, 304
    defined, 280
    organizing data, 283
    reading and analyzing graphs, 303–304
Statistics
    analyzing data, 303–304
    collecting data, 281–282
    data distributions, 305–307
    defined, 280
    measures of center, 284–290
    measures of spread and variability, 291–294
    organizing data, 283
    percents in, 54
    representing data, 295–302, 308
    statistical questions, 280 (*See also* Statistical questions)
Stevin, Simon, 108
Stone, Arthur, 276
Straight angle, 236

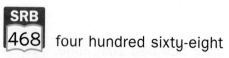

Index

Structure, in mathematics, 26–28
Subcategories, in hierarchies, 241, 243
Substitution, 214, 215, 216
Subtraction
   with base-10 blocks, 117
   counting-up subtraction, 123–124
   with decimals
      with base-10 blocks, 117
      methods of, 122, 124, 126
      practicing, 334, 350
      precision in, 128
   Distributive Property (of Multiplication over Subtraction), 205, 206, 231
   of fractions, 337
   with fractions, 176, 177, 179, 232, 337
   with mixed numbers, 181–182, 337
   order of operations, 203, 373
   trade-first subtraction, 121–122
   tying to addition, 219
   U.S. traditional subtraction, 125–127
   with whole numbers, 121, 123, 125, 127, 350
Subtraction of Fractions Property, 232
Subtraction Top-It, 350
Subtrahend, 127. *See also* Subtraction
Summer Olympics (real-world data), 363
Surface area, 263–264, 378
Surveys, 281, 282
Symmetrical data, 303, 305
Symmetry, 270

T

Tables
   bird counts, 281
   building height (real-world data), 359
   equivalent fractions, decimals, and percents, 377
   formulas, 378
   media time by youth, 282
   multiplication and division, 373
   personal references in measurement, 375
   place-value chart, 373
   population (real-world data), 357
   powers of 10, 99
   prefixes, 373
   for problem solving, 17, 27, 29, 31, 35, 222
   ratio/rate tables (See Ratio/rate tables)
   science and nature (real-world data), 366, 367, 368, 369, 370
   sports records (real-world data), 361, 362, 363, 364
   spreadsheets as, 229
   state facts (real-world data), 354–355
   unit conversions, 374
   Utah mileage chart (real-world data), 353
   "What's My Rule?", 199, 221, 223
   world land area (real-world data), 358
   world museums (real-world data), 360
Tablespoon (tbs), 374
Tabulating machine, 310
Tallest buildings of the world (real-world data), 359
Tallest mountains of the world (real-world data), 368
Tallest trees of the world (real-world data), 369
Tally charts, 283, 288
Tape diagrams, 45, 46, 47, 48, 62
Teaspoon (tsp), 374
Tenths, 108
Terawatt-hour, 370
Terminating decimals, 164
Terms, 206
Tessellations, 275
Theorems, 235
The whole, 155, 163
Thousandths, 108, 377
3-dimensional objects, 234, 244, 249
Time
   distance formula, 199, 213
   unit conversions, 71, 374
Toda, Takuo, 278
Ton (T), 67, 374, 375
Tools, 20–22
Topology, 235, 275
Tour de France records (real-world data), 362
Trade-first subtraction, 121–122
Transformations, 234, 271
Translations, 234
Trapezoid, 242, 243
Trees of the world (real-world data), 369
Trial and error, 214–215
Triangles
   area of, 13–16, 255–256, 260, 261, 378
   converting to parallelograms, 255
   dimensionality of, 234
   enlarging, 269
   equilateral
      defined, 240
      hexaflexagons and, 277
      hierarchy for, 241
      as regular polygons, 239
      in regular polyhedrons, 249
   hierarchy for, 241
   notation for, 240
   parts of, 240
   perimeter of, 128
   as polygons, 238, 239
   similarity of, 269
   theorem for, 235
   types of, 240
Triangular prism, 246, 248, 260
Triangular pyramid, 246, 247, 261

Index

Trihexaflexagon, 277
Tritetraflexagon, 276
Tuckerman, Bryant, 277
Turn-around rule for addition, 198. *See also* Commutative Property of Addition
Turns. *See* Rotations
2-dimensional objects, 234, 238, 244, 263. *See also* Angles; Lines; Line segments; Points; Polygons
2-finger width, 65

U

Uniform distribution, 306
Unit. *See* The whole
Unit conversions, 66, 67, 68, 69, 70–71, 128, 374
United States
   the census (photo essay), 309–314
   national parks (real-world data), 353
   population of (real-world data), 356, 357
   state facts (real-world data), 354–355
Unit fractions, 161, 174–175, 184
Unit percent, 59, 60
Unit rate
   defined, 49
   multiplicative comparisons with, 65
   problem solving with, 49, 50
   unit conversions and, 66, 67, 71
Unit ratios. *See also* Ratios
   defined, 49
   equivalent fractions and, 50
   map scales and, 79
   in money conversions, 43
   problem solving with, 50
   size-change factors and, 73–74
Universal Automated Computer (UNIVAC), 311
Unknowns. *See* Variables
Unsquaring, 90
Upper quartile (Q3), 292, 300, 301–302, 303
U.S. census (photo essay), 309–314
U.S. customary system, 67, 250, 374, 375
U.S. traditional addition, 120. *See also* Addition
U.S. traditional long division, 147–154. *See also* Division
U.S. traditional multiplication, 137–138. *See also* Multiplication
U.S. traditional subtraction, 125–127. *See also* Subtraction
Utah National Parks (real-world data), 353

V

Value, of digits, 97
Variability, 291, 293
Variable data, 280

Variables
   algebraic patterns with, 225
   defined, 198, 208
   in equations, 320–321, 327 (*See also* Equations)
   in expressions, 31, 200
   function of, 198, 207
   independent versus dependent, 223–224
   in inequalities, 4–8 (*See also* Inequalities)
   multiplication and, 201
   in number models, 19
   in number sentences, 207 (*See also* Number sentences)
   uses of, 198–199
   using rules, tables, and graphs to show relationships among, 222
Vertex (vertices)
   of an angle, 236
   defined, 236, 245
   of geometric solids, 245
   of a polygon, 238
   of a pyramid, 259
   of a quadrilateral, 242
   of a triangle, 240, 255
Vertical bar graph, 295
Viète, François, 198
Vinci, Leonardo da, 278
Volume
   of cubes, 378
   of cylinders, 224
   defined, 258
   fractional edge lengths, 262
   of geometric solids, 259
   personal references for, 66, 67, 375
   of prisms, 212, 260, 378
   of pyramids, 261, 378
   unit conversions, 374

W

Week (wk), 374
Weight
   measuring, 65
   personal references for, 67, 375
   unit conversions, 374
"What's My Rule?", 199, 221, 223
Whole numbers
   adding, 114, 118–120, 350
   defined, 89
   dividing
      extended division facts, 139
      fractions as, 89
      methods of, 143–144, 147–150
      practicing, 323, 350
   estimating, 323, 325, 332

multiplying
   extended multiplication facts, 132
   methods of, 133, 134, 136, 137
   by powers of 10, 129–130
   practicing, 325, 338, 349
place value of, 97, 373
properties of (See Properties of operations)
as rational numbers, 90
renaming as fractions, 89, 189, 329
subtracting, 121, 123, 125, 127, 350
Width, as height, 253
Wind tunnel, scale model of, 84
Wind turbines, scale model of, 85
Winter Olympics (real-world data), 363
World land area (real-world data), 358
World mountains (real-world data), 368
World trees (real-world data), 369
Wright Brothers, 84

x-axis, 90, 265, 266
x-coordinate, 90, 96, 265, 266, 267, 271

Yard (yd), 67, 68, 250, 374, 375
y-axis, 90, 265, 266, 365
y-coordinate, 90, 96, 265, 267, 271
Year (yr), 374

Zero
   absolute value and, 92
   Addition Property of, 231
   kinds of numbers and, 89, 90
   leading, in quotients, 148
   Opposites Property, 231
   position of, on number lines, 94
   role of
      in decimals, 111
      in dividing by powers of 10, 139, 140
      in front-end estimation, 113
      in multiplying by powers of 10, 129–131
      in rounding, 116